Nanotheranostics for Diagnosis and Therapy

Md. Abul Barkat • Farhan Jalees Ahmad •
Mohammad Akhlaquer Rahman •
Mohammad Azam Ansari
Editors

Nanotheranostics for Diagnosis and Therapy

Springer

Editors
Md. Abul Barkat
Department of Pharmaceutics
College of Pharmacy, University of Hafr Al-Batin
Hafr Al batin, Saudi Arabia

Farhan Jalees Ahmad
Department of Pharmaceutics
School of Pharmaceutical Education and Research, Jamia Hamdard
New Delhi, Delhi, India

Mohammad Akhlaquer Rahman
Department of Pharmaceutics and Industrial Pharmacy
College of Pharmacy, Taif University
Taif, Saudi Arabia

Mohammad Azam Ansari
Department of Epidemic Disease Research
Institute for Research and Medical Consultations, Imam Abdulrahman Bin Faisal University
Dammam, Saudi Arabia

ISBN 978-981-97-3114-5 ISBN 978-981-97-3115-2 (eBook)
https://doi.org/10.1007/978-981-97-3115-2

This Springer imprint is published by the registered company Springer Nature Singapore Pte Ltd.
The registered company address is: 152 Beach Road, #21-01/04 Gateway East, Singapore 189721, Singapore

Preface

In the contemporary era, the approach to disease treatment employing the available healthcare tools has been highly thwarted. With the involvement of complex etiology and progression of the disease, the reliance on the available therapies and diagnostic tools is considered to be quite less effective. In this regard, the modern healthcare system demands the availability of smart therapies and diagnostic tools to cater to healthcare needs. Recent advances in nanotechnology have spurred researchers worldwide to consider the great potential for using this technology to develop nanomedicine and nanotheranostics for the diagnosis and treatment of a variety of disease. Additionally, nanotheranostics enriched the field of nanomedicine by establishing a providential paradigm for these incurable diseases. These multimodal theranostic nanomedicines are capable of merging diagnostic and therapeutic agents into a single system on a tiny scale. Theranostic nanoparticles can assist in disease diagnosis, identifying its location, determining its stage, and providing information about the therapy response without or minimal harm to healthy cells/tissues.

Nanomaterial radiolabeling is a frequently used diagnostic technique in nuclear medicine imaging. Indeed, because of their clinical potential, several radiolabeled nanomaterials have been devised and developed for imaging cancer and other diseases. As multifunctional nanotheranostics, inorganic and organic nanoparticles such as polymer conjugates, dendrimers, micelles, liposomes, mesoporous silica nanoparticles, metal oxide, magnetic nanoparticles, quantum dots, nanocarbons, and nanoparticles of biodegradable polymers have been reported. Pharmaceutically, therapeutic and diagnostic agents are adsorbed, conjugated, entrapped, and encapsulated to achieve targeted co-delivery. The following is a concise overview of the main points covered in each chapter of the book.

Chapter 1 on the topic entitled "Introduction to Nanofabrication for Theranostics Application" concise overview of nanotheranostics, which is being explored as a promising technique for the treatment and diagnosis of several diseases.

Chapter 2 of Dr. Gupta on "Nanotheranostics: A Treasured Tailor for Disease Diagnosis and Treatment" covers a detailed outline about how nanotheranostics becomes a revolutionary force that connects the dots between diagnosis and therapy to produce.

Chapter 3 of Dr. Shah on the topic entitled as "Theranostics as a Novel Strategy for Targeted Delivery of Therapeutics: Transport and Fate Mechanism, Challenge, and Opportunities" covers various theranostics for targeted drug delivery and also outlines the current opportunities and challenges.

In Chap. 4, Dr. Chaudhary briefly explains about the "Nanotheranostics and Its Potential in Cancer Diagnostic and Its Management."

Dr. Kaushik in Chap. 5 reviews numerous "Bio-inspired Nanoparticles for Cancer Theranostics."

In Chap. 6, Dr. Goel provides an extensive note on "Targeted Theranostic Nanoparticles for Integrated Imaging and Therapy for the Management of Breast Cancer."

Dr. Redkar in Chap. 7 discusses in detail about the "Nanoscale Theranostics for Hepatocellular Carcinoma."

In Chap. 8, Dr. Rauf provides an extensive review on "Progress and Prospect of Nanocarriers: Design, Concept, and Recent Advances."

Dr. Garima in Chap. 9 outlines the "Theranostic Application of Nanomedicine in Neurodegenerative Diseases."

In Chap. 10, Dr. Hirulkar discusses the "Nanotheranostics: Clinical Status, Toxicity, Regulatory Consideration, and Future Prospects."

Dr. Israr in Chap. 11 provides a precise note on "Nanotheranostics in Prostate Cancer: Current Status and Future Potential."

In Chap. 12, Dr. Chauhan describes "Theranostics as an Emerging Integrated Imaging and Therapies in Precision Oncology."

This book, in this regard, endeavors to provide an updated account of recent advancements in the field of nanotechnology-based theranostics with special emphasis on their application in disease diagnosis and their treatment by ensuring the effective delivery of theranostic agents.

Lastly, the editors would like to express their gratitude to Springer and its team for offering a proficient platform for engaging with professionals in the area.

Hafr Al-Batin, Saudi Arabia — Md. Abul Barkat
New Delhi, Delhi, India — Farhan Jalees Ahmad
Taif, Saudi Arabia — Mohammad Akhlaquer Rahman
Dammam, Saudi Arabia — Mohammad Azam Ansari

Preface

In the contemporary era, the approach to disease treatment employing the available healthcare tools has been highly thwarted. With the involvement of complex etiology and progression of the disease, the reliance on the available therapies and diagnostic tools is considered to be quite less effective. In this regard, the modern healthcare system demands the availability of smart therapies and diagnostic tools to cater to healthcare needs. Recent advances in nanotechnology have spurred researchers worldwide to consider the great potential for using this technology to develop nanomedicine and nanotheranostics for the diagnosis and treatment of a variety of disease. Additionally, nanotheranostics enriched the field of nanomedicine by establishing a providential paradigm for these incurable diseases. These multimodal theranostic nanomedicines are capable of merging diagnostic and therapeutic agents into a single system on a tiny scale. Theranostic nanoparticles can assist in disease diagnosis, identifying its location, determining its stage, and providing information about the therapy response without or minimal harm to healthy cells/tissues.

Nanomaterial radiolabeling is a frequently used diagnostic technique in nuclear medicine imaging. Indeed, because of their clinical potential, several radiolabeled nanomaterials have been devised and developed for imaging cancer and other diseases. As multifunctional nanotheranostics, inorganic and organic nanoparticles such as polymer conjugates, dendrimers, micelles, liposomes, mesoporous silica nanoparticles, metal oxide, magnetic nanoparticles, quantum dots, nanocarbons, and nanoparticles of biodegradable polymers have been reported. Pharmaceutically, therapeutic and diagnostic agents are adsorbed, conjugated, entrapped, and encapsulated to achieve targeted co-delivery. The following is a concise overview of the main points covered in each chapter of the book.

Chapter 1 on the topic entitled "Introduction to Nanofabrication for Theranostics Application" concise overview of nanotheranostics, which is being explored as a promising technique for the treatment and diagnosis of several diseases.

Chapter 2 of Dr. Gupta on "Nanotheranostics: A Treasured Tailor for Disease Diagnosis and Treatment" covers a detailed outline about how nanotheranostics becomes a revolutionary force that connects the dots between diagnosis and therapy to produce.

Chapter 3 of Dr. Shah on the topic entitled as "Theranostics as a Novel Strategy for Targeted Delivery of Therapeutics: Transport and Fate Mechanism, Challenge, and Opportunities" covers various theranostics for targeted drug delivery and also outlines the current opportunities and challenges.

In Chap. 4, Dr. Chaudhary briefly explains about the "Nanotheranostics and Its Potential in Cancer Diagnostic and Its Management."

Dr. Kaushik in Chap. 5 reviews numerous "Bio-inspired Nanoparticles for Cancer Theranostics."

In Chap. 6, Dr. Goel provides an extensive note on "Targeted Theranostic Nanoparticles for Integrated Imaging and Therapy for the Management of Breast Cancer."

Dr. Redkar in Chap. 7 discusses in detail about the "Nanoscale Theranostics for Hepatocellular Carcinoma."

In Chap. 8, Dr. Rauf provides an extensive review on "Progress and Prospect of Nanocarriers: Design, Concept, and Recent Advances."

Dr. Garima in Chap. 9 outlines the "Theranostic Application of Nanomedicine in Neurodegenerative Diseases."

In Chap. 10, Dr. Hirulkar discusses the "Nanotheranostics: Clinical Status, Toxicity, Regulatory Consideration, and Future Prospects."

Dr. Israr in Chap. 11 provides a precise note on "Nanotheranostics in Prostate Cancer: Current Status and Future Potential."

In Chap. 12, Dr. Chauhan describes "Theranostics as an Emerging Integrated Imaging and Therapies in Precision Oncology."

This book, in this regard, endeavors to provide an updated account of recent advancements in the field of nanotechnology-based theranostics with special emphasis on their application in disease diagnosis and their treatment by ensuring the effective delivery of theranostic agents.

Lastly, the editors would like to express their gratitude to Springer and its team for offering a proficient platform for engaging with professionals in the area.

Hafr Al-Batin, Saudi Arabia — Md. Abul Barkat
New Delhi, Delhi, India — Farhan Jalees Ahmad
Taif, Saudi Arabia — Mohammad Akhlaquer Rahman
Dammam, Saudi Arabia — Mohammad Azam Ansari

Acknowledgments

We express our gratitude to Aishwarya Thyagarajan, the Production Editor (Books), for consistently liaising with the editors and promptly resolving any inquiries that arise during correspondence with the authors and publisher.

We express our gratitude to Springer and its staff for their assistance in the proofreading of the manuscript, leading to the ultimate publishing of the book.

We would like to extend our sincere gratitude to our family members for their comprehension and assistance.

Contents

Editors and Contributors

About the Editors

Md. Abul Barkat, PhD is currently serving as an Assistant Professor at the Department of Pharmaceutics, College of Pharmacy, University of Hafr Al-Batin, Kingdom of Saudi Arabia. He holds a Master's degree in Pharmaceutical Sciences from Jamia Hamdard University, New Delhi, India, and a PhD in Pharmaceutical Sciences (Pharmaceutics) from the Integral University Lucknow, India. With more than 12 years of teaching and research experience in the field of Pharmaceutical Science, his research interests include the development and optimization of herbal/synthetic-based nanostructured delivery systems, controlled release drug delivery systems, bio-enhanced drug delivery systems, nanomaterials and nanocomposites for a variety of conditions like burns, cancer, etc. To date, he has authored more than 75 research and review publications in various peer-reviewed journals, 15 book chapters, and 9 books.

Farhan Jalees Ahmad, PhD is currently Dean and Professor at the School of Pharmaceutical Education and Research, Jamia Hamdard, New Delhi, India, is an internationally known researcher in the area of Pharmaceutical Sciences. Having obtained his degree in M.Pharm. and Ph.D. (Pharmaceutics) from Jamia Hamdard, he continues to teach and leads a very productive research group which has been funded extensively by national and international funding agencies. The focus of his group has been in the area of nanomedicine, and many of his nanoproducts have been approved by the DCGI (Drugs Controller General of

India) and successfully transferred to the Defense Services for use in the armed forces. Nano DPI with a lung deposition of >60% was clinically proved in patients of COPD. He has published more than 480 publications, 12 book chapters, 11 books and has a US patent, 3 PCTs, and 24 Indian patents. With a total citation of 17,014 and H Index of 62, Professor Ahmad's work has been clearly well received by the scientific community.

Mohammad Akhlaquer Rahman, PhD is working as an Assistant Professor at the Department of Pharmaceutics, College of Pharmacy, Taif University, Kingdom of Saudi Arabia. He has 8 years of extensive experiences in teaching and research, published more than 40 research and review articles in referred and impact journals. He also contributed five book chapters with well-reputed international publishers. He attended international conference (Colloids and Nanomedicine 2012) in Amsterdam, Netherlands, with the financial support of Government of India. He received Best Research Award-2020 by VD Good™ Professional Association, India.

Mohammad Azam Ansari, PhD is currently working as Associate Professor in Epidemic Diseases Research Department at the Institute for Research and Medical Consultation, Imam Abdulrahman Bin Faisal University, Dammam, Saudi Arabia. He has earlier served as Assistant Professor in the Department of Applied Medical Sciences, Buraydah Private Colleges, Saudi Arabia (2014–2017). Dr. Ansari earned his Ph.D. in Microbiology (Medicine) from Aligarh Muslim University, Aligarh, India, in 2013. He has more than 8 years of research and teaching experience, and his research work mainly focused on nanoformulation and green synthesis of nanomaterials for biomedical applications such as antimicrobial, antibiofilm, anti-quorum sensing, anticancer, and nano drug delivery. He has published over 120 peer-reviewed articles in journals of international repute, 2 US patents, and received numerous grants and awards. He is also member of the editorial board and reviewer for a number of international and national journals.

Contributors

Farhan Jalees Ahmad Department of Pharmaceutics, School of Pharmaceutical Education and Research, Jamia Hamdard, New Delhi, India

Rumana Ahmad Department of Biochemistry, Era University, Lucknow, India

Abdulkareem Ali Alanezi Department of Pharmaceutics, College of Pharmacy, University of Hafr Al-Batin, Hafr Al-Batin, Saudi Arabia

Fahad Saad Alhodieb Department of Clinical Nutrition, College of Applied Health Sciences in Ar Rass, Qassim University, Ar Rass, Saudi Arabia

Mohammad Azam Ansari Department of Epidemic Disease Research, Institute for Research and Medical Consultations, Imam Abdulrahman Bin Faisal University, Dammam, Saudi Arabia

Syed Tauqeer Anwer Department of Biosciences, Jamia Millia Islamia, New Delhi, India

Pankaj Arora University Institute of Pharmaceutical Sciences and Research, Baba Farid University of Health Sciences, Faridkot, Punjab, India

Sumel Ashique Department of Pharmaceutical Sciences, Bengal College of Pharmaceutical Sciences and Research, Durgapur, West Bengal, India

Harshita Barkat Department of Pharmaceutics, College of Pharmacy, University of Hafr Al-Batin, Hafr Al-Batin, Saudi Arabia

Md. Abul Barkat Department of Pharmaceutics, College of Pharmacy, University of Hafr Al-Batin, Hafr Al-Batin, Saudi Arabia

Mithun Bhowmick Department of Pharmaceutical Sciences, Bengal College of Pharmaceutical Sciences and Research, Durgapur, West Bengal, India

Pratibha Bhowmick Department of Pharmaceutical Sciences, Bengal College of Pharmaceutical Sciences and Research, Durgapur, West Bengal, India

Pallavi M. Chaudhari Dr. D. Y. Patil College of Pharmacy, Pune, India

Sanjay Chauhan Institute of Pharmacy, Ram-Eesh Institute of Vocational and Technical Education, Greater Noida, Uttar Pradesh, India

Nishtha Chaurawal Department of Pharmacy, School of Chemical Sciences and Pharmacy, Central University of Rajasthan, Ajmer, Rajasthan, India

Pooja A. Chawla University Institute of Pharmaceutical Sciences and Research, Baba Farid University of Health Sciences, Faridkot, Punjab, India

Viney Chawla University Institute of Pharmaceutical Sciences and Research, Baba Farid University of Health Sciences, Faridkot, Punjab, India

Khalid Umar Fakhri Department of Biosciences, Jamia Millia Islamia, New Delhi, India

Garima Department of Biotechnology, DTU, New Delhi, India

Honey Goel University Institute of Pharmaceutical Sciences and Research, Baba Farid University of Health Sciences, Faridkot, Punjab, India

Neha Gupta Department of Pharmacy, School of Chemical Sciences and Pharmacy, Central University of Rajasthan, Ajmer, Rajasthan, India

Rutik Hirulkar Department of Pharmacy, School of Chemical Sciences and Pharmacy, Central University of Rajasthan, Ajmer, Rajasthan, India

Khalid Imtiyaz Department of Botany and Life Sciences, Shri Aurobindo College, University of Delhi, New Delhi, India

Juveriya Israr Institute of Biosciences and Technology, Shri Ramswaroop Memorial University, Barabanki, India

Department of Biotechnology, Era University, Lucknow, India

Monika Kaushik Department of Pharmacology, Amity Institute of Pharmacy, Amity University Gwalior, Gwalior, Madhya Pradesh, India

Sumayya Khan Department of Pharmacology, Faculty of Pharmacy, Maulana Azad University, Jodhpur, Rajasthan, India

Heya Khatoon Department of Pharmacy, Pandaveswar School of Pharmacy, Pandaveswar, West Bengal, India

Saadullah Khattak The 5th Affiliated Hospital of Wenzhou Medical University, Lishui, Zhejiang, China

Sanjesh Kumar Institute of Pharmaceutical Research, GLA University, Mathura, Uttar Pradesh, India

Sugandh Kumar School of Medicine, University of San Francisco California (UCSF), San Francisco, CA, USA

Anil Kumar Mavi Department of Botany and Life Sciences, Shri Aurobindo College, University of Delhi, New Delhi, India

Manojkumar Mukundrao Nitalikar Department of Pharmaceutics, Rajarambapu College of Pharmacy Kasegaon, Sangli, Maharashtra, India

Mohammad Oves Center of Excellence in Environmental Studies, King Abdulaziz University, Jeddah, Saudi Arabia

Radheshyam Pal Department of Pharmacy, Pandaveswar School of Pharmacy, Pandaveswar, West Bengal, India

Prachi Pannu Department of Biotechnology, DTU, New Delhi, India

Pooja Department of Biotechnology, DTU, New Delhi, India

Simran Preet Department of Biophysics, Panjab University, Chandigarh, India

Mohammad Akhlaquer Rahman Department of Pharmaceutics and Industrial Pharmacy, College of Pharmacy, Taif University, Taif, Saudi Arabia

Smriti Raina Department of Biotechnology, DTU, New Delhi, India

Mohd Ahmar Rauf Department of Surgery, Miller School of Medicine, University of Miami, Miami, FL, USA

Kaisar Raza Department of Pharmacy, School of Chemical Sciences and Pharmacy, Central University of Rajasthan, Ajmer, Rajasthan, India

Mayuresh Ramesh Redkar Department of Pharmaceutics, Yashwantrao Bhonsale College of Pharmacy Sawantwadi, Sindhudurg, Maharashtra, India

Department of Pharmaceutics, Rajarambapu College of Pharmacy Kasegaon, Sangli, Maharashtra, India

Riya University Institute of Pharmaceutical Sciences and Research, Baba Farid University of Health Sciences, Faridkot, Punjab, India

Moshahid Alam Rizvi Department of Biosciences, Jamia Millia Islamia, New Delhi, India

Subeel Shah Department of Pharmacy, School of Chemical Sciences and Pharmacy, Central University of Rajasthan, Ajmer, Rajasthan, India

Anjali Sharma Department of Biotechnology, DTU, New Delhi, India

Himanshu Sharma Teerthanker Mahaveer College of Pharmacy, Teerthanker Mahaveer University, Moradabad, Uttar Pradesh, India

Sahabjada Siddiqui Department of Biotechnology, Era University, Lucknow, India

Himanshu Singh Institute of Pharmacy, Ram-Eesh Institute of Vocational and Technical Education, Greater Noida, Uttar Pradesh, India

Mansi Singh Institute of Pharmaceutical Research, GLA University, Mathura, Uttar Pradesh, India

Richu Singla Multidisciplinary Research Unit, Guru Gobind Singh Medical College and Hospital, Faridkot, Punjab, India

Shailendra Kumar Sinha Department of Zoology and Environmental Science and Disaster Management, BBMKU, Dhanbad, Jharkhand, India

Mohamad Taleuzzaman Department of Pharmaceutical Chemistry, Faculty of Pharmacy, Maulana Azad University, Jodhpur, Rajasthan, India

Introduction to Nanofabrication for Theranostics Application

1

Md. Abul Barkat, Mohammad Akhlaquer Rahman, Mohammad Azam Ansari, and Farhan Jalees Ahmad

Abstract

Nanotheranostics is a treatment approach that integrates therapeutics and diagnostics using a single nanocarrier system. Its purpose is to monitor drug delivery, assess therapy response, and improve the drug efficacy and its safety. This novel fabrication in one system has garnered significant attention and has been acknowledged as a possible paradigm shift in addressing the limitations of traditional chemotherapy due to its ability to accomplish precise illness detection and treatment. Nanoparticulate systems are highly suitable for carrying theranostic agents due to their exceptional physicochemical properties. These properties include their nanoscale sizes, functional properties, extended blood circulation, ability to actively or passively target tumors, specific cellular uptake, and in certain cases, excellent optical properties that are ideal for simultaneous phototherapy and imaging. In general, the advancement of nanotechnology has made theranostics a tangible possibility, now undergoing the process of transitioning from laboratory research to practical use in medical settings. This chapter pro-

M. A. Barkat (✉)
Department of Pharmaceutics, College of Pharmacy, University of Hafr Al-Batin, Hafr Al-Batin, Saudi Arabia

M. A. Rahman
Department of Pharmaceutics and Industrial Pharmacy, College of Pharmacy, Taif University, Taif, Saudi Arabia

M. A. Ansari
Department of Epidemic Disease Research, Institute for Research and Medical Consultations, Imam Abdulrahman Bin Faisal University, Dammam, Saudi Arabia

F. J. Ahmad
Department of Pharmaceutics, School of Pharmaceutical Education and Research, Jamia Hamdard, New Delhi, India

M. A. Barkat et al. (eds.), *Nanotheranostics for Diagnosis and Therapy*,
https://doi.org/10.1007/978-981-97-3115-2_1

vides a concise overview of nanotheranostics, which is being explored as a promising technique for the treatment and diagnosis of several diseases.

Keywords

Nanotheranostics · Nanocarrier · Chemotherapy · Theranostic agents · Tumor targeting · Nanotechnology

1.1 Introduction

John Funkhouser, the CEO of Myocardial Solution Inc., introduced the word "Theranostic" in August 1998 in a press release (Funkhouser 2002). The word "theranostics" refers to drugs and procedures that are designed for both diagnosis and treatment simultaneously (Ahmed et al. 2012). Another definition of theranostics is the combination of diagnostic and therapy (Warner 2004). The modern era's advanced approach to theranostics is essentially the result of the Magic Bullet notion. In 1900, Paul Ehrlich, a German Nobel winner, introduced the idea of creating specific medications that may accurately target and eliminate areas of illness without harming healthy cells (Valent et al. 2016).

The initial clinical implementation of "theranostics" was proposed in 1946 (Seidlin et al. 1946) for patients with metastasized thyroid cancer using radioactive iodine therapy. This approach has been further studied and documented by Deb et al. (1996), Maresca et al. (2009), and Hillier et al. (2011). Subsequently, different therapies that use radioligands have become increasingly popular in early clinical trials, specifically for prostate cancer that expresses prostate-specific membrane antigen (PSMA). These therapies include radio-attached monoclonal antibody, small-molecule inhibitors of PSMA, and α- and β-emitting radioisotopes. The theranostics idea, which is at the forefront of medical advancements, departs from the traditional strategy of using a single treatment for all patients and instead embraces a comprehensive tailored medicine approach.

The objective of this therapy is to provide the appropriate treatment, tailored to the specific patient, at the optimal moment, while ensuring the administration of the correct dosage, resulting in a more focused and effective pharmacotherapy profile. When developing theranostic-based technologies for clinical translation, it is crucial to prioritize the following factors: achieving sufficient blood plasma circulation time, ensuring targeted delivery exclusively to cancerous tissues while avoiding accumulation in normal tissues and organs, preventing immune responses, and enabling simultaneous treatment and non-invasive monitoring to ensure successful drug delivery. In addition, it is desirable for the delivery method to be non-invasive, non-toxic, and biodegradable (Sneider et al. 2017).

1.2 Theranostics in Medicine

Theranostics, a field in medicine, encompasses several applications, which are delineated as follows (Kashyap et al. 2023; Yukawa et al. 2023; Andreou et al. 2017):

- It is a noninvasive molecular imaging technique to assess disease heterogeneity.
- Provides a more comprehensive understanding of the biodistribution and accumulation of the medication at the target location.
- Enhanced comprehension of the mechanism behind local medication release.
- Enhancing the release of drugs by using stimuli-responsive theranostics.
- Predicting drug responses and the corresponding undesirable effects using tailored therapy.

1.3 Nanotheranostics and Theranostic Nanomedicine

Nanotheranostics involves the application and advancement of nanomedicine techniques for advanced theranostics. This includes the utilization and improvement of various nanocarriers such as polymer conjugations, dendrimers, micelles, liposomes, metal and inorganic nanoparticles, carbon nanotubes, and nanoparticles of biodegradable polymers. These nanocarriers are used to deliver diagnostic and therapeutic agents in a sustained, controlled, and targeted manner, resulting in enhanced theranostic effects and reduced side effects. The objective is to identify and manage illnesses at their first phase, when they are very susceptible to being cured or, at the very least, treated. Nanotheranostics hold great potential for treating life-threatening diseases including cancer, cardiovascular disorders, and acquired immunodeficiency syndrome (AIDS). They provide the opportunity to simplify therapy, improve prognosis, conserve resources, and enhance the quality of life for patients (Kashyap et al. 2023).

Nanomedicine refers to the use of nanotechnology to diagnose, cure, and prevent illnesses at the cellular and molecular level. It can also be effective in developing theranostic candidates, which integrate diagnostic and therapy concurrently. This concept has been supported by several studies (Sumer and Gao 2008; Deveza et al. 2012; Janib et al. 2010). Theranostic nanomedicine is a highly advanced and versatile technology that can both diagnose and administer therapy to specific sick cells. It achieves this by using targeted ligands and biomarkers. Theranostic nanomedicine refers to the utilization of colloidal nanoparticles with diameters ranging from 10 to 1000 nm (Yu et al. 2012; Xie et al. 2010; Ye and Chen 2011). These materials are composed of large molecules or polymers that include diagnostic and therapeutic agents. The agents are either adsorbed, conjugated, encapsulated, or entrapped inside the materials. This allows for simultaneous diagnosis and therapy at the cellular and molecular level (Feng 2006; Muthu and Singh 2009; Muthu and Feng 2013). Theranostic nanomedicine surpasses other theranostics due to its advanced capabilities in a single platform. These capabilities include sustained/controlled release, targeted delivery, higher transport efficiency through endocytosis (Muthu and Singh 2009; Nirmala et al. 2023), stimulus-responsive agent release (smart delivery) (Muthu and Singh 2009), synergetic performance (such as combination therapy and siRNA co-delivery) (Zhao et al. 2013), multimodality diagnosis and/or therapies, and high-quality performances (such as oral delivery, evasion of multi-drug resistance (MDR) proteins, and autophagy inhibition). The incorporation of a solitary diagnostic or therapeutic agent into nanomedicine may not exhibit a high level of effectiveness, specificity, or sensitivity for its intended purposes (Mei et al. 2013; Smith and Smith 2012; Lammers et al. 2011).

Hence, the development of multimodality nanotheranostics involves utilizing the benefits of co-encapsulating several diagnostic and therapeutic modalities into targeted nanomedicine platforms (Tan et al. 2011). SiRNA can be included into theranostic nanomedicine to serve as an inhibitor of theranostic resistance. The utilization of siRNA-based theranostic nanomedicine has demonstrated significant advancements in both diagnosis and therapy, particularly as a multimodality treatment. Nanomedicine has been innovatively utilized for administering oral chemotherapy. A nanomedicine utilizing D-α-tocopheryl polyethylene glycol 1000 succinate (TPGS) has been developed for the oral administration of chemotherapeutic drugs such as docetaxel and paclitaxel. This nanomedicine has been shown to significantly improve the oral bioavailability of these medicines, as reported by Zhao in 2010 and Win in 2006 (Zhao and Feng 2010; Win and Feng 2006).

Future research will focus on the oral administration of theranostic nanomedicine, aiming to significantly advance the field of cancer theranostics and enhance the well-being of cancer patients. This will enable patients to get chemotherapy treatment in the comfort of their own homes, fulfilling their need for convenient and accessible healthcare. Autophagy, also known as autophagocytosis, is a cellular process that involves the breakdown of defective cellular components or foreign invaders within lysosomes. Autophagy also impacts nanomedicine following endocytosis, altering the intracellular pharmacokinetics of the nanomedicine, including its absorption, distribution, excretion, and metabolism, hence influencing its therapeutic effects. Hence, the incorporation of autophagy inhibitors into theranostic nanomedicine might augment the transportation of diagnostic and therapeutic substances. Advanced nanotheranostics is now being developed to simultaneously diagnose and treat illnesses with great efficiency at the cellular and molecular level (Muthu et al. 2014; Alhodieb et al. 2023).

Theranostic nanomedicine encompasses therapeutic substances such as hydrophobic drugs, proteins, peptides, and genetic elements. Theranostic nanomedicine employs various diagnostic agents alongside therapeutic agents. These diagnostic agents encompass optical imaging, which utilizes fluorescent dyes or quantum dots, magnetic resonance imaging (MRI) that relies on superparamagnetic metals like iron oxides, nuclear imaging that employs radionuclides, and computed tomography that utilizes heavy elements such as iodine. Quantum dots are crystalline semiconductors that generate a signal when activated by external sources of energy. Quantum dots are widely utilized nanomaterials in theranostic nanomedicine due to their notable benefits over organic dyes. In addition, quantum dots have strong emissions, higher absorption coefficients, high brightness levels, and they possess photostability (Santra 2012; Jahangir et al. 2019; Andreou et al. 2017; Yong et al. 2012). If an illness can be diagnosed at its early stage, it would be possible to cure or at least treat it effectively. Therefore, sophisticated theranostic platforms were developed to enable precise diagnosis, precise targeting, efficient and regulated delivery of materials (McCarthy et al. 2006). Theranostic nanomedicine offers several benefits for future generations, such as: (1) the ability to diagnose and track the dispersion of substances within cells and in living organisms; (2) the provision of enhanced treatment options; (3) Monitoring the therapeutic progress after the therapy (Ding & Wu 2012; Velikyan 2012). The diverse nanomedicine platforms are utilized for the specified applications, as shown in Table 1.1.

1.4 Nanotechnology-Assisted Cancer Theranostics

Nanomaterials possess a tiny size, which gives them a significant advantage in terms of having extensive surface areas and the ability to carry a big amount of pharmaceuticals. This enables the simultaneous delivery of many therapeutic medications and imaging agents. Furthermore, it is widely recognized that dysfunctional blood arteries in tumor tissues cause irregular molecular and fluid movement. As a result, nanoparticles can passively gather in tumors due to the well-established effects of increased permeability and retention (Kalyane et al. 2019; Nirmala et al. 2023). The incorporation of targeting ligands, including as folate, hyaluronic acid, transferrin, aptamers, antibodies, and peptides, onto nanoparticles has significantly enhanced their ability to target particular tumor tissues. This improvement is achieved by the specific recognition between the ligands and the receptors present on the surface of the tumor tissues (Guo et al. 2020; Barkat et al. 2020). Furthermore, the exceptional physicochemical features of many nanomaterials enable them to function as imaging agents independently, without the need for further loading of imaging agents.

Nanotechnology-assisted cancer theranostics offer several distinct advantages. These include the ability to passively or actively accumulate in tumor tissues through different mechanisms, as well as the excellent optical properties of certain nanoscale carriers that can directly contribute to targeted therapy and diagnosis (Jo et al. 2016). Additionally, these nanocarriers enable real-time imaging-guided therapy, and the combination of different therapeutic and diagnostic nanoparticles within a single system can produce synergistic therapeutic effects (Mi et al. 2016; Chen et al. 2017a).

1.5 Challenges Faced with Theranostic Regimen

The primary challenge linked to nanomedicine is the presence of tumor heterogeneity and an excessively simplistic pharmacokinetic model of the disease. According to this, it is hypothesized that the drug concentration within the tumor is solely determined by the medication present in the circulation for an extended duration (Dai et al. 2021; Chavda et al. 2023).

Table 1.1 Several sophisticated theranostic nanomedicine platforms are now being developed to enable simultaneous integration of diagnostic and therapeutic

Types of theranostic nanomedicine	Therapeutic/ diagnostic/targeting agent	Advancement	References
Drug-polymer conjugates	64Cu/RGD	Cancer imaging and radiochemo-therapy	Yuan et al. (2013)
Polymeric nanoparticles in conjugation with quantum dots	Docetaxel/folic acid	Co-delivery of docetaxel and quantum dots	Pan et al. (2010)
Solid lipid nanoparticles in conjugation with quantum dots	Paclitaxel/siRNA/ cRGD	Multimodal therapy	Shuhendler et al. (2012)
Dendrimers	Phthalocyanines/ LHRH	Delivery of single theranostic agent	Taratula et al. (2013)
Liposomes in conjugation with quantum dots	Docetaxel/folic acid	Co-delivery of docetaxel and quantum dots	Muthu et al. (2012)
Micelles in conjugation with iron oxide nanoparticles	Iron oxide nanoparticles	Delivery of single theranostic agent	Chandrasekharan et al. (2011)
Gold nanoparticles	DOX/CPLGLAG G peptide	Stimulus responsive drug release	Chen et al. (2013)
Carbon nanomaterials	Intrinsic property	Self photoluminescent and photothermal property	Daniel and Astruc (2004)
Gold nanoparticle clustered with graphene oxide	Fluorouracil	Detection and imaging and phototherapy, furthering into bioimaging-assisted cancer therapy	Roma-Rodrigues et al. (2019)
Albumin-nanoceria conjugated nanoparticles	Ceria	Inflammation control with guided imaging. Therapy shows potential as systemic arthritis treatment	McCarthy et al. (2010)
Dextran-coated magnetic fluorescent nanoparticle	Mesotetra (myhydroxyphenyl) chlorin	Imaging and therapy of atherosclerosis. Highly efficacious and sensitive imaging proved to be potent therapeutic platform	McCarthy et al. (2006)
Lipid latex nanoparticle system	Rosiglitazone	Imaging of atherosclerotic plaques and therapy. Preferable signal interception and therapeutic efficacy shown on models, leaving room for future clinical applications to be worked upon	Qin et al. (2016)

(continued)

Table 1.1 (continued)

Types of theranostic nanomedicine	Therapeutic/ diagnostic/targeting agent	Advancement	References
Modified liposomes of dual-ligand (anticarbonic anhydrase IX (anti-CA IX) antibody and CPP33)	Triptolide	Tumor penetration and tumor growth inhibition. Potential use as modified lipid vehicle for local targeted cancer drug delivery	Mukherjee et al. (2019)

This model failed to account for several parameters, including vascular density, blood vessel location inside the tumor, stromal cell thickness, and the presence of macrophages. Each of these characteristics has a unique impact on the distribution of drugs amongst different tumors, and it is important to consider each tumor as an individual organ while treating it (Ojha et al. 2022; Alizadeh et al. 2015).

Typically, tumors are classified as highly vascular structures, however there is considerable variability in vascularity across various tumor types (Junttila & de Sauvage 2013). For instance, squamous cell carcinoma has a lower number of blood vessels in comparison to colorectal cancer. The vascular density of this tumor may be evaluated using many imaging modalities such as MRI, CT, and others. In addition to the vasculature of the tumor, the permeability of the tumor vessels also has a significant impact. The determination is conducted in a laboratory environment by the utilization of Evan's blue dye. MRI can be employed in clinical settings to ascertain the permeability of a substance. A drug called gadofosveset, which binds to albumin, was recently authorized for the purpose of detecting tumor leakiness using contrast MRI (Jain 2013). Hence, a comprehensive characterization of the tumor vasculature and its permeability assists in attaining a more favorable prognosis.

Moreover, the implementation of theranostic nanomedicine in clinical settings faces substantial biological obstacles arising from the interaction between nanomaterials and biological systems (Gaspar 2007; Tinkle et al. 2014; Hua et al. 2018). When a nanomedicine comes into contact with biological substances, its potential toxicity or lack of compatibility can lead to the development of illnesses such as immunoreactions and inflammatory disorders. When nanoparticles come into contact with proteins in a biological system, they form a layer called a "protein corona" on the surface. The adsorption of this protein alters the stability, size, dispersibility, pharmacokinetics, biodistribution, and toxicity characteristics of the nanoparticles. Due to individual variations in patient response, regulatory clearance for theranostic nanomedicine may be challenging if safety is perceived as a uniform standard applicable to all patients (Dreifuss et al. 2015).

Several studies have indicated that nanotheranostics, which have significant therapeutic promise, have either poor or no diagnostic signs. The gold nanoparticles, measuring 20 nm in size, that were coated with anti-EGFR showed the greatest accumulation in tumors. On the other hand, the gold nanoparticles with a size of

50 nm produced the best results for CT imaging. The results stated above indicate that the use of theranostic nanomedicines as theranostic agents is limited by their tendency to disperse in tumors based on their size. Hence, it is imperative to engage in ongoing surveillance of patients for theranostics. Another significant obstacle to achieving scalable and cost-effective clinical use of theranostics nanomedicines is the challenging task of regulating and replicating synthesis processes (Accomasso et al. 2018). The feasibility of implementing extensive economic manufacturing poses a significant challenge for translation. Theranostics possess several dynamic functionalities and often need multiple phases of preparation, purification, and processing, resulting in challenges when scaling up. Consequently, most advanced theranostics are restricted to laboratory and preclinical investigations. Another significant concern that requires attention is the increasingly broad disparity between the research community and regulatory entities (Tinkle et al. 2014).

Various governmental regulations are being enforced to oversee the commercialization of nanomedicine, based on regulatory standards including quality control, formulation process, safety profile, and patent protection. The lack of well-defined regulatory and safety standards significantly hinders the prompt and effective implementation of theranostics in the market (Goyal et al. 2023). Nanoparticles used in commercial products have undergone established regulations. However, current standards may be inadequate and need further refinement to ensure the quality, reliability, and effectiveness of future nanotheranostics for human use. The majority of theranostics rely on optical imaging, which is of limited therapeutic significance because to its restricted ability to penetrate into tissues. PET and SPECT can be employed for investigation in both preclinical and clinical subjects, however, the routine application of these techniques for diagnostic and treatment monitoring is impeded by the risk of radiation exposure. Although still in the early stages of investigation, 19F MRI has great promise because to its little background interference, which allows for precise quantitative identification. Due to its restricted sensitivity and extensive usage of imaging chemicals, the biocompatibility, elimination, and toxicity of MRI must be meticulously evaluated (Ma et al. 2020). The prognostic approaches (planar scintigraphy, SPECT, and PET) have limited sensitivity and specificity compared to other detection modalities, which occasionally results in inaccurate patient choices. Utilizing imaging-based selection criteria may result in the exclusion of individuals who might potentially benefit from a certain therapy. The restrictions mostly originate from imaging system parameters, particularly spatial resolution and contrast resolution.

1.6 Concluding Remarks and Current Prospects

In conclusion, the fast advancement of theranostics based on nanotechnology has significantly accelerated the transformation of cancer treatment and detection. Ideal nanotheranostic systems should possess the following characteristics: (1) they should be non-toxic and biocompatible, (2) they should exhibit great stability and efficiency in loading drugs, (3) they should be easily prepared and modified, and (4) they should be capable of targeting tumors and effectively undergoing endocytosis (Truffi et al. 2016; Chen et al. 2017b). While the majority of nanoparticles do not exhibit apparent cytotoxicity, more research are important to ascertain safety in clinical applications. Furthermore, the process of preparing and ensuring the quality of nanoparticles remains intricate. In particular, the production of organic nanomaterials is often arduous and results in limited quantities. Furthermore, it is crucial to take into account the equilibrium between the effectiveness and dependability of nanotheranostic systems, notwithstanding the potential for introducing different agents in a single theranostic platform to offer various functionalities. Nevertheless, the nanosystem may be excessively intricate for mass production. However, present researchers are placing greater importance on these matters, and several efforts are underway to obtain more effective use of nanotheranostic platforms in cancer therapies. The advanced nanotheranostic represents a cutting-edge approach for future generations. It encompasses a wide range of features in nanomedicine, such as drug release that responds to stimuli like pH, temperature, magnetism, and ultrasound. It also includes synergistic performance, such as siRNA delivery and combination therapy, as well as a multi-modal approach and high-quality performance, such as oral delivery and autophagy inhibition. Additionally, additional research is required to investigate these principles using the existing clinical platform of nanomedicines, with the aim of developing adaptable and intelligent theranostic applications (Muthu and Wilson 2010; Bardhan et al. 2011). In the near future, theranostic nanomedicines loaded with quantum dots, gold nanoparticles, and iron oxide nanoparticles have the potential to serve as carriers to address these concerns. The iron oxide nanoparticles exhibit exceptional biosafety as they undergo degradation and metabolism within the serum Fe pool, leading to the formation of hemoglobin or integration into the metabolic pathway (Ho et al. 2011). Currently, there is a significant worry over the utilization of heavy metal-based quantum dots as theranostic probes for human applications. Studying the in vivo biodistribution and toxicology of quantum dots in animal systems, which may accurately reflect their toxicity in human usage, is a significant problem. We eagerly anticipate the development of biocompatible, non-immunogenic, ultra-small quantum dots with a size of less than 5.0 nm for loading in nanomedicine. These quantum dots should be eliminated by a renal clearance mechanism. This development is widely anticipated and has been mentioned in studies by Zhu et al. (2013) and Wu et al. (2013). Carbon nanotubes have the ability to induce harmful effects on living organisms. They generate oxidative stress by inducing lipid peroxidation, resulting in inflammation and cellular harm in vital organs including the lungs and liver (Ge et al. 2012; Yang et al. 2012). Furthermore, gold nanoparticles have been shown to cause cytotoxicity through oxidative stress (Mateo et al. 2014). However, it is worth noting that this cytotoxicity is not exclusive to gold nanoparticles, since other types of nanoparticles have also been observed to have similar effects (Setyawati et al. 2013; Ng et al. 2011). Hence, the safety of employing theranostic carbon nanotubes and gold nanoparticles for therapeutic purposes has become a matter of concern. There are several alternatives for using nanotheranostics to kill cancer cells. However, it is important not to overlook additional impacts of nanoparticles that are not poisonous to cells, such as causing blood vessels to become leaky and impairing cell movement during wound healing (Tay et al. 2014). Ultimately, the utilization of multimodal theranostic nanomedicines, which involve the integration of many diagnostic techniques and therapeutics, can offer significant benefits for theranostic applications. The utilization of nanomedicine, namely nanotheranostics, in a theranostic strategy that focuses on molecular biomarkers with sophisticated designs is expected to enhance the accuracy of cancer therapy.

References

Accomasso L, Cristallini C, Giachino C (2018) Risk assessment and risk minimization in nanomedicine: a need for predictive, alternative, and 3Rs strategies. Front Pharmacol 9:228

Ahmed N, Fessi H, Elaissari A (2012) Theranostic applications of nanoparticles in cancer. Drug Discov Today 17:928–934

Alhodieb FS, Rahman MA, Barkat MA, Alanezi AA, Barkat HA, Hadi HA, Harwansh RK, Mittal V (2023) Nanomedicine-driven therapeutic interventions of autophagy and stem cells in the management of Alzheimer's disease. Nanomedicine (Lond) 18(2):145–168. https://doi.org/10.2217/nnm-2022-0108

Alizadeh AA, Aranda V, Bardelli A, Blanpain C, Bock C, Borowski C et al (2015) Toward understanding and exploiting tumor heterogeneity. Nat Med 21:846–853

Andreou C, Pal S, Rotter L, Yang J, Kircher MF (2017) Molecular imaging in nanotechnology and theranostics. Mol Imaging Biol 19:363–372

Bardhan R, Lal S, Joshi A et al (2011) Theranostic nanoshells: from probe design to imaging and treatment of cancer. Acc Chem Res 44(10):936–946

Barkat HA, Das SS, Barkat MA, Beg S, Hadi HA (2020) Selective targeting of cancer signaling pathways with nanomedicines: challenges and progress. Future Oncol 16(35):2959–2980. https://doi.org/10.2217/fon-2020-0198

Chandrasekharan P, Maity D, Yong CX et al (2011) Vitamin E (D-alpha-tocopheryl-*co*-poly(ethylene glycol) 1000 succinate) micelles-superparamagnetic iron oxide nanoparticles for enhanced thermotherapy and MRI. Biomaterials 32(24):5663–5672

Chavda VP, Khadela A, Shah Y, Postwala H, Balar P, Vora L (2023) Current status of cancer nanotheranostics: emerging strategies for cancer management. Nanotheranostics 7(4):368–379. https://doi.org/10.7150/ntno.82263

Chen WH, Xu XD, Jia HZ et al (2013) Therapeutic nanomedicine based on dual-intelligent functionalized gold nanoparticles for cancer imaging and therapy in vivo. Biomaterials 34(34):8798–8807

Chen H, Zhang W, Zhu G, Xie J, Chen X (2017a) Rethinking cancer nanotheranostics. Nat Rev Mater 2:17024

Chen Y, Wu Y, Sun B, Liu S, Liu H (2017b) Two-dimensional nanomaterials for cancer nanotheranostics. Small 13:1603446

Dai H, Wang X, Shao J, Wang W, Mou X, Dong XJS (2021) NIR-II organic nanotheranostics for precision oncotherapy. Small 17:e2102646

Daniel MC, Astruc D (2004) Gold nanoparticles: assembly, supramolecular chemistry, quantum-size-related properties, and applications toward biology, catalysis, and nanotechnology. Chem Rev 104(1):293–346

Deb N, Goris M, Trisler K, Fowler S, Saal J, Ning S, Becker M, Marquez C, Knox S (1996) Treatment of hormone-refractory prostate cancer with 90Y-CYT-356 monoclonal antibody. Clin Cancer Res 2:1289–1297

Deveza L, Choi J, Yang F (2012) Therapeutic angiogenesis for treating cardiovascular diseases. Theranostics 2(8):801–814

Ding H, Wu F (2012) Image guided biodistribution and pharmacokinetic studies of theranostics. Theranostics 2(11):1040–1053

Dreifuss T, Betzer O, Shilo M, Popovtzer A, Motiei M, Popovtzer R (2015) A challenge for theranostics: is the optimal particle for therapy also optimal for diagnostics? Nanoscale 7:15175–15184

Feng SS (2006) New-concept chemotherapy by nanoparticles of biodegradable polymers: where are we now? Nanomedicine (Lond) 1(3):297–309

Funkhouser J (2002) Reinventing pharma: the theranostic revolution. Curr Drug Discov 08:17–19

Gaspar R (2007) Regulatory issues surrounding nanomedicines: setting the scene for the next generation of nanopharmaceuticals. Nanomedicine 2:143–147

Ge C, Meng L, Xu L et al (2012) Acute pulmonary and moderate cardiovascular responses of spontaneously hypertensive rats after exposure to single-wall carbon nanotubes. Nanotoxicology 6(5):526–542

Goyal AK, Ramchandani M, Basak T (2023) Recent advancements, challenges, and future prospects in usage of nanoformulation as theranostics in inflammatory diseases. J Nanotheranostics 4(1):106–126. https://doi.org/10.3390/jnt4010006

Guo X, Wei X, Chen Z, Zhang XB, Yang G, Zhou SB (2020) Multifunctional nanoplatforms for subcellular delivery of drugs in cancer therapy. Prog Mater Sci 107:24

Hillier S, Merkin R, Maresca K, Zimmerman C, Barrett J, Tesson M, Eckelman W, Mairs R, Joyal J, Babich J et al (2011) [131I]MIP-1375, a small molecule prostate-specific membrane antigen (PSMA) inhibitor for targeted therapy of prostate cancer (PCa). J Nucl Med 52:361

Ho D, Sun X, Sun S (2011) Monodisperse magnetic nanoparticles for theranostic applications. Acc Chem Res 44(10):875–882

Hua S, de Matos MBC, Metselaar JM, Storm G (2018) Current trends and challenges in the clinical translation of nanoparticulate nanomedicines: pathways for translational development and commercialization. Front Pharmacol 9:790

Jahangir MA, Gilani SJ, Muheem A, Jafar M, Aslam M, Ansari MT, Barkat MA (2019) Quantum dots: next generation of smart nano-systems. Pharm Nanotechnol 7(3):234–245. https://doi.org/10.2174/2211738507666190429113906

Jain R (2013) Measurements of tumor vascular leakiness using DCE in brain tumors: clinical applications. NMR Biomed 26:1042–1049

Janib SM, Moses AS, MacKay JA (2010) Imaging and drug delivery using theranostic nanoparticles. Adv Drug Deliv Rev 62(11):1052–1063

Jo SD, Ku SH, Won Y-Y, Kim SH, Kwon IC (2016) Targeted nanotheranostics for future personalized medicine: recent progress in cancer therapy. Theranostics 6:1362–1377

Junttila MR, de Sauvage FJ (2013) Influence of tumour micro-environment heterogeneity on therapeutic response. Nature 501:346–354

Kalyane D, Raval N, Maheshwari R, Tambe V, Kalia K, Tekade RK (2019) Employment of enhanced permeability and retention effect (EPR): nanoparticle-based precision tools for targeting of therapeutic and diagnostic agent in cancer. Mater Sci Eng C Mater Biol Appl 98:1252–1276

Kashyap BK, Singh VV, Solanki MK, Kumar A, Ruokolainen J, Kesari KK (2023) Smart nanomaterials in cancer theranostics: challenges and opportunities. ACS Omega 8(16):14290–14320. https://doi.org/10.1021/acsomega.2c07840

Lammers T, Aime S, Hennink WE et al (2011) Theranostic nanomedicine. Acc Chem Res 44(10):1029–1038

Ma B, Xu H, Zhuang W, Wang Y, Li G, Wang Y (2020) Reactive oxygen species responsive theranostic nanoplatform for two-photon aggregation-induced emission imaging and therapy of acute and chronic inflammation. ACS Nano 14:5862–5873

Maresca KP, Hillier SM, Femia FJ, Keith D, Barone C, Joyal JL, Zimmerman CN, Kozikowski AP, Barrett JA, Eckelman WC et al (2009) A series of halogenated heterodimeric inhibitors of prostate specific membrane antigen (PSMA) as radiolabeled probes for targeting prostate cancer. J Med Chem 52:347–357

Mateo D, Morales P, Avalos A et al (2014) Oxidative stress contributes to gold nanoparticle-induced cytotoxicity in human tumor cells. Toxicol Mech Methods 24(3):161–172

McCarthy JR, Jaffer FA, Weissleder R (2006) A macrophage-targeted theranostic nanoparticle for biomedical applications. Small 2(8–9):983–987

McCarthy JR, Korngold E, Weissleder R, Jaffer FA (2010) A light-activated theranostic nanoagent for targeted macrophage ablation in inflammatory atherosclerosis. Small 6:2041–2049

Mei L, Zhang Z, Zhao L et al (2013) Pharmaceutical nanotechnology for oral delivery of anticancer drugs. Adv Drug Deliv Rev 65(6):880–890

Mi Y, Shao Z, Vang J, Kaidar-Person O, Wang AZ (2016) Application of nanotechnology to cancer radiotherapy. Cancer Nanotechnol 7:11

Mukherjee A, Paul M, Mukherjee S (2019) Recent progress in the theranostics application of nanomedicine in lung cancer. Cancers 11:597

Muthu MS, Feng SS (2013) Theranostic liposomes for cancer diagnosis and treatment: current development and pre-clinical success. Expert Opin Drug Deliv 10(2):151–155

Muthu MS, Singh S (2009) Targeted nanomedicines: effective treatment modalities for cancer, AIDS and brain disorders. Nanomedicine (Lond) 4(1):105–118

Muthu MS, Wilson B (2010) Multifunctional radionanomedicine: a novel nanoplatform for cancer imaging and therapy. Nanomedicine (Lond) 5(2):169–171

Muthu MS, Kulkarni SA, Raju A et al (2012) Theranostic liposomes of TPGS coating for targeted co-delivery of docetaxel and quantum dots. Biomaterials 33(12):3494–3501

Muthu MS, Leong DT, Mei L, Feng SS (2014) Nanotheranostics—application and further development of nanomedicine strategies for advanced theranostics. Theranostics 4(6):660–677. https://doi.org/10.7150/thno.8698

Ng KW, Khoo PKS, Heng BC et al (2011) Cellular DNA damage response to zinc oxide nanoparticles hinges on the tumor suppressor p53 pathway. Biomaterials 32(32):8218–8225

Nirmala MJ, Kizhuveetil U, Johnson A, Balaji G, Nagarajan R, Muthuvijayan V (2023) Cancer nanomedicine: a review of nano-therapeutics and challenges ahead. RSC Adv 13(13):8606–8629. https://doi.org/10.1039/d2ra07863e

Ojha AK, Rajasekaran R, Pandey AK, Dutta A, Seesala VS, Das SK et al (2022) Nanotheranostics: nanoparticles applications, perspectives, and challenges. In: BioSensing, theranostics, and medical devices. Springer, pp 345–376

Pan J, Liu Y, Feng SS (2010) Multifunctional nanoparticles of biodegradable copolymer blend for cancer diagnosis and treatment. Nanomedicine (Lond) 5(3):347–360

Qin Y, Cui J, Zhang Y, Wang Y, Zhang X, Zheng H, Shu X, Fu B, Wu Y (2016) Integration of microfluidic injection analysis with carbon nanomaterials/gold nanowire arrays-based biosensors for glucose detection. Sci Bull 61:473–480

Roma-Rodrigues C, Pombo I, Raposo L, Pedrosa P, Fernandes AR, Baptista PV (2019) Nanotheranostics targeting the tumor microenvironment. Front Bioeng Biotechnol 7:197

Santra S (2012) The potential clinical impact of quantum dots. Nanomedicine (Lond) 7(5):623–626

Seidlin SM, Marinelli LD, Oshry E (1946) Radioactive iodine therapy. J Am Med Assoc 132:838–847

Setyawati MI, Tay CY, Leong DT (2013) Effect of zinc oxide nanomaterials-induced oxidative stress on the p53 pathway. Biomaterials 34(38):10133–10142

Shuhendler AJ, Prasad P, Leung M et al (2012) A novel solid lipid nanoparticle formulation for active targeting to tumor α(v) β(3) integrin receptors reveals cyclic RGD as a double-edged sword. Adv Healthc Mater 1(5):600–608

Smith BA, Smith BD (2012) Biomarkers and molecular probes for cell death imaging and targeted therapeutics. Bioconjug Chem 23(10):1989–2006

Sneider A, VanDyke D, Paliwal S, Rai P (2017) Remotely triggered nano-theranostics for cancer applications. Nanotheranostics 1:1–22

Sumer B, Gao J (2008) Theranostic nanomedicine for cancer. Nanomedicine (Lond) 3(2):137–140

Tan YF, Chandrasekharan P, Maity D et al (2011) Multimodal tumor imaging by iron oxides and quantum dots formulated in poly(lactic acid)-D-alpha-tocopheryl polyethylene glycol 1000 succinate nanoparticles. Biomaterials 32(11):2969–2978

Taratula O, Schumann C, Naleway MA et al (2013) A multifunctional theranostic platform based on phthalocyanine-loaded dendrimer for image guided drug delivery and photodynamic therapy. Mol Pharm 10(10):3946–3958
Tay CY, Cai PQ, Setyawati MI et al (2014) Nanoparticles strengthen intracellular tension and retard cellular migration. Nano Lett 14(1):83–88
Tinkle S, Mcneil SE, Mühlebach S, Bawa R, Borchard G, Barenholz YC, Tamarkin L, Desai N (2014) Nanomedicines: addressing the scientific and regulatory gap. Ann N Y Acad Sci 1313:35–56
Truffi M, Fiandra L, Sorrentino L, Monieri M, Corsi F, Mazzucchelli S (2016) Ferritin nanocages: a biological platform for drug delivery, imaging and theranostics in cancer. Pharmacol Res 107:57–65
Valent P, Groner B, Schumacher U, Superti-Furga G, Busslinger M, Kralovics R et al (2016) Paul Ehrlich (1854–1915) and His Contributions to the Foundation Birth of Translational Medicine. J Innate Immun 8(2):111–120
Velikyan I (2012) Molecular imaging and radiotherapy: theranostics for personalized patient management. Theranostics 2(5):424–426
Warner S (2004) Diagnostics + therapy = theranostics. Scientist 18:38–39
Win KY, Feng SS (2006) In vitro and in vivo studies on vitamin E TPGS-emulsified poly(D, L-lactic-*co*-glycolic acid) nanoparticles for paclitaxel formulation. Biomaterials 27(10):2285–2291
Wu X, Tian F, Zhao JX et al (2013) Evaluating pharmacokinetics and toxicity of luminescent quantum dots. Expert Opin Drug Metab Toxicol 9(10):1265–1277
Xie J, Lee S, Chen X (2010) Nanoparticle-based theranostic agents. Adv Drug Deliv Rev 62:1064–1079
Yang ST, Luo J, Zhou Q et al (2012) Pharmacokinetics, metabolism and toxicity of carbon nanotubes for biomedical purposes. Theranostics 2(3):271–282
Ye Y, Chen X (2011) Integrin targeting for tumor optical imaging. Theranostics 1:102–126
Yong KT, Wang Y, Roy I et al (2012) Preparation of quantum dot/drug nanoparticle formulations for traceable targeted delivery and therapy. Theranostics 2(7):681–694
Yu MK, Park J, Jon S (2012) Targeting strategies for multifunctional nanoparticles in cancer imaging and therapy. Theranostics 2(1):3–44
Yuan J, Zhang H, Kaur H et al (2013) Synthesis and characterization of theranostic poly(HPMA)-c(RGDyK)-DOTA-64Cu copolymer targeting tumor angiogenesis: tumor localization visualized by positron emission tomography. Mol Imaging 12(3):1–10
Yukawa H, Sato K, Baba Y (2023) Theranostics applications of quantum dots in regenerative medicine, cancer medicine, and infectious diseases. Adv Drug Deliv Rev 200:114863. https://doi.org/10.1016/j.addr.2023.114863
Zhao L, Feng SS (2010) Enhanced oral bioavailability of paclitaxel formulated in vitamin E-TPGS emulsified nanoparticles of biodegradable polymers: in vitro and in vivo studies. J Pharm Sci 99(8):3552–3560
Zhao J, Mi Y, Feng SS (2013) siRNA based nanomedicine. Nanomedicine (Lond) 8(6):859–862
Zhu Y, Hong H, Xu ZP et al (2013) Quantum dot-based nanoprobes for in vivo targeted imaging. Curr Mol Med 13(10):1549–1567

Nanotheranostics: A Treasured Tailor for Disease Diagnosis and Treatment

2

Neha Gupta, Nishtha Chaurawal, Fahad Saad Alhodieb, Md. Abul Barkat, Abdulkareem Ali Alanezi, Simran Preet, and Kaisar Raza

Abstract

Theranostics, a patient-centered approach to pharmacotherapy, is emerging as a target-centered, safe, and effective method. It involves both treatments and diagnostics. A relatively new area of medicine called "theranostics" combines focused therapy with targeted diagnostic testing. It offers a transition from traditional medicine to the more modern customized and precision medicine, emphasizing patient-centered treatment. It addresses personalized care based on each person's individuality, resulting in the appropriate medications for the patient at the proper time. Theranostics offer a successful treatment procedure that is customized and cost-effective. The three main components that form the basis of theranostics are biomarker profiling, proteomics, and pharmacogenetics. Theranostics is an intriguing field understood on many levels, with oncology receiving particular attention due to the significant roles played by carbon nano-

N. Gupta · N. Chaurawal · K. Raza (✉)
Department of Pharmacy, School of Chemical Sciences and Pharmacy, Central University of Rajasthan, Ajmer, Rajasthan, India
e-mail: drkaisar@curaj.ac.in

F. S. Alhodieb
Department of Clinical Nutrition, College of Applied Health Sciences in Ar Rass, Qassim University, Ar Rass, Saudi Arabia
e-mail: f.alhodieb@qu.edu.sa

M. A. Barkat · A. A. Alanezi
Department of Pharmaceutics, College of Pharmacy, University of Hafr Al-Batin, Al Jamiah, Hafr Al Batin, Saudi Arabia
e-mail: mdabarkat@uhb.edu.sa; aalanezi@uhb.edu.sa

S. Preet
Department of Biophysics, Punjab University, Chandigarh, UT, India
e-mail: simranpreet@pu.ac.in

M. A. Barkat et al. (eds.), *Nanotheranostics for Diagnosis and Therapy*,
https://doi.org/10.1007/978-981-97-3115-2_2

tubes, metallic nanoparticles, nanoparticles, and quantum dots in monitoring the effectiveness of therapy and treatment. To detect and cure diseases at the cellular and molecular level, theranostics medicine can attain systemic circulation, show host defenses, and deliver the medication and diagnostic agents to the targeted place. Thus, theranostics is a comprehensive shift in medicine from trial and error to customized and predictive medicine that improves the standard of pharmacological treatment.

Keywords

Diagnosis · Pharmacotherapy · Personalized medicine · Theranostics · Pharmacogenetics · Biomarker profiling · Carbon nanotubes · Nuclear medicine · Therapy

2.1 Introduction

Theranostics, the focus of this special issue of Molecular Pharmaceutics, where we go into the use of functional nanoscale agents in customized treatment (Shin et al. 2013). Using non-invasive techniques to visualize cellular (Kunjachan et al. 2015) and subcellular processes is known as molecular imaging (MacRitchie et al. 2020). In the emerging field of personalized medicine, molecular imaging is the observation, characterization, and quantification of biological processes at the cellular, subcellular, or even molecular level in living subjects that have grown significantly in importance. Utilizing molecular imaging probes in conjunction with conventional diagnostic imaging methods, molecular imaging assesses the expression of molecular markers suggestive of various disease stages (Chen and Chen 2010). The discipline has grown significantly in the last 20 years and offers a lot of promise for translational research. Working across disciplines to develop clinical imaging techniques for the advanced, perceptive detection, diagnosis, and treatment of disease at the molecular and cellular level with uncompromised specificity has been made possible by significant ascents in engineering, chemistry, molecular biology, and genetics (Pottoo et al. 2021; Rahman et al. 2023; Otsuka 2019; Friebe and Illanes 2023). The construction of a specified nanoscale structure for dual purposes, i.e., imaging and therapy, has been made possible by nanoparticles' unparalleled capac ity for drug administration and detection (theranostics). Theranostics applications are made possible by the nanoplatform's ability to include homing agents, contrast material, and medicines, overcoming biological and biophysical barriers. Dr. Snoke Svenson of Drug Delivery Solution LLC discusses his extensive expertise in introducing nanotechnology to the market in this viewpoint. He notes that more than 40 nanoformulations are under investigation at various stages of clinical development. What is interesting is that most of them use passive targeting strategies. The possible use of dendrimers in detecting and managing prostate cancer is covered by Dr. Xiankai Sun of the University of Texas Southwestern Medical Center. Although Sun and his associates' lab study focuses primarily on nuclear medicine, it also discusses pertinent chemical approaches for creating and synthesizing dendritic nanostructures. They also provide excellent examples of various targeting strategies (e.g., type 2 Transmembrane glycoprotein PSMA, integrin α, β3), diagnostic techniques (MRI, CT, PET, SPECT, optical, etc.), and treatments (chemotherapeutics, gene therapy, immunotherapy, radiation, etc.). In 2002, Funkhouser coined the term "theranostics," which describes any "material that combines the modalities of therapy and diagnostic imaging" into a single package or therapy, etc. (Kelkar and Reineke 2011).

2.1.1 Theranostics

The word "theranostics," which was first used by Funkhouser in 2002, refers to any "material that combines the modalities of therapy and diagnostic imaging" into a single package (Luk and Zhang 2014). The word "theranostics" was coined to define "the efforts to develop more specific, individualized therapies for disease, and to combine into a single system for patients," and the term refers to the simultaneous integration of therapy and diagnosis (Kevadiya et al. 2019). The goal of nanotheranostics is to improve the theranostics effects by applying and developing different types of nanocarriers, such as micelles, liposomes, metal and inorganic nanoparticles, carbon nanotubes, polymer conjugation, and nanoparticles of biodegradable polymers, for the sustained, controlled, and targeted co-delivery of therapeutic and diagnostic agents purpose is to diagnose and treat the disease at their earliest stage when the disease is most likely to curable or at least treatable (Hosseini et al. 2023).

Patient-centered care has always been the main focus and foundation for treating patients with drugs. However, the shift from universal medicine, or the "one medicine fits all" approach, to personalized medicine, which incorporates a particular method for individualized drug therapy based on individual differences in drug response, has emerged as a revolution (Elhag et al. 2022). This relates to the investigation of an incredible platform known as theranostics, which is a close relationship between diagnosis and therapeutics and may offer treatment protocols that are more tailored to each patient, increasing the likelihood of improved prognoses and bringing diagnosis and therapy closer together (Giammarile 2023). The goal is to provide safer and more targeted, effective pharmacotherapy by providing each patient with the "right" medication at the "right" dosage through the use of theranostics in medicine (Könik et al. 2021).

2.1.2 Nanotheranostics

The term "theranostics nanomedicine" addresses colloidal nanoparticles with a size range between 10 and 1000 nm (1 μm) (Bhattacharya et al. 2023). Nanotheranostics (NTs) comprise macromolecular materials or polymers in which therapeutic and diagnostic chemicals are concurrently absorbed, conjugated, entrapped, and encapsulated for cellular and molecular diagnosis and therapy (Fig. 2.1). Even for life-threatening illnesses like cancer, heart disease, diabetes and AIDS, it has promised for improving prognosis and making treatment considerably less complicated (Shrivastava et al. 2019). This will save money and improve patients' quality of life. Compared to other, theranostics nanomedicine can work better because of its advanced multiple capabilities, like drug targeted delivery, controlled release, promising transport capability like endocytosis, release of stimulant responsive agents (innovative delivery), synergistic performance (e.g., siRNA co-delivery, combination therapy), multimodality diagnosis and treatments, and quality performances (e.g., oral delivery, escape from multidrug resistance (MDR) protein, autophagy inhibition, etc.) (Muthu et al. 2014).

In nanomedicine, encapsulating a single medicinal or diagnostic agent could not have high sensitivity, specificity, or effectiveness for the intended uses. Thus, by co-encapsulating many therapeutic and diagnostic modalities in targeted nanomedicine platforms, multimodality nanotheranostics may be produced to leverage the benefits (Siafaka et al. 2021). It is interesting to note that siRNA can be used in theranostic resistance inhibition in nanomedicine for theranostics. Theranostic

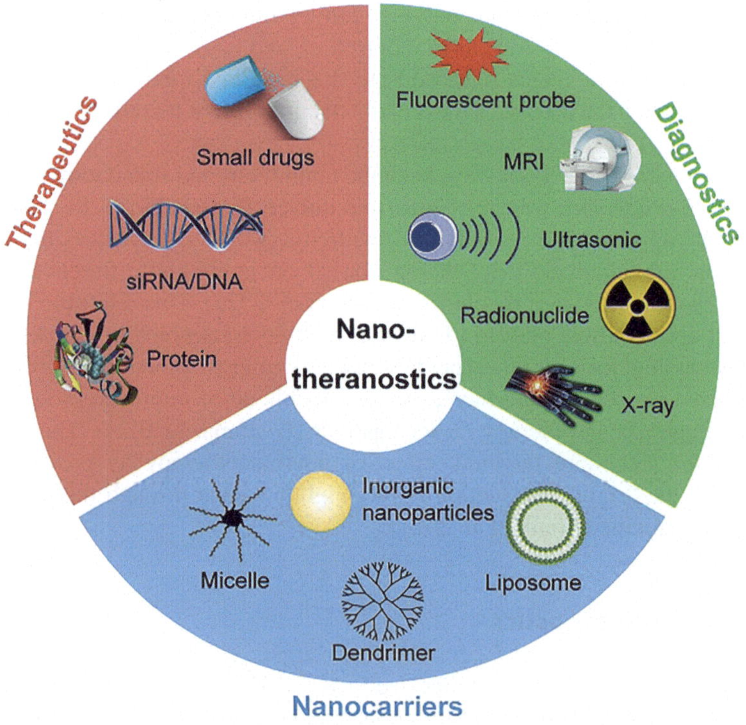

Fig. 2.1 Nanostructured-based theranostics system. [Reproduced under Creative Common Attribution- Non-commercial 4.0 International with Ivyspring International Publisher (Wang and Cui 2016)]

nanomedicine based on siRNA has significantly improved multimodality treatment and diagnostics (Chen et al. 2020).

Nanomedicine has also been created for oral chemotherapy, a unique approach. Our adumbrated D-α-tocopheryl polyethene glycol 1000 succinate (TPGS) based nanomedicine offers improved oral bioavailability of chemotherapy drugs like docetaxel and paclitaxel (Muthu et al. 2014).

2.2 Nanotheranostics Components

Mainly, nanotheranostics have three components:

1. Nanocarriers
   (a) Organic nanocarriers
   (b) Inorganic nanocarriers
2. Therapeutic agents
3. Diagnostic agents

2.2.1 Nanocarriers

Among all human ailments, cancer is one of the most concerning ones. In 2012, 8.2 million people died from this illness, and 14 million new cases were reported, according to the WHO World Cancer Report 2014 (Sung et al. 2021). It is anticipated that the number of cancer cases annually will reach 22 million during the next 20 years (Sathishkumar et al. 2023). Immunotherapy, chemotherapy, radiation therapy, and surgery are employed as cancer treatment modalities (Anand et al. 2023). To counteract the high rates of illness and mortality, however, effective new therapies are desperately needed (Lawn et al. 2010). Conventional chemotherapy has several drawbacks, such as excessive toxicity and insufficient cytostatic medication bio-distribution and pharmacokinetic profile (Ioele et al. 2022). Using nanomaterials to combine therapeutic and diagnostic chemicals into multifunctional nanoplatforms, biomedical substances can be transported and delivered across biological systems for the cure, treatment, and diagnosis, i.e., theranostics of cancer (Dessale et al. 2022).

2.2.1.1 Organic Nanocarriers

2.2.1.1.1 Polymer-Based Nanocarriers for Cancer Therapy

Diverse classes of nanomaterials can be used for cancer therapy as a theranostics system in nanoscale patterns. There are various types of nanocarriers, such as peptide NP, solid lipid NP, dendrimers, inorganic NP, carbon NP and nanotubes, nanodiamonds, nanoemulsions, etc. (Lôbo et al. 2021). However, the clinically available nanomaterials for human use are mostly liposomes and polymer-based nanoformulations. In reality, the first cancer medications that came into existence based on nanotechnology were pegylated liposomes containing the drug doxorubicin (Doxil) (Tenchov et al. 2021). These liposomes were approved for Kaposi's sarcoma cancer treatment of AIDS by the Food and Drug Administration (FDA) in 1995 and by the European Medicines Agency (EMA) in 1997 (where they are currently also indicated for the treatment of multiple myeloma, ovarian cancer, and metastatic breast cancer) (Fulton and Najahi-Missaoui 2023).

2.2.1.1.2 Polymeric Micelles

Polymeric micelles (PM) show considerable promise in improving the dosage range between a minimum effective therapeutic concentration and the minimum toxic concentration (therapeutic window) of lipophilic anticancer medicines like taxanes or platinates by serving as effective carriers for the regulated administration of poorly water-soluble medications (Xu et al. 2013). PM are nanoscale supramolecular structures with a 5–100 nm mean diameter. They comprise amphiphilic block copolymers and self-assemble in an aqueous medium at a polymer concentration called critical micelle concentration (CMC) (Cagel et al. 2017).

PM comprises a core-shell architecture, where the hydrophilic segment creates a coronal layer and a semi-solid core produced by a hydrophobic block of the copolymer. Although substitutes are being investigated, poly(ethylene glycol) (PEG) is the most common hydrophilic block in the copolymer structure when it comes to the polymers used to create PM (Barros et al. 2016). Because it prolongs the duration of blood circulation by preventing the mononuclear phagocyte system from identifying the carrier as a foreign substance, this polymer is very commonly employed in fabricating nanosystems (Fam et al. 2020).

On the other hand, a variety of polymers are utilized to produce the micellar core, with the ones with a more extended development history being poly(ethers), poly(esters), poly(amino acid), and *N*-(2-hydroxypropyl) meth acrylamide (HPMA) (Luque-Michel et al. 2017).

2.2.1.1.3 Polymer-Drug Conjugate

A group of nanoscale substances collectively referred to as polymer therapies have the characteristic of not having the bioactive ingredient encapsulated but rather attached to a polymeric water-soluble biocompatible carrier (Sairam et al. 2023). Ringsdorf originally put up a paradigm for macromolecular pro-drugs in the middle of the 1970s (Yang and Kopeček 2014). The pro-drug exhibits unique characteristics dictated by macromolecules and shows up in the drug-polymer conjugate's pharmacokinetic behavior (Larson and Ghandehari 2012).

Polymeric prodrugs, also known as polymer-drug conjugates, are widely recognized as a rapidly expanding and efficient method for enhancing the therapeutic use of pharmaceuticals (Alven et al. 2020). Drugs conjugated with polymers often have longer half-lives, are more stable, soluble in water, have reduced immunogenicity and antigenicity, and can target particular tissues or cells (Pasut and Veronese 2007).

In polymeric prodrugs and macromolecular prodrugs, polymers serve as carriers for the delivery of proteins, drugs, targeting moieties, and imaging agents (Khandare and Minko 2006). The effectiveness of several medications on the market for treating various diseases has already demonstrated the potential of polymer-drug conjugates (Junyaprasert and Thummarati 2023). Drug delivery methods known as polymeric pro-drugs are thought to work by releasing therapeutic drug molecules in lower doses from a polymer chain molecule over an extended period (Patil 2019). One example is PK1, the first PDC to move on to clinical trials in 1994. PK1 is a glycyl-phenylalanyl-leucyl-glycine linker-covalently attached HPMA copolymer to doxorubicin. Two phase II clinical studies are being studied to treat colorectal, lung, and breast cancer (Vasey et al. 1999).

2.2.1.1.4 Polymeric Nanoparticles

Polymeric nanoparticles (PNP) colloidal systems, which are submicron-sized and significantly more extensive than PM (50–500 nm), have revealed efficacy as effective carriers for the extended and continuous release of anticancer medications (Dristant et al. 2023). Several biocompatible polymers can be used to create these carriers. The kind of polymer employed may readily control the release of anticancer medications (Begines et al. 2020). PNP is often generated in two ways: either from pre-synthesized/existing polymer (e.g., by nanoprecipitation, emulsification/solvent evaporation, etc.) or from initial monomers that are polymerized (e.g., via emulsion polymerization) (Pulingam et al. 2022). The anticancer agent can be conjugated to the surface or core of the particle in polymeric nanocarriers, such as matrix systems where the anticancer agent is in a cavity surrounded by the polymer, i.e., called nanocapsule system dissolved or dispersed (nanospheres) or reservoir systems (Avramović et al. 2020).

When it comes to passive targeting, NP made of the biodegradable polymer poly(alkyl cyanoacrylate) (PACA) have been widely employed in drug delivery because of their capacity to encapsulate tiny hydrophobic molecules and increase the oral bioavailability of small molecular weight molecules of drugs because they cross cell membrane readily and absorbs into systemic circulation so increase the oral bioavailability. In reality, the doxorubicin trans drug is made by emulsion polymerization with anionic surfactants and is comprised of doxorubicin-loaded PIHCA (poly(isohexyl cyanoacrylate)) nanosphere formulation. It is currently the most sophisticated PNP under clinical investigation, but only for one indication.

Poly(esters) is another type of polymer that has also been utilized to manufacture PNP. This is the case study with Docetaxel-PNP, a formulation that consists of the medication docetaxel together with a combination of monovalent metal salts of PLA and amphiphilic diblock copolymers of monomethoxy PEG-PLA (Luque-Michel et al. 2017).

2.2.1.1.5 Chitosan Particles

The biodegradable cationic polymer known as chitosan (CS) is produced when chitin undergoes a partial deacetylation. Many studies have been done on using chitosan nanoparticles for therapeutic medication delivery in liver cancer. Ye et al. (2018) evaluation of the impact of a doxorubicin-loaded, chitosan-coated nanocarrier on liver cancer revealed remarkable suppressions of the disease cell development (Ye et al. 2018). To assess the HepG2 human liver cancer cell model in vitro, Loutfy et al. (2016) produced chitosan nanoparticles or CS-NPs. They found that CS-NPs had a relatively cytotoxic solid impact on liver cancer cells and proposed that CS-NPs would be an appropriate medication delivery option for liver cancer (Loutfy et al. 2016).

2.2.1.1.6 Dendrimers

One significant benefit of dendrimer-based drug delivery systems for anticancer treatment is their versatility. Dendrimers are a particular type of nanocarrier for drug delivery applications because of their physicochemical characteristics, which include their size and structure, aqueous solubility, monodispersity, and strong drug delivery capacity. Additionally, dendrimer possesses effective renal filtration capabilities (Kinberger et al. 2002).

The scientists created a multifunctional dendrimer-based NC coated with magnetite (Fe_3O_4) and polydopamine (PDA). According to reports, the manufactured nanocarriers are harmless and may efficiently carry chemotherapeutic and photothermal drugs to treat liver cancer cells at low nanoparticle concentrations (Jędrzak et al. 2019). According to Sharma et al. (2017), dendrimers' superior water solubility is caused by water-soluble functional groups on their surface. Drugs used to treat liver cancer are hydrophobic; this property is encapsulated in dendrimers' inner core and branches (Sharma et al. 2017).

2.2.1.1.7 Organic Nanofiber

Drug delivery for liver cancer has been achieved using organic nanofibers as vectors. The chemotherapeutic drug-containing nanofiber inhibits the liver cancer cells ability to metastasize and produce tumors.

An organic nanofiber that included the therapeutic gene miRNA-145 and the anticancer medication paclitaxel to observe how the transport of these two substances affected liver cancer. Research indicates that halting the growth of a liver tumor also stops it from spreading. In a mouse model Che et al. (2015) and Ji et al. (2018) developed doxorubicin-conjugated nanofibers that can act as a sustained release medication to stop tumor development in an SMMC7721 cell line (Ji et al. 2018). A doxorubicin-loaded nanofiber as a potential chemotherapeutic approach for treating hepatic cancer.

The resultant nanofiber had vigorous anti-tumor activity (Ruman et al. 2020).

2.2.1.2 Inorganic Nanoparticles

It is challenging to treat cancer with conventional medications, particularly multidrug-resistant cancers, due to several important issues, including poor pharmacokinetics and biodistributions, poor tumor selectivity, and the inability to pierce the extremely thick extracellular matrix.

Nanocarriers, with their distinct physical characteristics and biological capabilities, provide intriguing platforms to overcome MDR. Unlike traditional medications that diffuse broadly without targeting tropism, nanocarriers can preferentially accumulate in tumor tissues through passive or active targeting mechanisms. A range of nanocarriers, including lipid nanoparticles, polymer conjugates, liposomes, dendrimers, micelles, and inert nanoparticle carriers, as well as inorganic nanocarriers like silica nanoparticles, carbon-based nanocarriers, iron oxide nanoparticles, quantum dots (QDs), and gold nanoparticles, have shown unique advantages for drug delivery. These benefits include a high surface-to-volume ratio, which means when the surface area is more significant than volume, there is more plasma membrane relative to the inside of the cell, easily malleable size and shape, the ability to image the molecules, tunable pore structures, and superficial surface modification.

The "Trojan horse" method of drug delivery refers to the possibility that inorganic nanocarriers possess a stimuli-responsive release function that would selectively release the medicament in tumor microenvironments in response to external triggers/stimuli (e.g., light, magnetic fields, and hyperthermia) to prevent drug efflux transporters (in case of cancer mostly p-gp) from recognizing anticancer drugs. In the meantime, because inorganic nanocarriers are easily surface-modified, anticancer medicines and therapeutic genes (such as DNA or siRNA) might be loaded to generate synergistic therapeutic effects (Lin et al. 2016). Various types of inorganic nanocarriers are as follows:

2.2.1.2.1 Iron Oxide Nanoparticles

Iron oxide nanoparticles have significantly been studied for various clinical applications, such as CAs for MRIs, hyperthermia treatment, medication delivery, biosensing, and protein separation. These nanoparticles have exceptional biocompatibility and distinctive physical/surface characteristics. Specifically, the T2 relaxation period in tissues may be shortened by superparamagnetic iron oxide nanoparticles (SPIOs); ten varieties of SPIOs (superparamagnetic Iron oxide) have been authorized as T2 types of CAs for clinical MRI. Moreover, iron oxide nanoparticles can enhance drug internalization with cancer cells, allowing bioactive compounds to enter the cytoplasm through the endocytosis route without being recognized by pump transporters on the surface of multidrug-resistant cancer cells.

Doxorubicin was attached to polyethyleneimine (PEI) using pH-sensitive linkers, and the Doxorubicin-PEI combination was then deposited onto the surface of the iron oxide nanoparticle. The Doxorubicin could manage to get away. Doxorubicin -loaded nanoparticles can accomplish endosomal escape by using PEI's "proton sponge effect" to disrupt specific ABC transporters of glioma cells. Rat glioma C6 drug-resistant cells (C6-ADR) exhibit more significant intracellular drug accumulation through endosomes and ultimately reach the cytoplasm than free Doxorubicin (Lin et al. 2016).

2.2.1.3 Quantum Dots

Quantum dots, or QDs, are nanoscale semiconductor crystals with adjustable diameters ranging from 2 to 10 nm. High quantum yield, size-tunable light emission, strong chemical stability, and photostability are some of their distinctive optical qualities that mitigate the drawbacks of using organic dyes or other imaging modalities. Additionally, they can emit light in visible and near-infrared (NIR) wavelengths, which are advantageous for various biological applications. Regretfully, toxicity associated with QDs—particularly those based on Cd is a significant issue for biological applications. Because of this, it was essential to create Cd-free QDs, such as ZnO QDs, which have strong photoluminescence (PL) properties that are ideal for biological imaging while still being nontoxic and biodegradable (Abdelgalil et al. 2023).

2.2.1.4 Gold Nanoparticles

The colloidal gold suspension has a particle size of 1–100 nm, and its color changes from wine red to purple depending on the particle size—the nanometer-sized gold exhibits different properties from bulk gold. Gold nanoparticles have optical and electrochemical properties that strongly depend on their size, shape, interparticle distance, and surface chemistry. Other synthesis methods were developed for the preparation of various-sized AuNPs. The variation in particle size depends on the synthesis parameters like reaction temperature, pH, precursor concentration, and the ratio of precursor to reducing agent. The various sizes of AuNPs influence the Surface Plasmon Resonance (SPR) frequency. The high surface energy of gold causes the AuNPs to precipitate in fluids. By stabilizing the AuNPs and lowering the surface energy of gold, functionalizing molecules like polymers aid in modifying the surface of AuNPs and inhibit this self-assembly. Their surface chemistry is essential for using gold nanoparticles in biomedical applications (Patil et al. 2023).

2.2.1.5 Carbon-Based Nanocarriers

Since, carbon-based nanomaterials (CBNs) have unique structural dimensions and physicochemical features, they have collected much attention in various fields. Carbon nanotubes, graphene, mesoporous carbon, nanodiamonds, and fullerenes are the structural subtypes of CBNs. Higher drug-loading capacity, increased biocompatibility, and decreased immunogenicity have all been proven by these materials, which exhibit great optical activity and a significant multifunctional surface area. A great deal of research has been done on the creation of biocompatible scaffolds and nanomedicines using functionalized CBNs. Chemically altering CBNs to specific moieties (such as functional groups, compounds, and polymers) that have been identified as appropriate for use in biological applications. Due to their superior supramolecular π–π stacking, high adsorption capacity, and photothermal conversion ability, most of the materials under CBNs have been thoroughly investigated for cancer therapy (Debnath and Srivastava 2021). In mice with integrin $\alpha v\beta 3$-positive U87 human glioblastoma tumors, antibody-conjugated SWNTS produced a comparable strong photoacoustic contrast in vivo (Hahn et al. 2011). The Zhang group reported synthesizing CQDs that emitted either a green or red area by simply substituting glycerol for ethylene glycol as the synthetic solvent.

1,2,4-Triaminobenzene and urea were solvothermally treated in various solvents by Sahu and colleagues to generate CQDs that emitted distinct colors in the visible spectrum (Thi et al. 2023).

2.2.1.6 Mesoporous Silica Nanoparticles

In recent decades, chemists have been interested in nanomedicine using nanoparticles as a potential technique in the biomedical realm. In particular, since their initial synthesis in the early 1990s (Yanagisawa et al. 1990) and their first application in medication distribution, as documented by Vallet-Regi et al. (2001), mesoporous silica nanoparticles (MSNs) have drawn interest (Vallet-Regi et al. 2001).

Due to MSNs' unique qualities, which include their ability to contain medicinal substances in a small area, several organizations have created drug delivery systems based on MSNs. In addition, MSNs exhibit notable properties such as high silanol density, large pore volumes, tunable and narrow pore size distributions, and customizable porous architectures that provide facile dual-functionalization (inside and exterior). Several materials comprising molecular gates—gatekeepers or nano valves—sensitive to various stimuli have been developed due to additional reports of the first gated MSNs in the 2000s and the potential to create porous MSNs operating in aqueous solution. These molecular gates are grafted onto the MSN's exterior, obstructing cargo transport from the pores to the solution. Cargo is then released automatically in response to a variety of chemical, biological, or physical stimuli, which typically causes the molecular gate to separate and allow cargo to be released. Functionalized MSNs are effective, promising drug delivery carriers for targeted drug delivery, biologically compatible, enhance drug transportation across cell membranes and control release drug delivery. Numerous biomedical applications, notably tissue engineering, photothermal medication, photodynamic therapy (PDT), radiation treatment, treatment with ultrasound, antibiotic applications, and drug encapsulation and delivery for in vivo biological imaging, have demonstrated the immense potential of gated MSNs. Additionally, gated materials have been applied in communication and sensing applications. In vivo, models have shown the practical usage of gated MSNs for drug administration, improving the encapsulated medications' solubility and decreasing undesirable side effects (Lérida-Viso et al. 2023). Some specific theranostic nanomedicine system being developed for the simultaneous integration of treatment and diagnosis these are summarized in Table 2.1 (Muthu et al. 2014).

2.2.2 Therapeutic Agents

2.2.2.1 Role of Genetic Information

The human genome project, which focuses on genetic differences and predicted risk for many prevalent diseases, has supplied genetic information crucial to theranostics. Curiously, this has resulted in the expansion of disciplines like proteomics and pharmacogenetics.

2.2.2.1.1 Pharmacogenetics

Genotyping data are utilized to focus on optimum therapy in pharmacogenetics. This traditional therapeutics application studies inter-individual variation in DNA patterns associated with drug response, its release, absorption, and other definite pharmacokinetic data. Researchers have since broadened it into the field of pharmacogenomics (PGx), a name that was developed to encompass studies including

Table 2.1 Summary of specific advanced theranostic nanocarrier systems for disease diagnosis and their treatment

Theranostic nanomedicine	Types of material	Therapeutic agent	Diagnostic agent	Size	Targeting agent	Advancement	References
Drug polymer conjugates	HPMA [biodegradable *N*-(2-hydroxypropyl meth acrylamide)]	64cu (Copper-64)	64cu (Copper-64)	NA	RGD-peptide Drug conjugate	Malignant tumor imaging and chemotherapy through radiation	Taratula et al. (2013)
Polymeric nanoparticles	PLA-TPGS (polylactic acid–tocopheryl polyethylene glycol succinate)	Docetaxel (DTX)	Quantum dots (QDs)	~250 nm	Folic acid (vitamin B9)	Co-transportation of docetaxel and quantum dots	Pan et al. (2010)
Solid lipid nanoparticles (LNs)	Low-density lipoprotein (LDL), cholesterol	Paclitaxel/siRNA	Quantum dots (QDs)	~130 nm	cRGD-peptide	Multimodal therapy	Shuhendler et al. (2012)
Gold nanoparticles	Gold nanoparticles	Doxorubicin (DOX)	Gold nanoparticle	~55 nm	CPLGLAG G-peptide	Stimuli-responsive drug release	Chen et al. (2013)
Dendrimers	Polypropylenimine (PPE)	Phthalocyanines	Phthalocyanines	~62 nm	LHRH (Luteinizing hormone-releasing hormone)	Delivery of a single theranostic agent	Priester and ten Hagen (2023)
Liposomes	TPGS, phospholipid, cholesterol	DTX	Quantum dots (QDs)	~210 nm	Folic acid	Co-administration of docetaxel and quantum dots	Muthu et al. (2012)
Micelles	TPGS (tocopheryl polyethylene glycol succinate)	Iron oxide nanoparticles	Iron oxide nanoparticles	~178 nm	Passive targeting	Transportation of a single therapeutic agent	Chandrasekharan et al. (2011)

Theranostic nanomedicine	Types of material	Therapeutic agent	Diagnostic agent	Size	Targeting agent	Advancement	References
Carbon nanomaterials	SWCNTs (single-walled carbon nanotubes)	Intrinsic property	Intrinsic property	Length ~140 nm	Passive targeting	Itself photoluminescent and photothermal property	Robinson et al. (2010)

DNA changes, RNA, proteins, and any physiological process that may have therapeutic implications. Pharmacogenomics, or pharmacogenetics, studies how a person's genetic makeup influences medication reactions.

The word "pharmacogenomics" comes from the words "pharmacogenetics" (PGx), which is a fast-emerging discipline. Even before the first human genome was sequenced, in 2001, the first PGx dosage guidelines for antidepressant and psychiatric drugs were published. Comprehensive dosage recommendations and genotyping methodologies are now required due to the growing body of research and the desire to use PGx in clinical practice. The Dutch Pharmacogenetics Working Group (DPWG) was established in 2005 to create evidence-based PGx recommendations. 2011 saw funding for the Clinical Pharmacogenomics Implementation Consortium (CPIC). Currently, 21 genes and over 50 medications are covered by the PGx dosage guidelines published by the DPWG and CPIC together. "Pharmacology" (the study of the uses and the effect of medication) and "genomics" (the study of genes and their functions). Pharmacogenomics is the part of precision medicine.

2.2.2.1.2 Pharmacogenomics in the Drug Development Process

With the completion of the human genome project, technological advancements will soon make drug design, personalized medication therapy, genotype-based therapy, and illness prevention a reality. The field of pharmacogenomics has several applications in finding and developing new drugs, as well as in the titration of medication therapy according to a patient's genotype. Genetic polymorphism affects how much and to what degree genes are expressed, which can lead to individual differences in how drugs react to them and how they behave inside the body. These changes can also change phenotypic traits, making a person more susceptible to illness.

Two areas are listed below where pharmacogenomics significance and influence in drug development can be examined:

1. Structural pharmacogenomics
2. Functional pharmacogenomics

Whereas functional genomics evaluates the functional alterations brought about by structural variations in the genome, structural pharmacogenomics shows the structural differences in the genetic composition of various people. Changes in the kind of proteins generated due to genetic variations cause variations in specific bodily functions. Any illness or deformity may arise from this functional alteration. Consequently, structural and functional pharmacogenomics fields are precious in identifying and detecting genetic markers for illness and developing and optimizing pharmacological therapies for treating those diseases (Gupta and Jhawat 2017).

2.2.2.1.3 Proteomics

The ability of *Helicobacter pylori* to activate human neutrophils makes its Neutrophil Activator Protein (HP-NAP) a crucial virulence component neutrophil by releasing reactive oxygen species (ROS) and mediators. The HP-NAP is a potential vaccine target, therapeutic target, and *H. pylori* diagnostic marker. One-stage anionic exchange chromatography has been found by Shih et al. to be capable of purifying the recombinant HP-NAP produced in *B. subtilis* with a 91% recovery rate. The *Bacillus subtilis* DB104 expression pattern was explicitly expressed by the *Helicobacter pylori* neutrophil-activating protein (HP-NAP) to prevent lipopolysaccharide contamination. For the production of protein molecules, the nap gene was replicated into an immanent pRPA expression vector and then transformed into *Bacillus subtilis* DB104. When we compared the *Bacillus subtilis* DB104, the analysis of SDS-PAGE showed a remarkable increase in the production of a protein with a molecular weight of around 17 kDa in *B. subtilis* carrying pRPA-NAP. The 17 kDa protein was identified by immunoblotting analysis with an anti-HP-NAP antibody, MAb 16F4, as being HP-NAP and exclusively expressed in *B. subtilis* carrying pRPA-NAP (Shih et al. 2013). Human serum proteins, peptides, and other tiny constituents of low molecular weight have been linked to several pathological illnesses, including diabetes, cancer, cardiovascular disease, and infectious diseases. These problems are indicative of the underlying cell or tissue. In recent work, a proteomic pattern in serum was detected utilizing surface-enhanced laser/desorption ionization time-of-flight mass spectrometry (SELDI-TOF MS) and cutting-edge bioinformatic approaches to identify five species with molecular weights below 2500 Da. This allowed the differentiation of ovarian cancer from non-ovarian cancer disease (Tirumalai et al. 2003). In the early 1990s, GSK used gene-sequence databases to clone the gene-encoding production of LpPLA2 to investigate the potential involvement of Lp-PLA2 in atherosclerosis. Since then, research in the lab has shown that Lp-PLA2 produces inflammatory chemicals and that inhibiting Lp-PLA2 in animal models lowers atherosclerosis. GSK is moving medication candidates closer to the first round of human safety testing (Gilham 2002).

The integrated role of nanotheranostics and pharmacogenetics in diagnosis and treatment approaches in many genetic disorders are mentioned below in Fig. 2.2.

2.2.2.1.4 Pharmacogenomics Role in New Advances of Drug Discovery Through PBPK Modeling

Even though body weight is one of the many patient characteristics that chemotherapy dosage schemes in oncology always account for, the therapeutic protocols used and the dosage scaling that takes place is based on nomograms and how well the patient fits them rather than considering additional tumor characteristics. While there have only been a few attempts to use PBPK models for nanoformulations, recent presentations have shown the dynamism of such an approach and how specific parameters necessary for the PK profile of nanomedicines might be assessed in models.

By using PBPK modeling to better understand of the microenvironment complexity, phenotypic variation, and genetic heterogeneity of carcinomas, recent approaches in the drug development process that decrease attrition and boost productivity output for new anticancer therapies may be possible, such as:

Tamoxifen administration for patients with estrogen receptor-positive breast cancer is an illustration of customized cancer therapy. Tamoxifen is a prodrug that

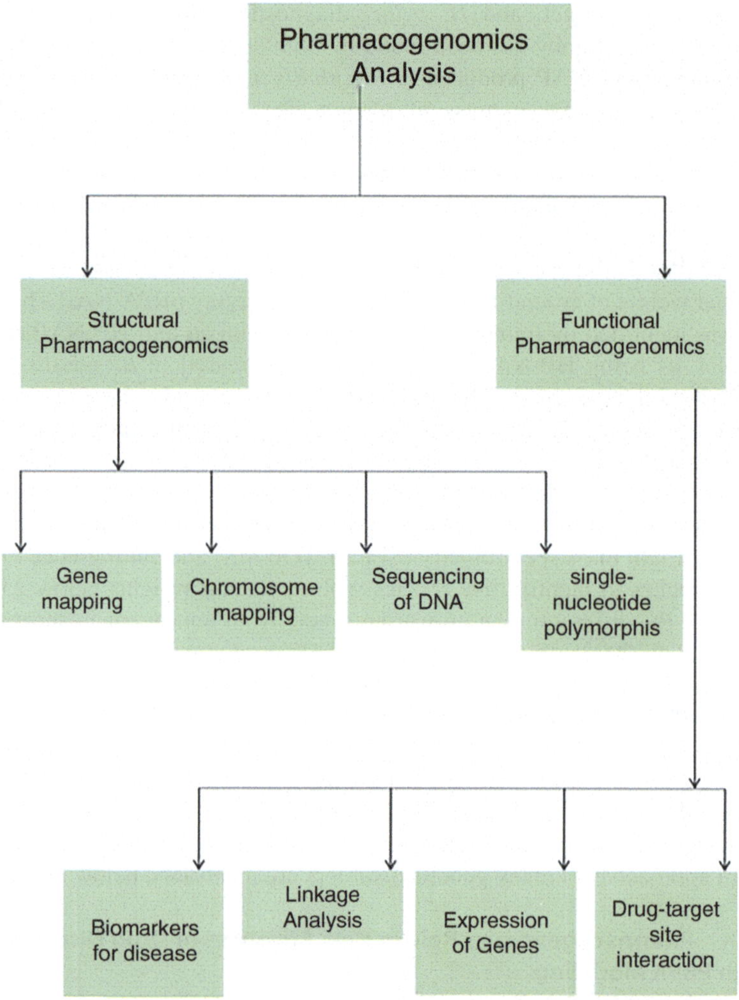

Fig. 2.2 The functional and structural aspects of pharmacogenomics. [Reproduced under Creative Common Attribution 3.0 International with Springer Nature (Chawla et al. 2021)]

forms metabolites with varying response effects through substantial hepatic and intestinal wall metabolism. Furthermore, it has been revealed that the rate-limiting process in synthesis of its primary active metabolite, 4-hydroxy-*N*-desmethyltamoxifen (endoxifen), is related to CYP2D6 activity and genetic variation. Therefore, understanding CYP2D6 PGx enables the prediction of each patient's unique pharmacological response and, in turn, the clinical result for women with breast cancer who are treated with tamoxifen.

Concurrent use of selective serotonin reuptake inhibitors has also been shown to have an impact on patient outcomes. One such drug is paroxetine, a potent CYP2D6 inhibitor. This clinical PGx information has been extensively used in tamoxifen-based PBPK approaches to provide a more specific clinical understanding of any essential in vitro–in vivo estimation of a drug's PK profile assessment. In order to determine whether co-delivery of endoxifen with tamoxifen might successfully solve genetic variability chemotherapeutic concerns and establish a prolonged pharmacodynamics profile in treated breast cancer patients, Tamoxifen-based PBPK modeling makes infrastructure development possible. As a result, a new era of immediate PTx in cancer therapeutics appears to be ushering in the development of multistate mechanistic PK/PD model-based approaches for individual therapeutic dosage schemes for Tamoxifen and other anticancer drugs, like temozolomide for glioblastoma (Vizirianakis et al. 2016).

2.2.2.2 Role of Drugs as Therapeutic Agents in Nanotheranostics

On the other hand, by creating nanocarriers with "on–off" theranostic functions, a customized analysis of the number of medications effectively transported to cancer locations may be produced. This justification has, therefore, sparked the creation of stimulus-responsive nanomaterials, opening up a wide range of possibilities to improve and regulate therapy dosage and repeatable medication delivery. Many stimulus-responsive nanomaterials have similar methods that entail building moieties, functional molecules, or linkers that are sensitive to chemicals (Ang et al. 2021).

We were formulating doxorubicin and multifunctional poly(amidoamine) (PAMAM) dendrimers as a unique platform for pH-trigger drug release and targeted delivery of drugs for cancer cell therapy. Using a pH-sensitive *cis*-aconityl linkage, we covalently conjugated doxorubicin onto the periphery of generation 5 (G5) PAMAM dendrimers that were partially acetylated and folic acid (FA) modified in this work to produce the G5.NHAc-FA-DOX conjugates (Gawali et al. 2019).

In another study, Kulhari and colleagues created micelles encasing the medication CTX for the treatment of prostate cancer using floriferous F68-*cis*-acotinyl linked-cabazitaxel (CTX) and F68-succinyl linked-CTX. The F68-*cis*-acotinyl linked-CTX obtained better apoptotic effects due to its increased pH sensitivity and regulated drug release. They also have reported the pH-stimulated drug release from other pH-sensitive chemical linkages. Functional groups like *b*-thiopropionate, orthoester, carbamate, cyclic acetal, hydrazone, ketal, imine, and benzoic-imines are among them (Jangid et al. 2020).

The creation of pH-labile magnetic nanocarriers for efficiently delivering the anticancer medication doxorubicin to tumor cells is reported here. Doxorubicin was covalently attached to the ascorbic acid-coated magnetic nanocarriers (AMNCs) through carbamate and hydrazone linkage. The drug-nanocarriers combination would reduce the quantity of drug that leaches into the circulation (pH 7.4) and facilitate drug release when the target cells internalize it (pH 5). It is significant to add that these pH-sensitive, colloidally stable nanocarriers selectively killed tumor cells instead of normal cells (Gawali et al. 2019). Here, we describe a highly effective photodynamic therapy (PDT) system made of a photosensitizer–MnO_2 nanosystem assembled by physic-sorption is the widely used photosensitizer chlorine e6 (Ce6) on MnO_2 nanosheets. This design uses MnO_2 nanosheets as an antioxidant and a carrier for Ce6 to lower intracellular GSH. It also demonstrates how effectively they quench Ce6's fluorescence to produce a turn-on fluorescence signal that can be used to track delivery efficiency (Fan et al. 2016).

This article describes a novel hydrogel that reacts to pH variations and near-infrared (NIR) light exposure: di benzaldehyde-functionalized polymer (DF-PEG)-polyaspartylhydrazide/black phosphorus nanosheets (BPNSs) hydrogel. When changes occur in environmental factors such as pH and near-infrared light (NIR), light increases rapidly, and the hydrogel becomes very responsive and releases its medication from the hydrogel system. Whole-cell studies revealed no appreciable cytotoxicity from the hydrogel itself. The crucial ability of hydrogel systems to form the solidified film when they inject individual components in vivo was one of their other characteristics. When considered collectively, these findings showed that the hydrogel holds considerable promise for the therapeutic management of cancer (Wu et al. 2019).

Li et al. provide two instances: In the first investigation, a diblock co-polymer that self-assembles onto NPs loaded with doxorubicin was produced by conjugating polylactic-*co*-glycolic acid (PLGA) with methoxyl-polyethylene glycol (MeO-PEG) using a Thioketal linker. In the second investigation, doxorubicin was co-loaded into NPs with α-tocopheryl succinate, which increases ROS production in cells, and MeO-PEG was substituted with PEG. When NPs were treated with 50 μM ROS, the cumulative release of doxorubicin from NPs was considerably greater after 6 h and 48 h. This release rose even further to over 80% after 48 h of incubating NPs with 100 μM ROS. The breakdown of the NPs, which increased their size and released doxorubicin, was ascribed to the cleavage of the Thioketal. linker, which increased the rate of doxorubicin release (Rinaldi et al. 2022).

2.2.2.3 Role of siRNA/DNA as Therapeutic Agents in Nanotheranostics

To accomplish the sequence post-transcriptional suppression of faulty genes in illnesses like cancer was a greater focus on the development of siRNA-based therapeutics after the discovery of RNA interference (RNAi) (Deng et al. 2014). All eukaryotic cells inhibit genes following transcription using the endogenous RNA interference (RNAi) process. RNAi might be induced by double-stranded RNAs (dsRNA), such as synthetic siRNA, endogenous microRNA (miRNA), as well as short hairpin RNA (shRNA) (Fellmann and Lowe 2014).

It is recognized that the framework metalloproteinase (MMP) family of enzymes, including MMP-2 and 9, plays a major role in advancing malignant tumor metastasis. Monomers of micelles containing MMP-sensitive linkers have been reported as a way to deliver drugs and siRNA to specific tumors given the presence of the overexpressed MMPs (Gonzalez-Avila et al. 2022).

A cell-surface binding peptide, a portion of the hydrophilic non-binding segment, and a siRNA binding segment containing thiols were the three essential properties of the blocked copolymers used to strengthen siRNA. This process produced multipurpose siRNA delivery micelles. Lysine amines conjugated with

2-iminothiolane (2IT) are found in the PEG-*b*-poly(lysine) (PEG-*b*-PLL) block copolymer that was used.

These cRGD-targeted tiny particles improved siRNA transport in both in vitro and in vivo. To focus on blood artery endothelial cells, mice had been programmed to integrate siRNAs targeting vascular endothelial growth factor receptor 2 (VEGFR2), whereas rodents targeted tumor mass with siRNAs targeting VEGF.

In vitro, the cRGD-modified, siRNA-loaded micelles demonstrated enhanced gene silencing capability, improved sub-cellular variance dispersion, and greater cell uptake. In mice, they also increased accumulation in the tumor mass and tumor-associated blood arteries following intravenous treatment. These were also effective in silencing the tumor's genes and halting epidermal HeLa–Luc tumors' growth after antiangiogenic siRNA therapy. In line with in vitro findings, no decrease in tumor development was shown when using naked siRNA, micelles devoid of the cRGD peptide, or those without 2IT-modified lysines (Jhaveri and Torchilin 2014).

For instance, there is a close relationship between the MDR of cancer cells and the Bcl-2 protein, one of the primary antiapoptotic defense proteins. Bcl-2-targeted siRNA can successfully overcome cancer cells' multidrug resistance (MDR) and increase the cancer cells' susceptibility to anticancer medications by knocking down the expression of the Bcl-2 protein in cancer cells.

Here, we describe the sequential administration of doxorubicin, an anticancer medication, and Bcl-2-targeted siRNA using PEI-functionalized graphene oxide (PEI-GO). We show that the PEI-GO is a great nanocarrier for delivering chemical medicines and siRNA to cancer cells. We also show that sequential administration of doxorubicin and siRNA by PEI-GO to cancer cells has a synergistic impact that significantly increases the efficacy of chemotherapy (Zhang et al. 2011). Faster and more effectively than smaller nanoparticles. Oishi et al. were the first to incorporate siRNA into gold nanoparticles; more recent developments use layer-by-layer assembly to construct supramolecular structures that permit the prolonged release of siRNA (Thakor and Gambhir 2013).

2.2.3 Diagnostic Agents

In recent years, the construction of several diagnostic tools, contrast agents, and drug delivery nanosystems has resulted from combining nanotools with biological sciences. Compared to larger monofunctional particles, multifunctional nanoscaled materials often exhibit promising physical activity, stability, and better biodistribution. The fields of radiology, nuclear medicine, and radiopharmacy employ various types of imaging techniques. Common imaging techniques in radiology include radiography, ultrasonography, computed tomography (CT), and magnetic resonance imaging (MRI). Numerous disorders, notably cancer, may now be diagnosed thanks to these techniques. Nevertheless, when it is apparent, the cancer cell will have around 1 billion cancer cells at a size of 1 cm^3. Thus, since the phenotypic alterations start early, the tumor area diagnosis would come too late. Early molecular identification of cancer is crucial for therapeutic intervention. As a result, research

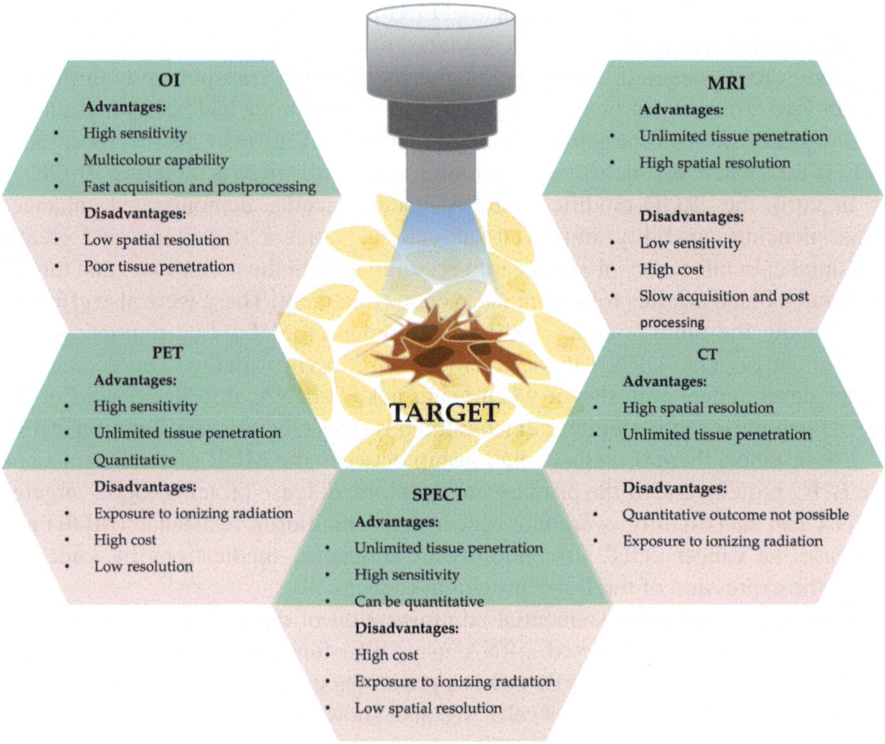

Fig. 2.3 Advantages and disadvantages of different imaging techniques. [Reproduced under Creative Common Attribution 4.0 International with MDPI (Murar et al. 2022)]

has shifted from nuclear medicine imaging to radiological imaging (Siafaka et al. 2021). Some important and advanced optical imaging-guided nanotheranostics towards personalized drug delivery are discussed below in Fig. 2.3 with their advantages and disadvantages as follows:

2.2.3.1 Optical Imaging

The efficacy of $NaGdF_4$@$CaCO_3$-PEG 250 (Aminolroayaei et al. 2021) core-shell nanoparticles in the detection of prostate tumors was investigated. Their findings indicate that the slow breakdown of $CaCO_3$ in the acidic tumor environment promotes the collision of $NaGdF_4$ with the surrounding aqueous media. The MRI signal and the water proton's comfort are improved due to this approach. Additionally, it produces CO_2 bubbles from the breakdown of $CaCO_3$, which provide powerful echoes for ultrasound imaging (Aminolroayaei et al. 2021). Zirconium-based metal organic frameworks (Zr-MOFs) reconstitute a vast family of water-soluble structures. The most popular Zr-MOFs are UiO-66, UiO-67, and UiO-68 are comprised of $[Zr_6O_6(OH_4)]^{12+}$ SBUs connected by terephthalate, 4,4′-biphenyl dicarboxylate and 4,4′-terphenyl dicarboxylate linkers, respectively, to form 12-connected structures having a fcu-a-topology (Margolis et al. 2007).

2.2.3.2 Dye-Doped Silica Nanoparticles

While adding dyes to NPs offers additional benefits, as previously mentioned, providing the loading of multiple dye molecules into a protective NP matrix allows for signal amplification, reduced chemical and photodegradation, improved contrast, and the ability to target specific biologically relevant sites. Dye alone is sufficient to increase signal-to-noise ratio (SNR), thereby improving the image contrast profile. In addition to the intended absorption profile, other factors such as fluorescence imaging or the inclusion of a therapeutic function, such as photodynamic therapy (PDT), may be considered when choosing the dye. Indium cyanide (ICG) dye is the most often employed for photo-acoustic imaging in molecular and nanoparticulate formulations. At around 780 nm, ICG's absorption peak falls within the physiologically significant near-infrared window, making it the only FDA-approved dye currently used for human applications. This allows for intense tissue imaging. PAT has been used to photograph blood and ICG-containing objects implanted in chicken breast tissue deeper than 5 cm. However, because of its short plasma half-life of up to 4 min and fast breakdown in aqueous conditions, ICG has limited in-vivo applicability. It has been demonstrated that encasing ICG in several NP matrices, such as calcium phosphate, poly(lactic-*co*-glycolic acid) (PLGA), and organically modified silica (ORMOSIL), increases its stability and blood circulation time, allowing for in vivo applications.

Among the NIR-emitting dyes that may be added to silica nanoparticles are polymethines (such as Cy5.5 and Cy7), indocyanine green (ICG), Alexa Fluor 750, and IRDye78. An enormous benefit of packing thousands of dye molecules onto a single silica nanoparticle is that the dye-loaded single NP is far brighter and more stable than the single-molecule version. The most common methods for creating dye-doped silica nanoparticles (NPs) are the sol–gel method (Stöber) or the microemulsion method, which involve just adding the dye or a modified form of it to the silica-forming solution (Hahn et al. 2011).

2.2.3.3 Photoacoustic Molecular Tomography (PMT)

Alexander G. Bell initially discovered the photoacoustic (PA) phenomena in 1880. Bell's experiment revealed that various materials created distinct tones when a spare/thin sheet of material exposed to a solar beam was abruptly disrupted by a revolving slotted disk (is shallow channels machines into the disc to aid in removing dust and gas). The result was an audible sound signal. However, the development of sensitive sensors and powerful light sources for spectroscopic analysis meant that the proper use of this approach had to wait over a century. It was reported in 1971 that PA spectroscopy could identify nitrogen oxide (NOx) pollution in air samples at a value of 0.01 parts per million (ppm). The PA approach was developed as an alternative to fluorescence measurement for the precise localization and determination of the excited state lifetime of oxygen-sensitive dyes, such as methylene blue (MB) and Pt(II) octaethylporphyrin (PtOEP) (Nie and Chen 2014).

2.2.3.4 AIEgens-Based Diagnostic Agents

It is possible to directly combine AIEgens with the right driving power or functionality to create nanomaterials with increased fluorescence emission. The supramolecular assemblies of AIEgens can give nanomaterials more controlled shapes and sizes that match their biological applications, in contrast to aggregates in mixed solvents. Three exemplary approaches to supramolecular assembly architecture-based theranostic nanomaterial fabrication have been documented: chemical bonding with amphiphilic polymers, coordination with metal ions, and modification using building blocks for non-covalent interactions. An AIE-based metallo-supramolecular assembly made of pyrocatechol violet and Zn^{2+}-cyclen based on TPE was created for the indicator-displacement assay in living cells. This allowed for the highly selective detection of molecules in living cells (Song et al. 2020). (QDs) (~10 nm), DOX/mPEGss-Tripp (117 nm), and M-NPAPF-Au (~120 nm) doxorubicin (DOX)-based. Interestingly, compared to AIE dots bigger than 25 nm, sub-10 nm AIE QDs were observed to improve tumor-targeting performance and lower liver retention. Better cellular uptake and NIR-II fluorescence tumor imaging were achieved by the AIE QDs constructed by microfluidics without additional target molecules or peptides (Cen et al. 2023).

Liu et al. developed and manufactured a dual-targeted enzyme-activatable bio-probe that included αvβ3 integrin and cathepsin B. The components of this bio-probe include hydrophilic units, αvβ3-attachable cRGD group, Cathepsin B-responsive GFLG peptide, and AIE fragment TPECM for both imaging and PDT. This bio-probe's high hydrophilicity allowed it to dissolve quickly in aqueous solution, and the free intramolecular movements generated excitation energy consumption, which allowed it to retain a fluorescence "off" state and low amount of ROS formation in an aqueous environment (Wang et al. 2018). TPEs and monomeric release AIEgens are very useful for imaging that has to be noninvasive since they only emit strongly when a dye binds to the target, preventing dye aggregation (Wang et al. 2023).

2.2.3.5 PET (Positron Emission Tomography)

Since PET imaging makes it possible to see and measure tumor characteristics like tumor metabolism and receptor expression that are only visible at the morphological level on traditional imaging, it is becoming increasingly important clinically in the management of cancer patients receiving radiation therapy. The most often used ligand for oncological PET imaging is 18FFDG, a glucose analogue. 42 inoperable patients with stage II–III nonsmall cell lung cancer (NSCLC) participated in a phase II study conducted by Kong et al. (Almuhaideb et al. 2011). To deal with a fixed risk of radiation-induced lung toxicity (grade >2), the patients were divided into groups of 39 (93%) and 3 (7%) who received tailored conformal radiotherapy, and the patients' doses were adaptively escalated to the residual tumor as determined by mid-treatment FDG-PET after an equivalent dose of 50 Gy in 2 Gy per fraction (EQD2) up to a physical total dose of 86 Gy (EQD2 of 92 Gy [alpha/beta = 10 for tumor]) in 30 daily fractions (Szejniuk et al. 2021). The clinical development of nanomedicines is at an early level. For instance, Cristal Therapeutics, a pharmaceutical firm, is now conducting a Phase I clinical study for CriPecR, a combination of docetaxel and the imaging agent zirconium-89, intended for PET imaging. To improve targeted therapy, this is being done to assess the biodistribution and tumor accumulation of the nanomedicine in solid tumors. Another business, Nanobiotix, is carrying out Phase I/II studies for NBTXR3R, a radio-enhancer that uses radiation-stimulated technology to eliminate tumors using hafnium oxide nanoparticles (Singh et al. 2020).

2.2.3.6 SPECT (Single Photon Emitted Computed Tomography)

For metastatic colorectal cancer and hepatocellular carcinoma, radioembolization or selective internal radiation therapy (SIRT) using 90Y microspheres is advised (Hanin et al. 2018). Due to the enlarged vasculature of tumors, radioactive microspheres are injected into the hepatic arteries in this procedure. Imaging after treatment is made possible by the production of Bremsstrahlung photons, formed when the atomic nuclei in tissue defect the beta particles released by 90Y (Kim et al. 2011). Following treatment, single photon emission computed tomography (SPECT) imaging is performed to evaluate the radionuclide's activity distribution qualitatively. Clinical data for five patients treated with SIRT using 90Y resin microspheres (SIR-Spheres) were obtained at the Royal Surrey NHS Foundation Trust in Guildford, UK, using the GE Optima 640 SPECT/CT system. In this investigation, the patient's data were anonymized (Ali Raza Naqvi and Babar Imran 2021). The 40-min SPECT acquisitions included 120 20-s projections, and the injected activity fell between 1.5 and 2.2 GBq. The range of the energy window was 75–225 keV. A general-purpose, medium-energy parallel-hole MEGP collimator was employed. The CT image was used as an anatomical picture in the HKEM and FHKEM after being obtained for attenuation estimation (Deidda et al. 2022).

2.2.3.7 MRI (Magnetic Resonance Imaging)

In this work, we developed and synthesized a multifunctional rattle-structured nanotheranostic for the storage of the chemical medicine doxorubicin and the hypoxia-sensitive enrichment of MnO_2. The inner core of the nanotheranostic was covered with hollow mesoporous silica. MnO_2 nanoscale films could be broken down into manganese ions (Mn^{2+}) in various acidic microenvironments brought on by hypoxia (Pi et al. 2023). These manganese ions were modified Tetraxetanum (DOTA) ligands for real-time T1-MRI by chelation, with doxorubicin release occurring on demand to achieve with the help of both normoxia and hypoxia-sensitive chemotherapy we can reduce or overcome hypoxia. Synergistic chemotherapy and hypoxia-driven T1-MRI by nanothermanostics is promising for intelligent diagnosis and tailored treatment (Song et al. 2018). Magnetic fluorescent nanoparticles for atherosclerosis with inflammatory theranostics were developed by McCarthy et al. Iron oxide nanomaterials coated with dextran were altered to use AlexaFluor 750 for near-infrared fluorescence and meso-tetra(mhydroxyphenyl) chlorin (THPC) for phototoxic therapy of inflammatory macrophages. Studies using near-infrared imaging on C57/BL6 mice showed that THPC-based nanomaterials may form aggregates in areas where macrophages and foam cells were abundant (Goyal et al. 2023).

2.2.3.8 Microfluidic Arrays

A microfluidic unit is characterized as a device that has one or more channels, at least one of which has a dimension less than 1 mm very small fluidic volumes in the nano liter and pico liter range may be controlled by the channels, which can have a width of less than a few micrometers (Vitorino et al. 2021). Almost every biological test operating at the molecular level may be integrated into a chip (lab-on-a-chip systems) thanks to microsystem technologies created for microfluidic chips (Sin et al. 2011). These methods allow the isolation and manipulation of live cells as well as the potential of toxicity tests, ELISA (enzyme-linked immunosorbent assays), PCR amplification, blood separation, and cytokine polymorphism genotyping. In 2006, Du et al. demonstrated that over 30% of the cancer cells were extracted from a combination of human cervical stromal cells (HCSC), cervical cancer cells (HCCC), and normal human glandular epithelial cells (HGEC) using an antibody-based microfluidics technique (Riehemann et al. 2009). There are some nanotheranostics formulations are mentioned below in Table 2.2 according to disease needs such as:

2.3 Challenges in Theranostics

Theranostic platforms built on nanotechnology provide chemotherapy unparalleled benefits for overcoming its long-standing drawbacks. Nanomaterials and nanotechnology-based approaches in oncology provide effective anticancer therapy, targeted medicine administration, early cancer diagnosis, and drug development (Kashyap et al. 2023). Gold nanorods are one of the theranostics agents that scientists have studied in great detail. Gold nanorods are accessible for both photo-acoustic imaging and photo-thermal treatment. They have good biocompatibility, superior heat production under laser irradiation, and facile functionalization. In actuality, however, the majority of gold nanorods that function as photothermal agents are too big; their length is around 40 nm, and their width is greater than 8 nm. Because of their bigger size, these gold nanorods are less biodegradable and easily excreted, raising questions about their long-term toxicity and eventually limiting their clinical translation (Yan et al. 2018). Nano-bio interaction is a significant obstacle to the clinical use of lipidic theranostic nanomedicine. Contrasting substances in lipidic theranostic nanomedicine may interact with biological material and induce problems like inflammation, immunoreactions, or related illnesses due to their incompatibility or potential toxicity. Several parameters, including the size, zeta potential, and solubility of the nanoformulation, significantly influence the degree of toxicity. Several nanomedicine formulations during research and clinical trials are intended for use in cytotoxics. In these cases, extending the therapeutic window might mitigate tolerability concerns or suboptimal target exposure, which

Table 2.2 Glimpse of nanoformulation for nanotheranostics according to disease needs

Cancer/disease	Types of nanomaterial/ composition	Physiochemical properties	Therapeutic/ imaging modality	References
Atherosclerosis	Gold NPs coated on silica/iron oxide shells	Size: 90–150 nm; stem cells bearing NPs are stinging and microinfusion into lesions	Targeting and therapy (plasmonic photo-thermal and stem cell therapy)	Kharlamov et al. (2015)
Myocardial infarction	Ultrasuperparamagnetic NP iron oxide (Ferumoxytol)	Size: 17–31 nm; 4–6 times enhanced phagocytosis; higher absorption by cells due to cardiac inflammation; possible visualization of cardiac muscle damage area	Targeting and imaging	Stirrat et al. (2017)
Type 1 diabetes	1. Gold NPs coupled with proinsulin peptide (c19-A3 GNP) 2. Ultrasuperparamagnetic NPs iron oxide (Ferumoxtan-10), dextran-coated	Size: 5 nm; intradermal transportation of drug by micro needles; negatively charged (−40 to −60 mV) Size: 30 nm; monocrystalline; with long-circulating properties; real time visualization of inflammatory lesions in vivo	Targeting and therapy Targeting and imaging	Tatovic et al. (2022)
Brain metastases	GdNPs (AGuIX)	Size: 4.5 nm; can be used in combination with fractionated stereotactic radiation in olig-brain metastases	Targeting and imaging (MRI)	Pan et al. (2010)

(continued)

Table 2.2 (continued)

Cancer/disease	Types of nanomaterial/ composition	Physiochemical properties	Therapeutic/ imaging modality	References
Gliosarcoma	Sphere shape GNPs carrying nucleic acids and drugs (NU-0129)	Size: 13 nm; can cross the BBB: Targeting of Bcl2L 12 gene to stop the growth of carcinomas	Targeting and therapy	Kumthekar et al. (2021)
Transthyretin-mediated (hATTR) amyloidosis	Lipid NPs patisiran (ALN-TTR02)	Size: 60–100 nm; liquid IV injection fends off the decay of left ventricles global longitudinal strain over 18 months	Targeting and therapy (to silence the gene)	Ioannou et al. (2023)
Thyroid cancer	Lymphotropic superparamagnetic NPs loaded with the drug (Ferumoxytol)	Size: 30 nm; IV injection (lyophilized iron oxide injected dose: 2.6 mg/kg of body weight); imaging with high sensitivity and specificity in detecting small lymph nodes (2 nm)	Imaging (through MRI) and therapy	Turkbey et al. (2020)

could impede the development of the treatment into a commercially successful product (Hare et al. 2017).

Theranostic nanomedicine offers enormous promise for enhancing human health by providing new perspectives on illness prevention, diagnosis, and therapy. Despite significant technological advancements in this area, nanotechnology is still in its infancy as a new paradigm for illness detection and treatment in clinics. The nano-bio interaction is one of the main obstacles to the adoption of theranostic nanomedicine in clinical settings. When a nanomedicine interacts with biological material, its possible toxicity or incompatibility may cause illnesses like inflammation, immuno-reaction, or similar conditions. The harmful impact is mainly determined by several factors, including the nanoformulations' size, ζ-potential, and solubility (Singh et al. 2020).

Nevertheless, because it is challenging to synthesize NP libraries with diverse characteristics quickly, accurately, and consistently, systematic parallel screening of the wide range of NP attributes becomes challenging. Microfluidic technologies have gained attention recently due to their ability to produce NP with a narrower size distribution, tuneable physical and chemical properties, and greater batch-to-batch reproducibility at a faster rate than traditional bulk techniques, which typically produce NPs with high polydispersity. Similarly, the production of monodisperse NPs with exact control over size, shape, chemical composition, drug loading, and surface features has been made possible using particle replication in non-wetting template (PRINT) technology (Shi et al. 2017).

According to data from pediatric oncology, mIBG is the only theranostic now available for systematic use for the imaging and therapy of neuroblastoma tumors that express the norepinephrine transporter (NET). The current gold standard for detecting neuroblastoma primary tumors and distant metastases, staging the illness, and evaluating the response to therapy is mIBG SPECT imaging. Despite its excellent specificity and sensitivity, mIBG SPECT has several disadvantages, such as long scan times, low picture quality, and thyroid damage from iodine (Goyal et al. 2023).

2.4 Conclusion

In conclusion, nanotheranostics, the combination of diagnostic and therapeutic functions at the nanoscale, improves our knowledge of illnesses and provides practitioners with robust instruments to act with unmatched precision. This unique combination reduces unintended harm to healthy tissues and expedites healthcare decision-making processes, signaling a paradigm change in the search for efficient and customized healthcare solutions. With its tailored approach that holds the key to opening new doors in the pursuit of better patient outcomes and a healthier future, nanotheranostics has the potential to completely transform how we diagnose and treat illnesses as it advances. Fundamentally, nanotheranostics has become a revolutionary force that connects the dots between diagnosis and therapy to produce.

References

Abdelgalil RM et al (2023) Engineered sericin-tagged layered double hydroxides for combined delivery of pemetrexed and ZnO quantum dots as biocompatible cancer nanotheranostics. ACS Omega 8(6):5655–5671. https://doi.org/10.1021/acsomega.2c07128

Ali Raza Naqvi S, Babar Imran M (2021) Single-photon emission computed tomography (SPECT) radiopharmaceuticals. In: Medical isotopes. IntechOpen. https://doi.org/10.5772/intechopen.93449

Almuhaideb A, Papathanasiou N, Bomanji J (2011) 18F-FDG PET/CT imaging in oncology. Ann Saudi Med 31(1):3–13. https://doi.org/10.4103/0256-4947.75771

Alven S et al (2020) Polymer-drug conjugate, a potential therapeutic to combat breast and lung cancer. Pharmaceutics 12(5):406. https://doi.org/10.3390/pharmaceutics12050406

Aminolroayaei F et al (2021) Recent nanotheranostics applications for cancer therapy and diagnosis: a review. IET Nanobiotechnol 15(3):247–256. https://doi.org/10.1049/nbt2.12021

Anand U et al (2023) Cancer chemotherapy and beyond: current status, drug candidates, associated risks and progress in targeted therapeutics. Genes Dis 10(4):1367–1401. https://doi.org/10.1016/j.gendis.2022.02.007

Ang MJY et al (2021) Emerging strategies in developing multifunctional nanomaterials for cancer nanotheranostics. Adv Drug Deliv Rev 178:113907. https://doi.org/10.1016/j.addr.2021.113907

Avramović N et al (2020) Polymeric nanocarriers of drug delivery systems in cancer therapy. Pharmaceutics 12(4):298. https://doi.org/10.3390/pharmaceutics12040298
Barros SM et al (2016) A review of solute encapsulating nanoparticles used as delivery systems with emphasis on branched amphipathic peptide capsules. Arch Biochem Biophys 596:22–42. https://doi.org/10.1016/j.abb.2016.02.027
Begines B et al (2020) Polymeric nanoparticles for drug delivery: recent developments and future prospects. Nanomaterials 10(7):1403. https://doi.org/10.3390/nano10071403
Bhattacharya T et al (2023) Advancement in biopolymer assisted cancer theranostics. ACS Appl Bio Mater 6(10):3959–3983. https://doi.org/10.1021/acsabm.3c00458
Cagel M et al (2017) Polymeric mixed micelles as nanomedicines: achievements and perspectives. Eur J Pharm Biopharm 113:211–228. https://doi.org/10.1016/j.ejpb.2016.12.019
Cen P et al (2023) Aggregation-induced emission luminogens for in vivo molecular imaging and theranostics in cancer. Aggregate 4(5):352. https://doi.org/10.1002/agt2.352
Chandrasekharan P et al (2011) Vitamin E (D-alpha-tocopheryl-*co*-poly(ethylene glycol) 1000 succinate) micelles-superparamagnetic iron oxide nanoparticles for enhanced thermotherapy and MRI. Biomaterials 32(24):5663–5672. https://doi.org/10.1016/j.biomaterials.2011.04.037
Chawla R et al (2021) Integrated role of nanotechnology and pharmacogenetics in diagnosis and treatment of diseases. Pharmacogenetics 11:97643. https://doi.org/10.5772/INTECHOPEN.97643
Che H-L et al (2015) Simultaneous drug and gene delivery from the biodegradable poly(ε-caprolactone) nanofibers for the treatment of liver cancer. J Nanosci Nanotechnol 15(10):7971–7975. https://doi.org/10.1166/jnn.2015.11233
Chen K, Chen X (2010) Design and development of molecular imaging probes. Curr Top Med Chem 10(12):1227–1236. https://doi.org/10.2174/156802610791384225
Chen W-H et al (2013) Therapeutic nanomedicine based on dual-intelligent functionalized gold nanoparticles for cancer imaging and therapy in vivo. Biomaterials 34(34):8798–8807. https://doi.org/10.1016/j.biomaterials.2013.07.084
Chen Z et al (2020) Theranostic small interfering RNA nanoparticles in cancer precision nanomedicine. WIREs Nanomed Nanobiotechnol 12(2):e1595. https://doi.org/10.1002/wnan.1595
Debnath SK, Srivastava R (2021) Drug delivery with carbon-based nanomaterials as versatile nanocarriers: progress and prospects. Front Nanotechnol 3:1–22. https://doi.org/10.3389/fnano.2021.644564
Deidda D et al (2022) Hybrid kernelised expectation maximisation for bremsstrahlung SPECT reconstruction in SIRT with 90Y micro-spheres. EJNMMI Phys 9(1):25. https://doi.org/10.1186/s40658-022-00452-4
Deng Y et al (2014) Therapeutic potentials of gene silencing by RNA interference: principles, challenges, and new strategies. Gene 538(2):217–227. https://doi.org/10.1016/j.gene.2013.12.019
Dessale M, Mengistu G, Mengist HM (2022) Nanotechnology: a promising approach for cancer diagnosis, therapeutics and theragnosis. Int J Nanomed 17:3735–3749. https://doi.org/10.2147/IJN.S378074
Dristant U et al (2023) An overview of polymeric nanoparticles-based drug delivery system in cancer treatment. Technol Cancer Res Treat 22:153303382311520. https://doi.org/10.1177/15330338231152083
Elhag DA et al (2022) Inflammatory bowel disease treatments and predictive biomarkers of therapeutic response. Int J Mol Sci 23(13):6966. https://doi.org/10.3390/ijms23136966
Fam SY et al (2020) Stealth coating of nanoparticles in drug-delivery systems. Nanomaterials (Basel) 10(4):787. https://doi.org/10.3390/nano10040787
Fan H et al (2016) A smart photosensitizer-manganese dioxide nanosystem for enhanced photodynamic therapy by reducing glutathione levels in cancer cells. Angew Chem Int Ed 55(18):5477–5482. https://doi.org/10.1002/anie.201510748
Fellmann C, Lowe SW (2014) Stable RNA interference rules for silencing. Nat Cell Biol 16(1):10–18. https://doi.org/10.1038/ncb2895

Friebe M, Illanes A (2023) Advancements in medical imaging and image-guided procedures: a potential—or rather likely—paradigm shift in diagnosis and therapy: understand disruption and take advantage of it! Appl Sci 13(16):9218. https://doi.org/10.3390/app13169218
Fulton MD, Najahi-Missaoui W (2023) Liposomes in cancer therapy: how did we start and where are we now. Int J Mol Sci 24(7):6615. https://doi.org/10.3390/ijms24076615
Gawali SL, Barick KC, Shetake NG, Rajan V, Pandey BN et al (2019) pH-labile magnetic nanocarriers for intracellular drug delivery to tumor cells. ACS Omega 4(7):11728–11736. https://doi.org/10.1021/acsomega.9b01062
Giammarile F (2023) Theranostic in nuclear medicine—the paradigm of NET. Hell J Nucl Med 26:42–43
Gilham I (2002) Theranostic an emerging tool in drug discovery and commercialisation. Drug Discov World 6:17–23
Gonzalez-Avila G et al (2022) Nanotechnology and matrix metalloproteinases in cancer diagnosis and treatment. Front Mol Biosci 9:918789. https://doi.org/10.3389/fmolb.2022.918789
Goyal AK, Ramchandani M, Basak T (2023) Recent advancements, challenges, and future prospects in usage of nanoformulation as theranostics in inflammatory diseases. J Nanotheranostics 4(1):106–126. https://doi.org/10.3390/jnt4010006
Gupta S, Jhawat V (2017) Quality by design (QbD) approach of pharmacogenomics in drug designing and formulation development for optimization of drug delivery systems. J Control Release 245:15–26. https://doi.org/10.1016/j.jconrel.2016.11.018
Hahn MA et al (2011) Nanoparticles as contrast agents for in-vivo bioimaging: current status and future perspectives. Anal Bioanal Chem 399(1):3–27. https://doi.org/10.1007/s00216-010-4207-5
Hanin FK, Yussof NA, Sabirin J (2018) Selective internal radiation therapy using yttrium-90 radioembolization for hepatocellular carcinoma. J Global Oncol 4(Supplement 2):81s–81s. https://doi.org/10.1200/jgo.18.13200
Hare JI et al (2017) Challenges and strategies in anti-cancer nanomedicine development: an industry perspective. Adv Drug Deliv Rev 108:25–38. https://doi.org/10.1016/j.addr.2016.04.025
Hosseini SM et al (2023) Theranostic polymeric nanoparticles as a new approach in cancer therapy and diagnosis: a review. Mater Today Chem 29:101400. https://doi.org/10.1016/j.mtchem.2023.101400
Ioannou A, Fontana M, Gillmore JD (2023) Patisiran for the treatment of transthyretin-mediated amyloidosis with cardiomyopathy'. Heart Int 17(1):27. https://doi.org/10.17925/HI.2023.17.1.27
Ioele G et al (2022) Anticancer drugs: recent strategies to improve stability profile, pharmacokinetic and pharmacodynamic properties. Molecules 27(17):5436. https://doi.org/10.3390/molecules27175436
Jangid AK et al (2020) A nanoscale, biocompatible and amphiphilic prodrug of cabazitaxel with improved anticancer efficacy against 3D spheroids of prostate cancer cells. Mater Adv 1(4):738–748. https://doi.org/10.1039/D0MA00189A
Jędrzak A et al (2019) Dendrimer based theranostic nanostructures for combined chemo- and photothermal therapy of liver cancer cells in vitro. Colloids Surf B Biointerfaces 173:698–708. https://doi.org/10.1016/j.colsurfb.2018.10.045
Jhaveri AM, Torchilin VP (2014) Multifunctional polymeric micelles for delivery of drugs and siRNA. Front Pharmacol 5:1–26. https://doi.org/10.3389/fphar.2014.00077
Ji Y et al (2018) Drug-bearing supramolecular MMP inhibitor nanofibers for inhibition of metastasis and growth of liver cancer. Adv Sci 5(8):1700867. https://doi.org/10.1002/advs.201700867
Junyaprasert VB, Thummarati P (2023) Innovative design of targeted nanoparticles: polymer–drug conjugates for enhanced cancer therapy. Pharmaceutics 15(9):2216. https://doi.org/10.3390/pharmaceutics15092216
Kashyap BK et al (2023) Smart nanomaterials in cancer theranostics: challenges and opportunities. ACS Omega 8(16):14290–14320. https://doi.org/10.1021/acsomega.2c07840
Kelkar SS, Reineke TM (2011) Theranostics: combining imaging and therapy. Bioconjug Chem 22(10):1879–1903. https://doi.org/10.1021/bc200151q

Kevadiya BD et al (2019) Neurotheranostics as personalized medicines. Adv Drug Deliv Rev 148:252–289. https://doi.org/10.1016/j.addr.2018.10.011

Khandare J, Minko T (2006) Polymer–drug conjugates: progress in polymeric prodrugs. Prog Polym Sci 31(4):359–397. https://doi.org/10.1016/j.progpolymsci.2005.09.004

Kharlamov AN et al (2015) Silica–gold nanoparticles for atheroprotective management of plaques: results of the NANOM-FIM trial. Nanoscale 7(17):8003–8015. https://doi.org/10.1039/C5NR01050K

Kim YC et al (2011) Usefulness of bremsstrahlung images after intra-arterial Y-90 resin microphere radioembolization for hepatic tumors. Nucl Med Mol Imaging 45(1):59–67. https://doi.org/10.1007/s13139-010-0066-4

Kinberger GA, Cai W, Goodman M (2002) Collagen mimetic dendrimers. J Am Chem Soc 124(51):15162–15163. https://doi.org/10.1021/ja0212031

Könik A et al (2021) Theranostics: the role of quantitative nuclear medicine imaging. Semin Radiat Oncol 31(1):28–36. https://doi.org/10.1016/j.semradonc.2020.07.003

Kumthekar P et al (2021) A first-in-human phase 0 clinical study of RNA interference-based spherical nucleic acids in patients with recurrent glioblastoma. Sci Transl Med 13(584):3945. https://doi.org/10.1126/scitranslmed.abb3945

Kunjachan S et al (2015) Noninvasive imaging of nanomedicines and nanotheranostics: principles, progress, and prospects. Chem Rev 115(19):10907–10937. https://doi.org/10.1021/cr500314d

Larson N, Ghandehari H (2012) Polymeric conjugates for drug delivery. Chem Mater 24(5):840–853. https://doi.org/10.1021/cm2031569

Lawn SD, Harries AD, Wood R (2010) Strategies to reduce early morbidity and mortality in adults receiving antiretroviral therapy in resource-limited settings. Curr Opin HIV AIDS 5(1):18–26. https://doi.org/10.1097/COH.0b013e328333850f

Lérida-Viso A et al (2023) Biosafety of mesoporous silica nanoparticles; towards clinical translation. Adv Drug Deliv Rev 201:115049. https://doi.org/10.1016/j.addr.2023.115049

Lin G et al (2016) Inorganic nanocarriers overcoming multidrug resistance for cancer theranostics. Adv Sci 3(11):1–14. https://doi.org/10.1002/advs.201600134

Lôbo GCNB et al (2021) Nanocarriers used in drug delivery to enhance immune system in cancer therapy. Pharmaceutics 13(8):1167. https://doi.org/10.3390/pharmaceutics13081167

Loutfy SA et al (2016) Synthesis, characterization and cytotoxic evaluation of chitosan nanoparticles: in vitro liver cancer model. Adv Nat Sci Nanosci Nanotechnol 7(3):035008. https://doi.org/10.1088/2043-6262/7/3/035008

Luk BT, Zhang L (2014) Current advances in polymer-based nanotheranostics for cancer treatment and diagnosis. ACS Appl Mater Interfaces 6(24):21859–21873. https://doi.org/10.1021/am5036225

Luque-Michel E et al (2017) Clinical advances of nanocarrier-based cancer therapy and diagnostics. Expert Opin Drug Deliv 14(1):75–92. https://doi.org/10.1080/17425247.2016.1205585

MacRitchie N et al (2020) Molecular imaging of inflammation—current and emerging technologies for diagnosis and treatment. Pharmacol Ther 211:107550. https://doi.org/10.1016/j.pharmthera.2020.107550

Margolis DJA et al (2007) Molecular imaging techniques in body imaging. Radiology 245(2):333–356

Murar M, Albertazzi L, Pujals S (2022) Advanced optical imaging-guided nanotheranostics towards personalized cancer drug delivery. Nanomaterials 12(3):399. https://doi.org/10.3390/nano12030399

Muthu MS et al (2012) Theranostic liposomes of TPGS coating for targeted co-delivery of docetaxel and quantum dots. Biomaterials 33(12):3494–3501. https://doi.org/10.1016/j.biomaterials.2012.01.036

Muthu MS, Leong DT, Mei L, Feng S-S (2014) Nanotheranostics—application and further development of nanomedicine strategies for advanced theranostics. Theranostics 4(6):660–677. https://doi.org/10.7150/thno.8698

Nie L, Chen X (2014) Structural and functional photoacoustic molecular tomography aided by emerging contrast agents. Chem Soc Rev 43(20):7132–7170. https://doi.org/10.1039/c4cs00086b

Otsuka H (2019) Clinical imaging technology and the diagnosis in patient-centered interdisciplinary care. J Med Investig 66(1.2):31–34. https://doi.org/10.2152/jmi.66.31

Pan J, Liu Y, Feng S-S (2010) Multifunctional nanoparticles of biodegradable copolymer blend for cancer diagnosis and treatment. Nanomedicine 5(3):347–360. https://doi.org/10.2217/nnm.10.13

Pasut G, Veronese FM (2007) Polymer–drug conjugation, recent achievements and general strategies. Prog Polym Sci 32(8–9):933–961. https://doi.org/10.1016/j.progpolymsci.2007.05.008

Patil J (2019) A review on polymer drug conjugate-what, why and how? Int J Pharm Sci Res 6(11):4611–4621. https://doi.org/10.13040/IJPSR.0975-8232.6(11).4611-21

Patil T et al (2023) Gold nanoparticles: synthesis methods, functionalization and biological applications. J Clust Sci 34(2):705–725. https://doi.org/10.1007/s10876-022-02287-6

Pi F et al (2023) Alleviating the hypoxic tumor microenvironment with MnO_2-coated CeO_2 nanoplatform for magnetic resonance imaging guided radiotherapy. J Nanobiotechnol 21(1):1–14. https://doi.org/10.1186/s12951-023-01850-1

Pottoo FH, Barkat MA, Harshita AMA, Javed MN, Sajid Jamal QM, Kamal MA (2021) Nanotechnological based miRNA intervention in the therapeutic management of neuroblastoma. Semin Cancer Biol 69:100–108. https://doi.org/10.1016/j.semcancer.2019.09.017

Priester MI, ten Hagen TLM (2023) Image-guided drug delivery in nanosystem-based cancer therapies. Adv Drug Deliv Rev 192:114621. https://doi.org/10.1016/j.addr.2022.114621

Pulingam T et al (2022) Exploring various techniques for the chemical and biological synthesis of polymeric nanoparticles. Nanomaterials (Basel) 12(3):576. https://doi.org/10.3390/nano12030576

Rahman M, Almalki WH, Afzal O, Alfawaz Altamimi AS, Najib Ullah SNM, Abul Barkat M, Beg S (2023) Chiral-engineered supraparticles: emerging tools for drug delivery. Drug Discov Today 28(1):103420. https://doi.org/10.1016/j.drudis.2022.103420

Riehemann K et al (2009) Nanomedicine—challenge and perspectives. Angew Chem Int Ed 48(5):872–897. https://doi.org/10.1002/anie.200802585

Rinaldi A et al (2022) Applications of the ROS-responsive Thioketal linker for the production of smart nanomedicines. Polymers (Basel) 14(4):687. https://doi.org/10.3390/polym14040687

Robinson JT et al (2010) High performance in vivo near-IR (>1 μm) imaging and photothermal cancer therapy with carbon nanotubes. Nano Res 3(11):779–793. https://doi.org/10.1007/s12274-010-0045-1

Ruman U et al (2020) Nanocarrier-based therapeutics and theranostics drug delivery systems for next generation of liver cancer nanodrug modalities. Int J Nanomed 15:1437–1456. https://doi.org/10.2147/IJN.S236927

Sairam AB et al (2023) Toxicity of polymeric nanodrugs as drug carriers. ACS Chem Health Saf 30(5):236–250. https://doi.org/10.1021/acs.chas.3c00008

Sathishkumar K et al (2023) Cancer incidence estimates for 2022 & projection for 2025: result from National Cancer Registry Programme, India. Indian J Med Res 156:598–607. https://doi.org/10.4103/ijmr.ijmr_1821_22

Sharma AK et al (2017) Dendrimer nanoarchitectures for cancer diagnosis and anticancer drug delivery. Drug Discov Today 22(2):314–326. https://doi.org/10.1016/j.drudis.2016.09.013

Shi J et al (2017) Cancer nanomedicine: progress, challenges and opportunities. Nat Rev Cancer 17(1):20–37. https://doi.org/10.1038/nrc.2016.108

Shih KS et al (2013) One-step chromatographic purification of helicobacter pylori neutrophil-activating protein expressed in *Bacillus subtilis*. PLoS One 8(4):e60786. https://doi.org/10.1371/journal.pone.0060786

Shin SJ, Beech JR, Kelly KA (2013) Targeted nanoparticles in imaging: paving the way for personalized medicine in the battle against cancer. Integr Biol 5(1):29–42. https://doi.org/10.1039/c2ib20047c

Shrivastava S et al (2019) A review on theranostics: an approach to targeted diagnosis and therapy. Asian J Pharm Res Dev 7(2):63–69. https://doi.org/10.22270/ajprd.v7i2.463

Shuhendler AJ et al (2012) A novel solid lipid nanoparticle formulation for active targeting to tumor α v β 3 integrin receptors reveals cyclic RGD as a double-edged sword. Adv Healthc Mater 1(5):600–608. https://doi.org/10.1002/adhm.201200006

Siafaka PI et al (2021) Current update on nanoplatforms as therapeutic and diagnostic tools: a review for the materials used as nanotheranostics and imaging modalities. Asian J Pharm Sci 16(1):24–46. https://doi.org/10.1016/j.ajps.2020.03.003

Sin MLY et al (2011) System integration—a major step toward lab on a chip. J Biol Eng 5:6. https://doi.org/10.1186/1754-1611-5-6

Singh D, Dilnawaz F, Sahoo SK (2020) Challenges of moving theranostic nanomedicine into the clinic. Nanomedicine 15(2):111–114. https://doi.org/10.2217/nnm-2019-0401

Song R et al (2018) A multifunctional nanotheranostic for the intelligent MRI diagnosis and synergistic treatment of hypoxic tumor. Biomaterials 175:123–133. https://doi.org/10.1016/j.biomaterials.2018.05.018

Song N et al (2020) Nanomaterials with supramolecular assembly based on AIE luminogens for theranostic applications. Adv Mater 32(49):1–27. https://doi.org/10.1002/adma.202004208

Stirrat CG et al (2017) Ferumoxytol-enhanced magnetic resonance imaging assessing inflammation after myocardial infarction. Heart 103(19):1528–1535. https://doi.org/10.1136/heartjnl-2016-311018

Sung H et al (2021) Global cancer statistics 2020: GLOBOCAN estimates of incidence and mortality worldwide for 36 cancers in 185 countries. CA Cancer J Clin 71(3):209–249. https://doi.org/10.3322/caac.21660

Szejniuk WM et al (2021) High-dose thoracic radiation therapy for non-small cell lung cancer: a novel grading scale of radiation-induced lung injury for symptomatic radiation pneumonitis. Radiat Oncol 16(1):1–11. https://doi.org/10.1186/s13014-021-01857-8

Taratula O et al (2013) A multifunctional theranostic platform based on phthalocyanine-loaded dendrimer for image-guided drug delivery and photodynamic therapy. Mol Pharm 10(10):3946–3958. https://doi.org/10.1021/mp400397t

Tatovic D et al (2022) Safety of the use of gold nanoparticles conjugated with proinsulin peptide and administered by hollow microneedles as an immunotherapy in type 1 diabetes. Immunother Adv 2(1):ltac002. https://doi.org/10.1093/immadv/ltac002

Tenchov R et al (2021) Lipid nanoparticles—from liposomes to mRNA vaccine delivery, a landscape of research diversity and advancement. ACS Nano 15(11):16982–17015. https://doi.org/10.1021/acsnano.1c04996

Thakor AS, Gambhir SS (2013) Nanooncology: the future of cancer diagnosis and therapy. CA Cancer J Clin 63(6):395–418. https://doi.org/10.3322/caac.21199

Thi TD et al (2023) Control the solubility of carbon quantum dots by solvent engineering. HPU2 J Sci Nat Sci Technol 2(3):51–58

Tirumalai RS et al (2003) Characterization of the low molecular weight human serum proteome. Mol Cell Proteom 2(10):1096–1103. https://doi.org/10.1074/mcp.M300031-MCP200

Turkbey B et al (2020) Ferumoxytol-enhanced MR lymphography for detection of metastatic lymph nodes in genitourinary malignancies: a prospective study. Am J Roentgenol 214(1):105–113. https://doi.org/10.2214/AJR.19.21264

Vallet-Regi M et al (2001) A new property of MCM-41: drug delivery system. Chem Mater 13(2):308–311. https://doi.org/10.1021/cm0011559

Vasey PA et al (1999) Phase I clinical and pharmacokinetic study of PK1 [*N*-(2-hydroxypropyl) methacrylamide copolymer doxorubicin]: first member of a new class of chemotherapeutic agents-drug-polymer conjugates. Cancer Research Campaign Phase I/II Committee. Clin Cancer Res 5(1):83–94

Vitorino R et al (2021) Microfluidics for peptidomics, proteomics, and cell analysis. Nanomaterials (Basel) 11(5):1118. https://doi.org/10.3390/nano11051118

Vizirianakis IS et al (2016) Enabling personalized cancer medicine decisions: the challenging pharmacological approach of PBPK models for nanomedicine and pharmacogenomics (review). Oncol Rep 35(4):1891–1904. https://doi.org/10.3892/or.2016.4575

Wang J, Cui H (2016) Nanostructure-based theranostic systems. Theranostics 6(9):1274–1276. https://doi.org/10.7150/THNO.16479

Wang D et al (2018) Theranostics based on AIEgens. Theranostics 8(18):4925–4956. https://doi.org/10.7150/thno.27787

Wang H et al (2023) Aggregation-induced emission (AIE), life and health. ACS Nano 17:14347. https://doi.org/10.1021/acsnano.3c03925

Wu RS et al (2019) pH-sensitive black phosphorous–incorporated hydrogel as novel implant for cancer treatment. J Pharm Sci 108(8):2542–2551. https://doi.org/10.1016/j.xphs.2019.03.003

Xu W, Ling P, Zhang T (2013) Polymeric micelles, a promising drug delivery system to enhance bioavailability of poorly water-soluble drugs. J Drug Deliv 2013:1–15. https://doi.org/10.1155/2013/340315

Yan C et al (2018) Concentration effect on large scale synthesis of high quality small gold nanorods and their potential role in cancer theranostics. Mater Sci Eng C 87:120–127. https://doi.org/10.1016/j.msec.2018.02.021

Yanagisawa T et al (1990) The preparation of alkyltrimethylammonium–kanemite complexes and their conversion to microporous materials. Bull Chem Soc Jpn 63(4):988–992. https://doi.org/10.1246/bcsj.63.988

Yang J, Kopeček J (2014) Macromolecular therapeutics. J Control Release 190:288–303. https://doi.org/10.1016/j.jconrel.2014.04.013

Ye B et al (2018) Chitosan-coated doxorubicin nano-particles drug delivery system inhibits cell growth of liver cancer via p53/PRC1 pathway. Biochem Biophys Res Commun 495(1):414–420. https://doi.org/10.1016/j.bbrc.2017.10.156

Zhang L et al (2011) Enhanced chemotherapy efficacy by sequential delivery of siRNA and anticancer drugs using PEI-grafted graphene oxide. Small 7(4):460–464. https://doi.org/10.1002/smll.201001522

Theranostics as a Novel Strategy for Targeted Delivery of Therapeutics: Transport and Fate Mechanism, Challenge, and Opportunities

3

Subeel Shah, Nishtha Chaurawal, Fahad Saad Alhodieb, Md. Abul Barkat, Simran Preet, and Kaisar Raza

Abstract

The target is tagged using a diagnostic radioisotope in nuclear medicine, and therapy is administered when the target's emergence is confirmed. Using I-131 (RAI) to treat thyroid cancer, theranostic notion was first used in 1964. The nation's Bhabha Atomic Research Centre (BARC) began to sell additional theranostic radiotracers throughout time. Peptides such as DOTATATE and PSMA are easily accessible in kit form inside the country, while Lu-177 is now manufactured in India. The word "theranostic" refers to a therapeutic approach that incorporates medication with imaging to treat a variety of diseases in a targeted or individualized manner. In order to identify and cure diseases at the cellular and molecular scales, theranostic medicine can attain systemic circulation, elude host defenses, and deliver the medication and diagnostic agents to the targeted place. Target site delivery of therapeutic treatments can be facilitated by ther-

---

S. Shah · N. Chaurawal · K. Raza (✉)
Department of Pharmacy, School of Chemical Sciences and Pharmacy, Central University of Rajasthan, Ajmer, Rajasthan, India
e-mail: drkaisar@curaj.ac.in

F. S. Alhodieb
Department of Clinical Nutrition, College of Applied Health Sciences in Ar Rass, Qassim University, Ar Rass, Saudi Arabia
e-mail: f.alhodieb@qu.edu.sa

M. A. Barkat
Department of Pharmaceutics, College of Pharmacy, University of Hafr Al-Batin, Al Jamiah, Hafr Al Batin, Saudi Arabia
e-mail: mdabarkat@uhb.edu.sa

S. Preet
Department of Biophysics, Panjab University, Chandigarh, UT, India
e-mail: simranpreet@pu.ac.in

M. A. Barkat et al. (eds.), *Nanotheranostics for Diagnosis and Therapy*,
https://doi.org/10.1007/978-981-97-3115-2_3

anostics possessing enhanced pharmacokinetic properties, increasing drug efficacy and safety and reducing toxicity. The present chapter deals with various theranostics for targeted drug delivery. It also outlines the current opportunities and challenges.

Keywords

Theranostics · Targeted drug delivery · Nanomaterial · Nanoparticles · Diagnostics

3.1 Introduction

3.1.1 Overview of Theranostics

Theranostics (portmanteau) is a combination of diagnosis and treatment. A portmanteau is a term that results from combining two or more words to form a new word (Pene et al. 2009). In August 1998 news release from Cardiovascular Diagnostics, US consultant John Funkhouser first used the term "theranostics" to refer to a substance that enables the simultaneous diagnosis, treatment, and monitoring of a health issue (Idée et al. 2013). Theranostics combines these properties into a single "package," as compared to developing and using different components for these two purposes. This has the potential to eliminate undesired differences in biodistribution and selectivity that presently exist between additional imaging and therapeutic agents. The capacity to visualize and track sick tissue, drug delivery kinetics, and medication efficacy is the ultimate objective of the theragnostic discipline. The eventual objective is to be able to adjust therapy and dosage with previously unachievable control. The science of nanomedicine can advance toward a time of more efficient and individualized treatment techniques by personalizing medicine rather than taking a "one size fits all" approach (Kelkar and Reineke 2011).

3.1.2 Significance of Theranostics in Drug Delivery

Theranostics are a fast-growing discipline that blends customized medicine with the special potential provided by nanotechnology to deliver therapy to targeted tissues, specifically improving therapeutic efficacy and minimizing off-target consequences. Theranostic techniques integrate non-invasive imaging techniques with targeted administration of drug therapy, which may stem from various physicochemical and physiological principles. There are several therapeutic approaches for theranostics as illustrated below in the Fig. 3.1 and they can be manufactured to have low renal clearance, decreased immunogenicity, and reduced antigenicity, as well as excellent delivery qualities (Hapuarachchige and Artemov 2020). It has been demonstrated that nanoformulations may successfully cure cancer. Nanocarriers can carry many therapeutic medicines to tumor locations, hence improving treatment efficacy.

Nanoformulations such as liposomes, metallic nanoparticles, carbon nanotubes, quantum dots and others can be employed in theranostics. These compounds

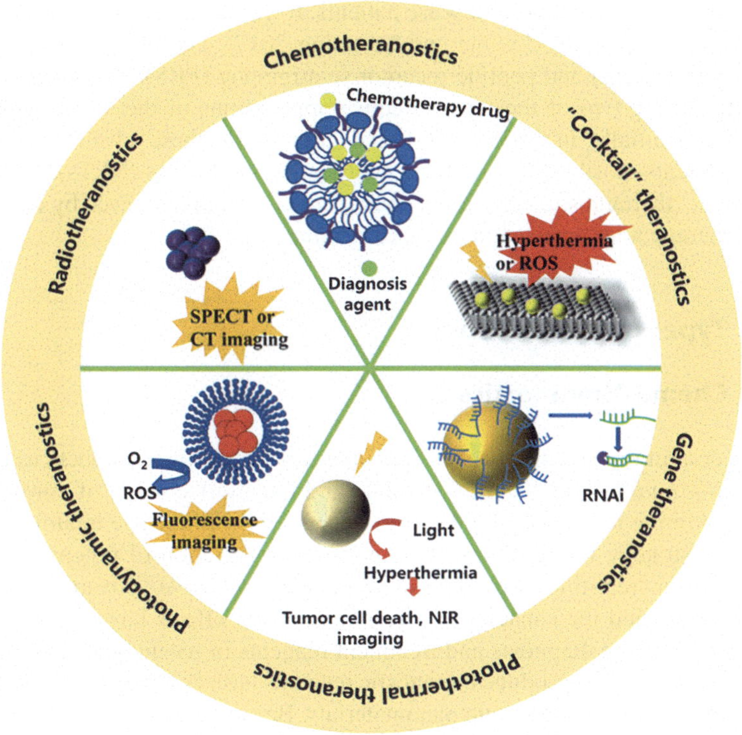

Fig. 3.1 Therapeutic approaches for theranostics. [Reproduced under Creative Common Attribution 4.0 International License with China Anti-Cancer Association (Xue et al. 2021)]

utilized in imaging, therapeutics, and targeting require more thorough study and customized medications for their respective uses in order to achieve optimal theranostic applications and use. With ongoing technological advancements, research, and invention, the field of theranostics has a bright future. These developments might improve theranostics nanoparticles' solubility, excretion, longer circulation times, and greater compatibility with more efficient delivery vehicles (Vats et al. 2017).

3.1.3 Historical Perspective

The phrases "therapy" and "diagnostic" are the roots of the term "therapeutic medicine." It describes a process that consists of a diagnostic test, the outcome of which provides an extremely high likelihood of a successful outcome when applying a certain therapy. It was created in 1998 by PharmaNetics director John Funkhouser, who, as part of research on the business's development approach, created a test for tracking the effectiveness of a novel anticoagulant medication (Kunikowska 2021). Sodium iodide (131I) was licensed by the US Food and Drug Administration (FDA) in 1951 for treatment in thyroid disease patients. It was the first radiopharmaceutical to receive FDA approval (Levine and Krenning 2017). Peptide receptor radionuclide therapy (PRRT) and peptide receptor scintigraphy (PRS) for cancer imaging and treatment are two of the most effective applications of theranostic notion in radiotherapy. Initially, in the late 1980s and early 1990s, these advancements were used for patients mostly suffering from neuroendocrine tumors (NETs). In the new discipline of "theragnostic," a tailored course of treatment is achieved by combining targeted therapy with precise diagnosis (Krenning et al. 1994).

3.2 Types of Theranostics

3.2.1 Chemo-Theranostics

More successful therapy approaches are being developed by researchers as a result of the rising number of cancerous patients and the shortcomings of conventional therapies. Aiming to specifically destroy cancer cells with precise treatment might be a useful strategy for resolving present issues. Integration of precise treatment features such as real-time drug release monitoring, selective activation, imaging-guided therapy, and the capacity to assess therapeutic effects have drawn a lot of interest to integrated diagnosis and treatment reagents in recent years. An ongoing effort has been made to produce theranostic reagents based on polymers, small molecules, and organic and inorganic nanomaterials. Because of the advanced development of chemotherapeutic medications and their ease of estimate and dependability, organic small-molecule theranostic reagents based on chemotherapy have attracted a lot of interest. These resources provide a helpful idea for tailored therapeutic approaches and relies on studies regarding chemotherapy and its related results. Numerous nanotheranostic substances have been produced as a result of their size characteristics' enhanced permeation and retention (EPR) impact (Rong et al. 2022). Due to the theranostic agents' quick growth and ongoing emergence, Kim's team produced fluorogenic reaction-based prodrug conjugates as targeted cancer theranostics more thorough overview of the latest advancements in small-molecule theranostic agents based on chemotherapy in 2018 (Lee et al. 2018).

Reactive oxygen species (ROS) overproduction in cancer cells has been extensively documented. Focusing on the quinazolinone derivative BQA, Gao et al. created a dual-responsive AIE probe that is H_2O_2 and self-assembled. It was then decorated with a tetra peptide (GGFF) and an aryl boronated immolate linker. The hydroxyl group may be easily restored by the probe's reaction with H_2O_2. The internal hydrogen bond was broken in the soluble state and below the critical assembly concentration (CAC), leading to a poor emission. The intramolecular hydrogen bond may be preserved in the aggregate state and above the critical assembly concentration (CAC), as evidenced by the significant emission that was seen. Cancer cells showed a high fluorescence intensity based on the dual response to H_2O_2 and self-assembly. By using the intracellular H_2O_2 threshold and nonlinear fluorescence intensification, this redox self-assembly method may successfully distinguish cancer cells from healthy cells (Gao and Tang 2020).

3.2.2 Radio-Theranostics

Featuring numerous advancements and new prospects, radiotheranostics are perhaps the most clinically sophisticated theranostics application. Radiotheranostics relies heavily on imaging of the same target region to choose patients for radio-targeted therapies; as a result, treatment and imaging are tightly related. In addition, radiotheranostics enters the mainstream of cancer treatments, and it is going through a transformation. Stabilizing end-stage diseases that are resistant to conventional therapies and enhancing the quality of life for these patient groups have been the primary objectives of radiotheranostic indications. Individuals with neuroendocrine tumors, thyroid and prostate malignancies that would otherwise be incurable have shown better results from early clinical trials (Luster et al. 2008). When metastatic thyroid cancer was treated with radioactive iodine, the disease's overall survival rate increased from low to around 85% (Herrmann et al. 2020).

Either a therapeutic or a diagnostic version employs intense radiopharmaceuticals, ideally with quantitative PET/CT employing positron emitters. For each patient, the tumor reason and stage are confirmed using the diagnostic radiopharmaceutical. On the other hand, it confirms that the radiopharmaceutical in question is a particular and target-specific chemical, making it the best choice for a particular patient. For the therapeutic version of the theranostic drug, precise diagnostic imaging also makes quantitative dosimetric data—expressed in terms of the maximal radiation dosage to the target tissue and the tolerated dose to healthy organs—possible. This is administered with the best radioactivity possible using the right radiopharmaceutical for each patient.

Nuclear medicine and radio-pharmacy use radiopharmaceuticals as key tools, which are accessible in both therapeutic and diagnostic forms (often for PET/CT). The procedure involves replacing the diagnostic radionuclide with its therapeutic equivalent in order to convert the derivative from diagnostic to therapeutic (Roesch and Martin 2023). A recently discovered bifunctional chelator called TRAP (triazacyclononance–triphosphinate) was created by Notni et al. It allowed the conjugation of three targeting vectors while exhibiting better 68Ga-labeling properties than DOTA and NOTA (milder reaction conditions, broader pH range during labelling, and higher specific activities while maintaining high complex stability) (Notni et al. 2010).

Because of their benign nature and easy chemistry to facilitate surface alterations, silica nanoparticles (NPs) are appealing candidates for radio-nanomedicine. Benezra et al. created multimodal PEGylated silica NPs coupled with a targeting peptide $\alpha v\beta 3$ integrin arginine–glycine–aspartic (RGD) and a prolonged half-life of nanoparticle. Comparing the created NPs to previous inorganic NPs platforms, a mix of structural, optical, and biological features was present. The applicability of this NPs system was evaluated using melanoma models that were both tiny (mouse) and big (mini pig) (Desai et al. 2022).

3.2.3 Immuno-Theranostics

In recent years, nuclear medicine has attracted more interest and seems to be heading toward rapid achievement, particularly with regard to theranostic operations. Using nuclear medicine and molecular imaging techniques to see a disease site in its whole-body manifestation before starting a comprehensive therapy targeted at the same target seems to be a clear and promising application of the theranostic principle (Strosberg et al. 2017). These significant accomplishments also made nuclear medicine appear as one of the most promising collaborators in cancer patients' oncological therapies. Nonetheless, intensive research in immunological oncology (IO) has led to the development of numerous checkpoint inhibitor medicines and other immune-modulating anticancer medications, coinciding with the advancement of targeted radioisotope therapy (Decazes and Bohn 2020).

Numerous collaborative teams have published a number of high-caliber studies and trials that use nuclear imaging techniques to non-invasively see and measure the targets of IO treatments. De Vries et al.'s first-in-human investigation revealed that tumor patients' absorption of zirconium-89-labeled atezolizumab (anti-PD-L1) was generally strong but varied (Bensch et al. 2018). Furthermore, it has the potential to cure certain individuals with cancer completely; immunotherapy is the most effective and relevant current cancer treatment. As a result, it would appear sensible to introduce the conceptual phrase "Immuno-theranostics" and put "Immuno" before "This method." This means, in our opinion, not only the imaging validation of the target, which is the identification of an IO target but also the modulation of an IO target to increase its accessibility for immunotherapy, which would ultimately result in a higher percentage of patients being cured by IO. In the context of the antagonistic TME and its plasticity, we firmly believe that immuno-theranostics will play a significant part in solving the mystery of ineffectual IO therapy and that true synergism is possible. Inevitably, immunotherapeutic is far more complicated than traditional theranostics and requires a deeper biological comprehension of a dynamic, diverse target that is local, systemic, and possibly changing over time (Schaefer et al. 2020).

3.2.4 Combined Quantum Dots Immuno-Theranostics Nanomedicines

In order to facilitate photothermal therapy (PTT) by hydrogen bonding and electrostatic assembly, Qian et al. created quantum dots (CDs) with coatings made of polymers of mesoporous silica nanoparticle frameworks (CD@MSNs). CDs have been shown using laser imaging to emit multiple wavelengths at 488, 552, and 636 nm in the green, yellow, and red regions. In complex physiological settings, this nanoplatform may successfully avert the breaking down or inactivation of tumor antigen proteins, hence enhancing immune response (Luo et al. 2018).

In order to image both in vivo and in vitro, a number of biological adaptable and harmless bio-imaging agents, such as graphene quantum dots (GQDs), were studied. To enable photoimmunotherapy guided by fluorescence and magnetic resonance imaging, Wu et al. created Gd3+/Cy3 scanning sensors, immunoadjuvant polycationic polymer/CpG oligodeoxy nucleotide nanoparticles (PC), and polydopamine coated Graphene quantum dots photosensitive nanomaterials (GCpD) (Nasser et al. 2023).

3.2.5 Gene-Theranostics

A novel discipline called gene theranostics—where a nano-bioconjugate is utilized to both diagnose and cure disease—has emerged as a result of the integration of gene therapy, NPs, and medical imaging. Binding typically occurs between a nanoparticle, which supplies the imaging signal, and a vector conveying genetic information. Gene transduction effectiveness and contrast imaging are both improved by the synergistic impact of this probe's synthesis.

The application of nanomaterials in gene theranostics investigations is a unique technique resulting from the coupling of this treatment with the theranostics idea. By introducing therapeutic genes using integrative or non-integrative vectors, gene therapy is utilized to cure monogenic hereditary or inherited diseases. Safely altered viral vectors, known as "integrative vectors," allow a therapeutic transgene to be inserted into the target cell's DNA. The therapeutic transgene is not persistently expressed in the target cell when using non-integrative vectors (Herranz et al. 2011).

Researchers Räty et al. established the utilization of magnetic nanoparticles, enabling the manufacture of an inert nano bioconjugate in gene therapy, specifically for non-invasive MRI imaging of baculovirus biodistribution in vivo. Commercial ultrasmall superparamagnetic iron oxide nanoparticles (USPIO) were utilized as the NPs (Räty et al. 2006).

Despite being employed for gene delivery and imaging, gold nanoparticles have received significantly less attention in magnetic resonance imaging (MRI), mostly because the NPs must undergo a second alteration in order to be active. According to Kamei et al. nanoparticles can be used to enhance gene therapy, and gold nanoparticles can increase the potential of a number of imaging methods (Kamei et al. 2009). The thermogenic potential of magnetic graphene nanoparticles functionalized with chitosan was investigated by Wang et al. It was shown that by combining the delivery of genes and chemotherapeutic drugs, magnetic graphene nanoparticles coated with chitosan displayed strong theranostic properties, cancers in mice, and MR imaging in the tumors (Wang et al. 2013).

3.2.6 Phototherapy-Theranostics

Some common characteristics of therapeutic medications are tumor targeting, imaging, and treatment. Theranostic drugs that possess photosensitizers that emit fluorescence and are phototoxic can be utilized as both therapeutic and imaging agents since they can kill tumor cells in addition to imaging them. Photothermal treatment (PTT) or photodynamic therapy (PDT) can be used to fulfil the therapeutic role of photosensitizer. Reactive oxygen species (ROS) are used by PDT to harm cancerous cells, whereas PPT uses light-induced hyperthermia to destroy cancerous cells. As a result, photosensitizers embellished with tumor targeting units combine the three theranostic agent activities of imaging, treatment, and tumor targeting. Several photosensitizers also exhibit tumor-targeting properties inherent in their structure. It is possible to accomplish tumor targeting, imaging, and treatment through photosensitizers despite the need for further modifying groups. As a result, photosensitizers have been essential to creating theranostic treatments (Gao et al. 2023).

It is now widely known that tumor cells vary from typical cells in the following specific ways: Additionally, tumor cells are often under oxidative stress. Certain receptors on the surface of tumor cells and internal enzymes are significantly expressed (Baig et al. 2019). Porphyrins and derivatives of phthalocyanines share a skeleton, making them the first generation of photosensitizers. At the moment, the creation of theranostic agents makes extensive use of all of these. Theranostic agent 1, created by Wang et al. was selective to FR over-expressed KB cells and less so to FR-negative A549 cells. It was created by conjugating FA to zinc tetraaminophthalocyanine (Wang et al. 2014). Because of their favorable photophysical characteristics and comparatively high stability even in aqueous conditions, BODIPY (Fluorinated Boron-Dipyrromethene) derivatives are also interesting photosensitizers in the development of anticancer theranostic drugs. Ke et al. conjugated FA with oligoethylene glycol as a linker of varying lengths to create two theranostic agents, Two folate-conjugated diiododistyryl boron dipyrromethene (Ke et al. 2013).

Because of their intense fluorescence, photosensitizers with aggregation-induced emission (AIE) characteristics have been investigated for the development of theranostic agents. Hu et al. also created the theranostic agent 20, which has a peptide specific to cancer cells and an AIE fluorophore acting as a photosensitizer (Hu et al. 2014).

3.3 Theranostic Agents in Targeted Drug Delivery

Throughout the pharmaceutical sector, theranostic is a new name for goods with both medicinal and diagnostic qualities. Creating theranostic nanomedicines, which combine therapeutic and diagnostic effects into a single formulation, is highly beneficial for addressing global public health concerns. To have the greatest potential effect, the medicine is delivered to the targeted organ via nanoparticles (NPs). In addition, they lengthen the duration that a medication remains in the circulation near the tumor, which makes them advantageous even for passive targeted drug delivery. Targeted theranostic nanomedicines have a potential future in medication delivery. Targeted medication delivery using theranostic nanomedicines has the potential to address several issues that are typically associated with ineffective cancer treatments and diagnoses (Lodhi et al. 2021).

Clinical therapies face a significant problem in the effective and site-specific administration of therapeutic medicines. Improving treatments' bioavailability and pharmacokinetic qualities through the increased permeability and retention (EPR) effect is one method by which nano-sized carriers, including liposomes, micelles, and polymeric nanoparticles, have been studied. Targeting ligand conjugation onto nanocarriers may provide selective distribution and enhance the situation even more (Yu et al. 2010).

Depending on their functions and physical positions, the carrier and surface modifiers of a theranostic nanoparticle can be divided into a minimum of three parts: the biomedical payload, the carrier, and the surface modifier. The carrier should sufficiently protect the biological payloads under physiological conditions during delivery to the desired target site. The modifiers are attached to the surface of the carrier. They are expected to give theranostic nanomaterials additional characteristics, such as long circulation time, barrier-penetrating ability, and target-specific binding ability (Lim et al. 2015). In the field of theranostics, the nanocarriers as shown in Fig. 3.2 have a dual function that allows for targeted drug delivery and diagnostic imaging.

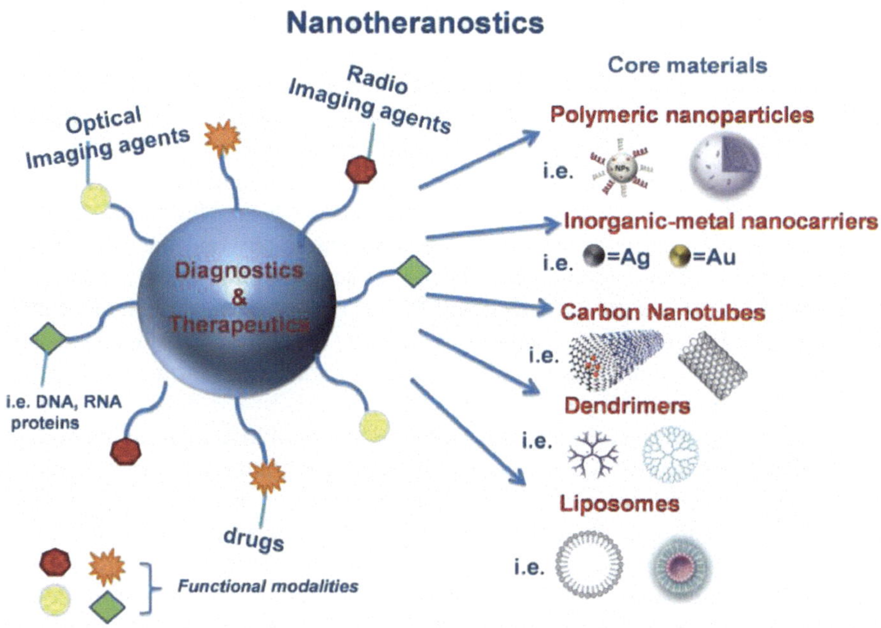

Fig. 3.2 Various nanocarriers used for theranostics. [Reproduced under Creative Common Attribution-Non Commercial-NoDerivs 4.0 International with Elsevier (Siafaka et al. 2021)]

3.3.1 Image-Based Theranostics

Theranostics are a fast-growing discipline that blends customized medicine with the special potential provided by nanotechnology to deliver therapy to targeted tissues, specifically, improving treatment efficacy and reducing off-target consequences. Imaging diagnostics and treatment are two applications for theranostic probes. Imaging and therapeutic molecules should be components of a single platform that is customized with distinct subunits for specific recognition of target molecules, image markers, and therapeutic chemicals. In order for the application to be genuinely theranostic (Langbein et al. 2019).

Numerous different types of imaging, including optical (bioluminescence or fluorescence), nuclear single photon emission computed tomography (SPECT) and positron emission tomography (PET) photoacoustic, magnetic particle, ultrasound, and magnetic resonance imaging methods, are employed in the field of theranostics (Penet et al. 2012). The imaging outcomes by using the pretargeting component, (1) confirm the appearance of receptors on the surface are present at the intended site, (2) accurately timing the second drug delivery component's administration depending on the pretargeting component's higher specific retention at the target site and subsequent clearance from other parts of the body. The expression of receptors may be differences between primary tumor and distant metastatic sites, making them difficult to biopsy. A "go" signal for the injection of the anticancer drug carrier is therefore provided by pictures from the pretargeting component, which confirm receptor expression. Equally significant is a visualization of the drug carrier, which can establish that the carcinogenic component was delivered and retained (Arranz and Ripoll 2015).

A theranostic system's biological imaging component would offer vital diagnostic data on the existence and anatomical position of cellular targets targeted for therapeutic drug delivery. Physicians practicing nuclear medicine are familiar with this idea, which originated in the middle of the 1940s when radioiodine was used as maybe the first imaging-based molecular theranostic agent. Nowadays, Modern molecular imaging, utilizing positron emission tomography and the 18F-labeled glucose analogue FDG, has completely changed therapies, most notably in the field of cancer. Fludeoxyglucose-18 (FDG) and PET is now often used interchangeably with molecular image scanning, and its use in patient care for cancer staging, restaging, and tracking treatment response for several reimbursable reasons has become essential (Lee and Li 2011).

Theranostic systems were created for the co-delivery of MRI contrast agents and therapeutic agents in order to lessen side effects, increase bioavailability, get around complicated delivery obstacles, and evaluate therapeutic responses; theranostics MRI contrast agents are T1-MRI and T2-MRI (Zhang et al. 2022). To solve this problem and provide real-time information on the distribution route and related treatment responses, many MRI-traceable, tailored therapeutics have been developed. One important biological barrier for the transport of drugs to the brain is the blood–brain barrier (BBB), which shields the brain from dangerous viruses. Using nonspecific or active magnetic targets, Israel et al. have described a number of methods to breach or improve permeability across the BBB using magnetic iron oxide nanoparticles (David et al. 2017).

According to MRI scans, the BBB at the glioblastoma (GBM) location was shown to be compromised, and an ex-vivo biodistribution research found that PLGA nanoparticles may concentrate in the mouse brain after magnetic therapy. In an attempt to get around the BBB, magnetic resonance-guided focused ultrasound (MRgFUS) has also been tried. This method functions by temporarily disrupting the blood–brain barrier (BBB) using low-energy ultrasound, and it can accurately focus systemically given medicine to certain CNS regions (Abrahao et al. 2019).

Studies conducted in vivo showed improved medication administration at the tumor location, and dual MRI-near infrared fluorescence (NIRF) imaging might help accomplish precise tumor resection. In a different investigation, polypeptide nanoparticles containing SPIONs and the anticancer drug cisplatin were created for stimuli-responsive MRI-guided theranostics (Feng et al. 2018). A single targeted molecule, such as an antibody or peptide, can be used by theranostics to facilitate both therapeutic and diagnostic imaging. Radiometals for targeted radiopharmaceutical treatment (TRT) and positron emission tomography (PET) imaging have been included in recent theranostics. DOTA-octreotate (dotatate) can be radiolabeled with either Ga-68 (68Ga-dotatate; NetSpot) or Lu-177 (177Lu-dotatate; Lutathera), which are FDA-approved for PET imaging and TRT, respectively, using a single chelator. This demonstrates the theranostic medications' potential for translation (Foster et al. 2021).

3.3.2 Nanoparticle-Based Theranostics

After years of independent research, the field of nanoparticle-based treatment and imaging has advanced to a point where knowledge of them has allowed for the development of NP-based theranostics, or nanoplatforms that combine imaging and therapeutic capabilities. This is similar to classical theranostics in several ways, but with a stronger emphasis on "codelivery." It expands on the prior paradigm by permitting imaging to be carried out throughout a treatment regimen addition to before or after. Several nanoparticles are already imaging agents, which makes it easy up of agents in affected regions (Cai and Chen 2008). When coated appropriately, iron oxide nanoparticle-based theranostic agents (IONP) may be combined with therapeutic molecules easily. The Zhang group linked the anticancer medication methotrexate (MTX) onto an aminated IONP surface. Studies conducted in vitro showed that the particles internalized into cells and gathered in lysosomes, where the presence of proteases and the low pH caused the drug molecules to be released. The binding of paclitaxel (PTX) to IONP surfaces via a phosphodiester moiety was reported by Hwu et al. The fields of nanoparticle-based treatment and imaging are having difficulty moving forward with clinical trials, and as the offspring of these two fields, nanoparticle-based theranostics is still in its infancy. Theranostics exhibits benefits and drawbacks as listed in Table 3.1. However, the need for

Table 3.1 Pros and cons of theranostics

Type of nanotheranostics	Pros	Cons	References
Mesoporous silica nanoparticles	The large expanse of surface Biological compatibility Attainable porosity Surface reactive and accessible functionality	In vivo-toxicity	Xu et al. (2023))
Liposomes	Adaptability in formulation, modification, and synthesis Convenient conjugation and integration using targeting agents Quick cellular absorption and well describe cell internalization process	Poor drug loading performance Fluctuation from batch to batch in manufacturing High cost of manufacture Low solubility	Daraee et al. (2016))
Gold nanoparticles	Sizeable surface area Lower hydrodynamic mean size Appropriate for photodynamic therapy Surface modification is conveniently	High expenses for production on a wide scale Biocompatibility issue Absence of uniform procedures for translating into the clinical	Kesharwani et al. (2023))
Quantum dots	High photo- and chemical stability Feasibility of using synergistic applications in therapy and diagnosis	Hazardousness Challenging surface alteration Particle accumulation and deterioration	Azam et al. (2021))
Carbon nanotubes	Large surface area High penetration capacity to biological barriers Suitable for photodynamic therapy	Nonbiodegradable Lack of a standard protocol to prepare high-purity	Shao et al. (2013))

individualized treatment and the push from nanotechnology advancements have already turned nanoparticle-based theranostics into a hotspot for study (Xie et al. 2010).

3.3.2.1 Liposomes

The application of therapeutic and diagnostic agents is essential for the early diagnosis and management of illness. The conventional approaches to creating these agents frequently have harmful side effects or inadequate agent delivery to the target cells or tissues. A lot of work has been put into creating efficient nanocarriers in recent years in an attempt to overcome these constraints. Using therapeutic and diagnostic chemicals in one system to provide simultaneous treatment and detection is another successful strategy. Efficient theranostic systems that possess high stability and targeting ability and do not interact with one another between various therapeutic and diagnostic chemicals inside the system are necessary to accomplish these aims. Liposomes seem to be one of the most promising nanocarrier systems for therapeutic applications among those studied so far (Xing et al. 2016).

The application of liposomes as a diagnostic tool emphasizes the use of diverse imaging modalities and the identification of diverse molecular target types. Fluorescence is one of the many popular and frequently utilized medical imaging techniques that have been carried out using liposomes as nanocarriers. In order to understand the biological processes of live cells or tissues, it enables the viewing of biomolecule location, gene expression, and enzyme activity. Enhanced pharmacokinetic liposomes can aid in the delivery of fluorescent imaging chemicals to the intended location. In order to create functionalized-quantum dot–liposome (f-QD–L) hybrid nanoparticles for cancer imaging, Kostarelos and colleagues encapsulated PEG-coated quantum dots (QDs) into the aqueous phase of DOPC-supported liposomes (Al-Jamal et al. 2008).

Investigations employing liposomes for theranostic purposes are becoming more and more common. A photosensitizer, hexadecyl amine chlorin e6 (hCe6)-loaded liposome, and a hypoxia-activated prodrug, AQ4N, was described by Feng et al.

PET imaging is possible for liposomes, and they can be used for photodynamic treatment (PDT) when exposed to 660 nm light-emitting diodes (LEDs). Additionally, in tumor-bearing animals, Poly(9,9-dioctylfluorene-2,7-diylco-benzothiadiazole) (PFBT) as a fluorescent probe and anticancer medications (doxorubicin and folate)-loaded liposome demonstrated efficient imaging and therapeutic effectiveness. Liposomal preparations containing the photosensitizer Ce6, the hypoxia-activated prodrug Tirapazamine (TPZ), and a gene probe were created by Zhang et al. These preparations provide synergistic photodynamic chemotherapy. Additionally, liposome NPs coated in gold were reported by Rengan et al. for PTT (Lee and Im 2019).

Regarding brain-targeted medication administration and imaging, glutathione-reduced gold nanoparticles (AuGSH) and DTX can be combined to create transferrin (Tf)-receptor-targeted gold-based theranostic liposomes. Glycoprotein transferrin is utilized in receptor-mediated endocytosis of nanoparticles to improve medication absorption for cancer treatment, including DTX, and are appropriate for both imaging and therapeutic applications because of their appealing qualities, which include biosafety, stability, easy customization, and high surface plasmon uptake. They may be employed in a variety of imaging modalities, including photoacoustic and CT imaging (Fernandes 2023).

3.3.2.2 Exosomes

Exosomes' exceptional biodistribution and biocompatibility have drawn attention to their use as medication or nucleic acid delivery vehicles. They are able to pass through the BBB and are membrane-permeable (Zhuang et al. 2011). In a non-invasive way, exosomes are good indicators for cancer detection. Unlike traditional drug delivery techniques, exosomes are non-hazardous, non-immunogenic, and exhibit a higher rate of penetration through tumor cells. Lack of targeted toxicity as well as immunogenicity can be mitigated by using exosomes as carriers, which can also improve low bioavailability (Gulati et al. 2022).

Liquid biopsy could be used to diagnose both benign and malignant liver cancers using exosomes. According to Wang et al. exosomes are essential for the detection of hepatocellular carcinoma (HCC). According to Zhang et al. the hepatic microenvironment is regulated by exosome-delivered receptors for epidermal growth factor (EGnjFR), which encourages the liver metastasis of stomach cancer. Exosome proteins produced from serum have been identified by Zhao et al. as possible candidate biomarkers for hepatocellular carcinoma (HCC). Exosomes may serve as an indirect indicator of the liver's diseased condition. As a result, they can be applied to the assessment and diagnosis of liver disease (Luo et al. 2022).

3.3.2.3 Dendrimer

The dendrimer molecule is symmetrical and almost spherical, with repeated structures resembling polymers. The majority of polymers do not dissolve in water. If necessary, though, the dendrimers can be rendered water-soluble. In nanomedicine, the dendrimer serves as a flexible drug delivery nano vehicle. Its adaptable structure enables a broad range of compounds to be conjugated to the outside while maintaining the ability to transport various medications by creating covalent or non-covalent connections inside the core (Kesharwani et al. 2014). A nanocomposite consisting of an FA-conjugated dendrimer containing Technetium-99m (99mTc) and Mn(II) has been shown recently by Luo et al. for targeted treatment and SPECT and MRI detection. This technique used MRI to give high spatial resolution and SPECT to provide information about the tumor's physiological state. Moreover, it could be less hazardous because Mn(II) was used rather than Gd(III). Another helpful theranostic strategy to get over MRI's limited sensitivity would be to combine it with fluorescence imaging (Ray et al. 2018).

The first gadolinium-incorporated PAMAM dendrimer for MRI imaging was created by Wiener et al. The relaxivity, which is correlated with the strength of the MRI signal, was raised by a single PAMAM dendrimer containing many Gd chelates. The relaxivity of macromolecular Gd(III) chelates can be significantly improved by delaying their molecular tumbling because the relaxivity of the chelates influences the water exchange rate. For MRI to be successful, high relaxivity and regulated contrast agent pharmacokinetics are essential. The development of several macromolecular contrast agents based on dendrimers has included conjugating Gd(III) chelates to the dendrimer periphery. Many nano globular (POSS core) dendrimers allow the conjugation of Gd(III) chelates, just like the PAMAM dendrimer does. The main variables governing in vivo contrast enhancement in MRI are the pharmacokinetics and renal excretion properties of the nano-globular contrast agents (Sk and Kojima 2015).

3.3.3 Metallic Nanomaterials for Theranostics

The past few years has grown substantially in interest in the application of nanotechnology to both the prevention and treatment of cancer, particularly in the development of novel diagnostic and therapeutic tools. Numerous attempts have been made to integrate monitoring and therapeutic features into a single, efficient nanomedicine system. This idea, known as "theranostics," is a nanosystem with the ability to assess therapy responses, diagnose problems, and successfully spread. Therapeutic methods and systems have significant promise for monitoring and improving nanomedicine-mediated targeting of drugs. By combining clinical capabilities with molecular imaging technologies, treatments can be helpful in medication selection, treatment planning, goal response control, and follow-up organizing based on the particular molecular characteristics of the disorder. The use of metallic nanoparticles in bio-applications, including biophysical research, imaging, cancer treatment, and medical diagnostics, has grown in popularity recently (Mostafavi et al. 2021).

3.3.3.1 Tungsten-Based Nanoparticles

This is used to create a novel theranostics nanoplatform using gold nanoparticles (Au NPs) and up conversion nanoparticles (UCNPs) loaded into tungsten trioxide (WO_3). The nanosystem functioned as an effective radio-sensitizer for radio chemotherapy against cancer because of its enhanced oxygen-generation effects and doxorubicin loading capability. The use of tungsten disulfide (WS_2)-based nanomaterials as PTT agents was covered by Niknam et al. The high expression of H_2O_2 that results in cell membrane rupture, mitochondrial failure, reactive oxygen species (ROS) generation, and oxidative stress could be catalyzed by the nanocomposite in conjunction with X-ray irradiation. The outcomes demonstrated that local RT/PTT could effectively stop tumors metastasis, destroy nearby tumors, and stop cancers from coming back. To improve RT results, the nanocomposite may also simultaneously produce high temperatures when exposed to near-infrared radiation (Niknam et al. 2022).

3.3.3.2 Iron-Based Nanoparticles

Iron is the most studied and applied metal in cancer therapy. Superparamagnetic iron oxide nanoparticles are a significant class of these materials, and because of their magnetic and biocompatibility qualities, they have been thoroughly studied for use in cancer treatment. The effectiveness of iron oxide nanoparticles was assessed in vitro and in vivo using MRI by Sun et al. after synthesizing them conjugated with the therapeutic drug methotrexate and the targeted ligand chlorotoxin. These multifunctional nanoparticles have the potential to be used in cancer detection and therapy (a process known as thermogenic application), as evidenced by their results, which revealed enhanced cytotoxicity in tumor cells (Sun et al. 2008).

Research revealed that silver-based nanoparticles (AgNPs) enter cells via endocytosis and localize inside the cytoplasmic perinuclear region and endo lysosomal compartment. AgNPs cause reactive oxygen species (ROS) and have an impact on cell respiration. AgNPs cause oxidative stress, DNA damage, apoptosis, and mitochondrial damage to cancer cells, among other detrimental effects on cells. Research revealed that vascular endothelial growth factor, which is involved in angiogenesis, can be affected by AgNPs. Combining therapy and diagnosis is called theranostics. Due to its anticancer properties, biosynthesized AgNPs is useful as a targeted drug delivery and bioimaging carrier, among other theranostic uses. It is an economical, secure, easy, and environmentally responsible method (Sakthi Devi et al. 2022). AgNPs have the ability to inhibit lung cancer cells migration and invasion in vitro, according to researchers. They propose that a key factor is nanoparticle size, with 13 nm AgNPs being the most efficient. The impact begins to diminish at 45 nm and disappears entirely for AgNPs that are 92 nm in size (Takáč et al. 2023).

3.3.3.3 Copper-Based Nanoparticles

To achieve the combination therapy of malignancies, Cu-based nanocomposites with exceptional characteristics were successfully created. Due to their special physical and chemical characteristics, Cu-based NMs have emerged as one of the areas of focus for cancer theranostics research. In the last 5 years, over 15,000 peer-reviewed reports on cu-based NMs for imaging and cancer therapy have been published (Zhong et al. 2022).

It was demonstrated by Zhou et al. that polyethene glycol (PEG)-coated copper(II) sulfide nanoparticles (PEG-CuS NPs) with a peak absorption of 1064 nm could be used effectively as a mediator for photo thermolysis of cancer cells along with a contrast substance for in vivo imaging of the 4T1 breast tumors vasculature. Ding et al. suggested the aqueous production of PEGylated copper sulfide particles with a controlled size between 3 and 7 nm to lessen the shortcomings of earlier research. Particles with a greater potential for tumor imaging, especially those smaller than 5 nm, were shown to have this ability in later preclinical investigations (Ren et al. 2019).

3.3.3.4 Gold-Based Nanoparticles

Nowadays, gold nanoparticles (AuNPs) are extensively used in phototherapy and bio-imaging because of their very sensitive optical and electrical characteristics that may be adjusted (the surface plasmon resonance). Multifunctional AuNPs, a novel idea known as "theranostics," may have therapeutic and diagnostic features that may be combined into a single system, enabling therapy and diagnosis at the same time and tracking therapeutic outcomes. Due to their advantageous physical characteristics and customized surface functionalization, gold nanoparticles (AuNPs) offer a promising platform for the development of theranostics (Guo et al. 2017).

The distinct and sensitive nanoplatforms that GNPs offer for imaging, diagnosis, and treatment are made possible by their physicochemical characteristics. Localized surface plasmon resonance is the most important and potent optical property among these GNPs features, and it is utilized in surface-enhanced processes, including surface-enhanced fluorescence, surface-enhanced SERS, and other plasmonic performances (Gao et al. 2021). The effectiveness of gold nanoparticles in radiosensitizing cancer cells was demonstrated by Unak et al. They created targeted

AuNPs by attaching an antibody called anti-metadherin (Anti-MTDH), which binds to a receptor that is overexpressed in breast cancer cells (Akhter et al. 2012).

3.3.4 Carbon-Based Nanomaterial for Theranostics

The application of nanostructured elements for the establishment of cancer theranostic substances is gaining a lot of attention due to recent developments in the field of nano-biotechnology. Many kinds of nanostructured materials have been created and researched for cancer theranostic applications during the past 20 years. As a result, hundreds of cancer-related research publications have been published. Because of their special physicochemical qualities, carbon-based nanostructured materials (such as CNTs, graphene, fullerenes, and nanodiamonds) have been thoroughly studied for cancer theranostics along with the potential to create extremely effective multi-function diagnostic as well as therapeutic agents (Augustine et al. 2017).

Because of its capacity to carry therapeutic molecules and enable visual inspection of cells and tissues—both essential for the diagnosis and treatment of sick and damaged tissues—carbon-based nanomaterials, or CBNs, have demonstrated a huge influence in the biomedical area. The compounds that makeup CBN are fullerenes, carbon nanotubes (CNTs), graphene (G), and its derivatives, such as graphene oxide (GO), nanodiamonds (NDs), and carbon-based quantum dots (CQDs) (Patel et al. 2019).

3.3.4.1 Carbon-Nanotubes

Whereas the term "carbon nanomaterials" has gained widespread usage in recent times, its origins may be traced back to the 1950s. Meanwhile, Sumio Iijima's 1991 synthesis and structural analysis of carbon nanotubes (CNTs) marked the beginning of the first real systematic study of carbon nanomaterials. Since then, carbon nanotubes (CNTs) have been extensively researched, primarily for use in electronics and high-strength composite materials. The sp2-carbon hollow cylindrical structure of CNTs may be controlled in terms of length, diameter, and wall count. Based on their multifunctional qualities, CNTs are being intensively investigated in nanomedicine in fields including drug delivery and imaging, frequently concurrently in a theranostic approach as shown in Fig. 3.3. Through the employment of photoacoustic imaging (PA) technology in cell imaging, Avti et al. directly identified, located, and evaluated the number of carbon nanotubes (CNTs) in various tissue specimens. The strong NIR absorbance of CNTs in these specimens made the signals of these samples clear and consistent. Additionally, Wang and associates reported on a particular type of silica-coated, gold nanorods functionalized multi-wall carbon nanotubes (MWNTs) modified by RGD that had the potential to actively target in-vivo PA imaging of stomach cancer (Tang et al. 2021).

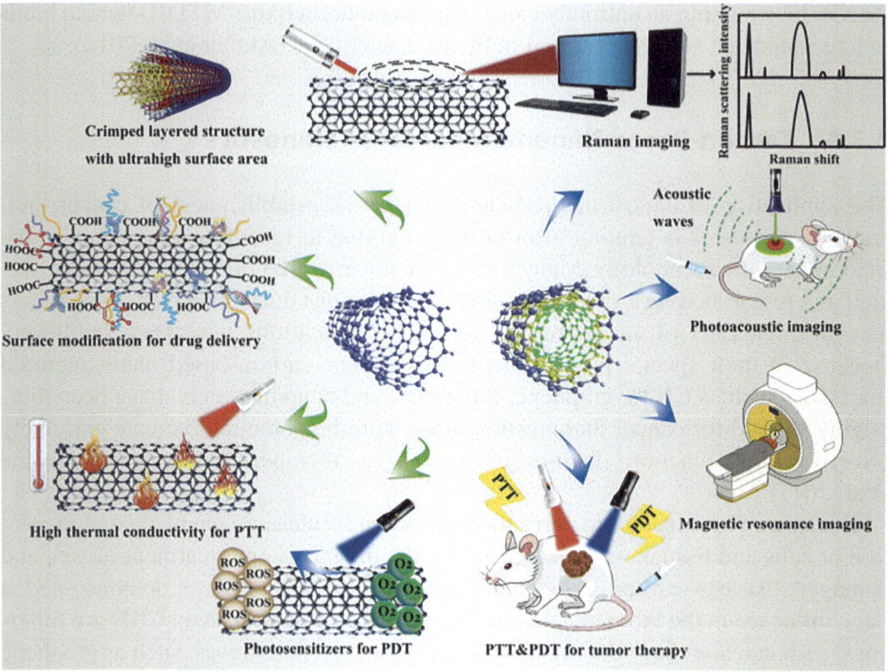

Fig. 3.3 Overview of the diverse applications of CNTs in cancer theranostics. [Reproduced under Creative Common Attribution 4.0 International with Springer Nature (Tang et al. 2021)]

3.3.4.2 Quantum-Dots

Carbon quantum dots, or CQDs, are a potential family of imaging agents and drug carriers with low toxicity, good biocompatibility, optical characteristics, and drug-loading capacity for a range of biological applications. Nevertheless, the majority of CQDs that have been previously documented have nonselective interactions with both normal cells and tumor cells. CQDs can target specific tumors with therapeutic drugs by imaging them. This discovery might point to a viable method for precisely imaging tumors and delivering treatment to them (Li et al. 2020).

To treat tumors that overexpress epidermal growth factor receptor (EGFR) EGFR, a quantum-dot PEGylated poly(lactic acid) micelle (QD-PLA-PEG) coupled with anti-EGFR nanobody, 7D12, was developed. Because the indium phosphate core/zinc sulfide shell quantum dot core of the micelles showed stable NIR fluorescence, optical imaging could be used to track the micelles' location both in vivo and in vitro. Therefore, the nanobody (Nb)-conjugated QD-PLA-PEG micelles demonstrated a much greater tumor-targeting ability than the nontargeted ones in an MDA-MB-468 breast cancer xenograft mouse model. For tumors over-expressing EGFR, nb conjugated QD-PLA-PEG micelles can potentially be used as a targeted theranostic approach (Wang et al. 2017).

3.3.4.3 Graphene-Based Nanoparticles

Many nanomaterials have been extensively investigated recently for possible theranostic uses. Graphene, a relatively new nanomaterial, has generated a lot of attention in the scientific community and the general public. Because of its unique physicochemical properties, which include electrical, optical, thermal, and mechanical properties, graphene is being investigated for many possible uses. Since graphene's exact structure varies widely depending on preparation techniques and degree of oxidation, there has been discussion about it over the years. There are an increasing number of medical applications for graphene, including drug delivery, diagnostics, tissue engineering, and gene transfection, all with the goal of using it as a theranostic material (Orecchioni et al. 2015).

Miao et al. created cholesteryl hyaluronic acid (HA) conjugated graphene oxide (GO) loaded with doxorubicin for both in vitro and in vivo targeted administration of doxorubicin to (KB cell line) epidermal cancer cells. According to in vitro research, the targeted nanocomposite increased cell mortality by 40.3% when compared to the non-targeted one. In a similar vein, in tumors-bearing athymic mice, subcutaneous injection of the targeted nanocomposites led to a 14.1% reduction in tumors volume in just 24 days. n a different investigation, mice with subcutaneous HeLa tumors received doxorubicin via an adipic acid dihydrazide (ADH) linkage that connected HA and GO to create a HA-tagged GO-based targeted drug delivery system. In comparison to non-targeted nanocomposites and free doxorubicin, the targeted nanocomposites demonstrated 17% and 12% higher tumor suppression rates after 16 days, respectively (Roy and Jaiswal 2017).

3.3.5 Miscellaneous

3.3.5.1 Peptide-Based Theranostic Agents

Peptides are formed mainly from 50 amino acids and have tremendous biocompatibility and biodegradability properties; it ensures that peptides are good bioactive properties as well as targeted binding, stimulus-response and therapy. These properties show that peptides are better components for biomedical devices in clinical applications and applicable in the field of biomedical devices as site-specific targeting and theranostics agents as described in below Table 3.2 (Rong et al. 2020).

3.3.5.2 Aptamer-Based Theranostic Agents

A diverse range of molecular targets can be recognized by aptamers, which are single-stranded RNA and DNA-based oligonucleotides that fold into three-dimensional forms. A growing number of targeted probes for in vivo molecular imaging and treatment are being made from aptamers because of their chemical flexibility and similarities to antibodies in terms of selectivity and binding, having tremendous bioactive properties, which makes them used for various therapy and diagnosis purposes (Bohrmann et al. 2022). Some aptamers, like (EpCAM) Epithelial cell adhesion molecule aptamer, bind with epithelial adhesion molecules (EpCAM) as well as aptamer sgc8, which binds to protein tyrosine kinase-7

Table 3.2 Peptide nanoparticle conjugates for cancer theranostics

Peptide	Nanoparticle	Conjugate	Indications	References
Trans-activator of transcription peptide	Gold nanoparticles	Gold-s-coordinate bond	Transfection assay, gene interference, skin penetration, in vivo anti-tumor efficacy	Yakati et al. (2022))
KDKPPR-peptide	Gold nanorod	Maleimide-thiol chemistry	Molecular affinity, cytotoxicity phototoxicity	Youssef et al. (2019))
Angiopep and activable CPP	PEG-PCL nanoparticles	EDC/NHS chemistry	Cellular uptake, cancer cell apoptosis, in vivo imaging antiglioma effect	Tangthong et al. (2021))
ATWLPPR peptide	Silica nanoparticles	EDC/NHS chemistry	Molecular affinity MRI, PET, CT photosensitivity (anti-tumor efficacy), tumor tissue selectivity	Bechet et al. (2015))
T7 peptide	Gold-iron oxide NP	CD-adamantane interaction	Drug loading/release, cellular uptake, anti-tumor efficacy, in vivo imaging	Sukumar et al. (2019))
iRGD peptide	Liposomes	Maleimide-thiol chemistry	Drug release, cellular uptake, cytotoxicity, ROS detection, molecular imaging	Yan et al. (2016))

(PTK-7), have been isolated for a variety of target molecules in recent years. These molecules include organic and inorganic molecules, peptides, proteins, nucleic acids, bacteria, etc. When compared to other targeted ligands aptamers have a number of advantageous characteristics, including as small size, rapid tissue penetration, low or even non-immunogenicity, a high degree of chemical stability and binding affinity, and variable chemical modification (Gao et al. 2022).

3.4 Theranostics Enhanced Pharmacokinetics for Effective Drug Delivery

The word "theranostics" was introduced to describe scientific advancements that have led to the development of more targeted and customized treatments for a range of diseases. Additionally, these advancements have resulted in the combination of therapeutic and diagnostic applications into one agent, which has the potential to create a novel therapeutic paradigm that encompasses drug delivery, diagnosis, and response monitoring (Jeelani et al. 2014).

A definition of theranostic nanomedicine is "nanomedicine that combines therapeutics and diagnostics." In fact, certain nanomedicines possess inherent therapeutic and diagnostic qualities. They serve as platforms or nanomedicines that are self-theranostic. The targeting moiety of an advanced thermodynamic nanomedicine coupled with a specific target may identify, bind, and internalize the target by a specific method, such as receptor-mediated endocytosis.

Comparing nanomedicine to traditional medicinal products, there are some attractive advantages. Initially, conventional protocols may be utilized to prepare the nanocarrier readily using nanotechnology to deliver intended functions and attain precise targeting. Nanocarriers are, secondly, capable of enhancing the pharmacokinetics and improving the biodistribution of currently available treatment moieties to the intended tissues, which may enhance effectiveness (Sonali et al. 2018).

Several systems are used for theranostic purposes, including:

- Radioisotopes
- Genetic materials (DNA sequences, RNA, and proteins)
- Quantum dots
- Dendrimers
- Liposomes
- Magnetic nanoparticles
- Fullerenes and nanotubes
- Gold nanoparticles
- Microbubbles
- Porous nanoparticles

Targeting platforms that co-encapsulate several therapeutic and diagnostic modalities can help build multimodality theranostics, which can leverage the benefits (Dhurjad et al. 2020).

Pharmacokinetics refers to the study of how the body processes a drug, including its absorption, distribution, metabolism, and excretion (ADME). Theranostics can contribute to the improvement of pharmacokinetics in several ways:

3.4.1 Targeted Drug Delivery

Synthesis and characterization of a targeted theranostic hyperbranched polymer were accomplished with success. When compared to the untargeted counterpart, the Hsp70 targeted aptamer added to the polymer can offer noticeably improved accumulation within xenograft tumors and theranostic hyperbranched polymer show greater cellular uptake and higher accumulation in the tumor tissue but also had longer retention at the site of the tumor presumably because of the ligand–receptor interaction and target specificity (Zhao et al. 2017).

For several kinds of drug delivery systems, however, a pharmacokinetic monitoring strategy is required to maximize the results of these approaches. A theranostic device that allows for pharmacokinetic monitoring, as well as medication administration, would be extremely helpful in medical operations. A bubble liposome has frequently been utilized as a drug molecule carrier and as a contrast agent for imaging. Ultrasound-based drug administration and imaging have been the subject of significant study (Tsunoi et al. 2015).

3.4.2 Monitoring of Drug Response

Instantaneous monitoring of patients' response to therapy is possible with the use of theranostic methods. Based on the distinctive pharmacokinetic profile of each patient, this information can be utilized to change the patient's therapy or modify the dosage of drugs.

Because they enable individualized treatment, theranostic nanoparticles have the potential to significantly enhance cancer management. Even though a lot of theranostic nano-constructs have shown promise in preclinical research, their poor therapeutic effectiveness and restricted targeting capacity continue to hinder clinical translation. We describe the application of new ultrasmall porous silica nanoparticles (UPSN) that exhibit improved in vivo pharmacokinetics, including escape from the reticuloendothelial system (RES) organs and significant target tissue accumulation (12% ID/g in the tumor). UPSN is conjugated with the isotope pair 90/86Y, allowing for internal radiation in addition to noninvasive imaging (Ferreira et al. 2021).

3.4.3 Personalized Treatment

Personalized medicine has the ability to enhance pharmacokinetics and minimize adverse effects by utilizing a patient's unique genetic and molecular profile. Better pharmacokinetic results can be achieved by medical professionals tailoring the choice of therapeutic agents and optimizing medication dosages based on the patient's particular features (Dincer et al. 2017).

A promising approach to treating tumors might involve the use of theranostic probes to simultaneously activate therapy based on the expression level of certain tumor-related biomarkers, as well as optical detection and imaging of those biomarkers. Here, a novel theranostic nanoprobe called an up-conversion nano-beacon (UCNPs-MB/Dox) is suggested for ratiometrically detection and visualization. The theranostic agents of nano-beacons based on radiometric up-conversion fluorescence have the potential to identify the biomarkers present on tumors and are promising theranostics for non-invasive cancer (Ding et al. 2016).

3.5 Challenges and Future Prospects

The implementation of nanocarrier-based theranostics in clinical settings is hampered primarily by the nano-bio interaction. Nanotechnology for theranostics purposes may be harmful or incompatible when it interacts with biological material, which might result in diseases like inflammation, immunoreaction, or related disorders. One major barrier to theranostics being used in clinical settings is the need to have a synthesis method that is regulated and reproducible. The primary problems in producing nanoparticles on a wide scale are low yield, irregular physical and chemical characteristics, and poor batch-to-batch reproducibility. To overcome the biological barrier of nanotheranostics, it is important to comprehend the link between patient biology and disease heterogeneity and nanomedicine. This is because the fundamental reason for the failures observed in the clinical trial translation of promise theranostics.

In the long run, theranostic technologies have the potential to completely change the healthcare industry if they continue to progress. In order to maximize the therapeutic effectiveness of theranostic drugs, ongoing study endeavors have to concentrate on improving their design as well as their engineering. Theranostics might be made even more accurate and effective by integrating cutting-edge technology like artificial intelligence and nanomedicine. Furthermore, investigating new biomarkers and molecular targets for theranostic applications would broaden the range of illnesses that may be successfully treated with this strategy. Researchers must work together as the field of theranostics develops in order to transfer these discoveries from the lab to clinical practice, eventually improving the safety of patients and bringing in a new age of customized medicine.

3.6 Conclusion

In conclusion, theranostics offers a versatile method that combines therapeutic and diagnostic functions using nanotechnology, making it a potential paradigm in the fields of drug targeting and pharmacokinetics improvement. Unprecedented precision and customized medicine opportunities are provided by the capacity to concurrently diagnose, track, and treat illnesses at the molecular level. Real-time evaluation of therapy success is made possible by the synergistic combination of medicinal chemicals with imaging agents, which enables doctors to customize treatments depending on unique patient responses. Additionally, the ability to maximize systemic toxicity and minimize off-target effects while optimizing pharmacokinetics through tailored drug delivery systems improves therapeutic results.

References

Abrahao A, Meng Y, Llinas M, Huang Y, Hamani C, Mainprize T, Aubert I, Heyn C, Black SE, Hynynen K, Lipsman N, Zinman L (2019) First-in-human trial of blood–brain barrier opening in amyotrophic lateral sclerosis using MR-guided focused ultrasound. Nat Commun 10:1–9. https://doi.org/10.1038/s41467-019-12426-9

Akhter S, Ahmad MZ, Ahmad FJ, Storm G, Kok RJ (2012) Gold nanoparticles in theranostic oncology: current state-of-the-art. Expert Opin Drug Deliv 9:1225–1243. https://doi.org/10.1517/17425247.2012.716824

Al-Jamal WT, Al-Jamal KT, Bomans PH, Frederik PM, Kostarelos K (2008) Functionalized-quantum-dot-liposome hybrids as multimodal nanopartides for cancer. Small 4:1406–1415. https://doi.org/10.1002/smll.200701043

Arranz A, Ripoll J (2015) Advances in optical imaging for pharmacological studies. Front Pharmacol 6:1–7. https://doi.org/10.3389/fphar.2015.00189

Augustine S, Singh J, Srivastava M, Sharma M, Das A, Malhotra BD (2017) Recent advances in carbon based nanosystems for cancer theranostics. Biomater Sci 5:901–952. https://doi.org/10.1039/c7bm00008a

Azam N, Najabat Ali M, Javaid Khan T (2021) Carbon quantum dots for biomedical applications: review and analysis. Front Mater 8:1–21. https://doi.org/10.3389/fmats.2021.700403

Baig MH, Adil M, Khan R, Dhadi S, Ahmad K, Rabbani G, Bashir T, Imran MA, Husain FM, Lee EJ, Kamal MA, Choi I (2019) Enzyme targeting strategies for prevention and treatment of cancer: implications for cancer therapy. Semin Cancer Biol 56:1–11. https://doi.org/10.1016/j.semcancer.2017.12.003

Bechet D, Auger F, Couleaud P, Marty E, Ravasi L, Durieux N, Bonnet C, Plénat F, Frochot C, Mordon S, Tillement O, Vanderesse R, Lux F, Perriat P, Guillemin F, Barberi-Heyob M (2015) Multifunctional ultrasmall nanoplatforms for vascular-targeted interstitial photodynamic therapy of brain tumors guided by real-time MRI. Nanomed Nanotechnol Biol Med 11:657–670. https://doi.org/10.1016/j.nano.2014.12.007

Bensch F, van der Veen EL, Lub-de Hooge MN, Jorritsma-Smit A, Boellaard R, Kok IC, Oosting SF, Schröder CP, Hiltermann TJN, van der Wekken AJ, Groen HJM, Kwee TC, Elias SG, Gietema JA, Bohorquez SS, de Crespigny A, Williams SP, Mancao C, Brouwers AH, Fine BM, de Vries EGE (2018) 89Zr-atezolizumab imaging as a non-invasive approach to assess clinical response to PD-L1 blockade in cancer. Nat Med 24:1852–1858. https://doi.org/10.1038/s41591-018-0255-8

Bohrmann L, Burghardt T, Haynes C, Saatchi K, Häfeli UO (2022) Aptamers used for molecular imaging and theranostics—recent developments. Theranostics 12:4010–4050. https://doi.org/10.7150/THNO.72949

Cai W, Chen X (2008) Multimodality molecular imaging of tumor angiogenesis. J Nucl Med 49:113S. https://doi.org/10.2967/jnumed.107.045922

Daraee H, Etemadi A, Kouhi M, Alimirzalu S, Akbarzadeh A (2016) Application of liposomes in medicine and drug delivery. Artif Cells Nanomed Biotechnol 44:381–391. https://doi.org/10.3109/21691401.2014.953633

David HC, Amani MN-J, Nancy EA (2017) 乳鼠心肌提取 HHS public access. Physiol Behav 176:139–148. https://doi.org/10.1016/j.jconrel.2020.01.009.Magnetic

Decazes P, Bohn P (2020) Immunotherapy by immune checkpoint inhibitors and nuclear medicine imaging: current and future applications. Cancers (Basel) 12:371. https://doi.org/10.3390/cancers12020371

Desai P, Rimal R, Sahnoun SEM, Mottaghy FM, Möller M, Morgenroth A, Singh S (2022) Radiolabeled nanocarriers as theranostics—advancement from peptides to nanocarriers. Small 18:1–19. https://doi.org/10.1002/smll.202200673

Dhurjad L, Sagle D, Deshmukh A, Narkhede M (2020) Cellulose nanocrystals: a versatile nanoplatform. Asian J Pharm Res Dev 8:92–96

Dincer C, Bruch R, Wirth S, Schumann S, Urban GA (2017) Biosensors and personalized drug therapy: what does the future hold? Expert Rev Precis Med Drug Dev 2:303–305. https://doi.org/10.1080/23808993.2017.1395691

Ding Q, Zhan Q, Zhou X, Zhang T, Xing D (2016) Theranostic upconversion nanobeacons for tumor mRNA ratiometric fluorescence detection and imaging-monitored drug delivery. Small 12:5944–5953. https://doi.org/10.1002/smll.201601724

Feng L, Xie R, Wang C, Gai S, He F, Yang D, Yang P, Lin J (2018) Magnetic targeting, tumor microenvironment-responsive intelligent nanocatalysts for enhanced tumor ablation. ACS Nano 12:11000–11012. https://doi.org/10.1021/acsnano.8b05042

Fernandes DA (2023) Liposomes for cancer theranostics. Pharmaceutics 15:2448. https://doi.org/10.3390/pharmaceutics15102448

Ferreira CA, Goel S, Ehlerding EB, Rosenkrans ZT, Jiang D, Sun T, Aluicio-Sarduy E, Engle JW, Ni D, Cai W (2021) Ultrasmall porous silica nanoparticles with enhanced pharmacokinetics for cancer theranostics. Nano Lett 21:4692–4699. https://doi.org/10.1021/acs.nanolett.1c00895

Foster A, Nigam S, Tatum DS, Raphael I, Xu J, Kumar R, Plakseychuk E, Latoche JD, Vincze S, Li B, Giri R, McCarl LH, Edinger R, Ak M, Peddagangireddy V, Foley LM, Hitchens TK, Colen RR, Pollack IF, Panigrahy A, Magda D, Anderson CJ, Edwards WB, Kohanbash G (2021) Novel theranostic agent for PET imaging and targeted radiopharmaceutical therapy of tumour-infiltrating immune cells in glioma. EBioMedicine 71:1–10. https://doi.org/10.1016/j.ebiom.2021.103571

Gao M, Tang BZ (2020) AIE-based cancer theranostics. Coord Chem Rev 402:213076. https://doi.org/10.1016/j.ccr.2019.213076

Gao Q, Zhang J, Gao J, Zhang Z, Zhu H, Wang D (2021) Gold nanoparticles in cancer theranostics. Front Bioeng Biotechnol 9:1–20. https://doi.org/10.3389/fbioe.2021.647905

Gao F, Yin J, Chen Y, Guo C, Hu H, Su J (2022) Recent advances in aptamer-based targeted drug delivery systems for cancer therapy. Front Bioeng Biotechnol 10:1–9. https://doi.org/10.3389/fbioe.2022.972933

Gao J, Jiang H, Chen P, Zhang R, Liu N (2023) Photosensitizer-based small molecule theranostic agents for tumor-targeted monitoring and phototherapy. Bioorg Chem 136:106554. https://doi.org/10.1016/j.bioorg.2023.106554

Gulati R, Nandi D, Sarkar K, Venkataraman P, Ramkumar KM, Ranjan P, Janardhanan R (2022) Exosomes as theranostic targets: implications for the clinical prognosis of aggressive cancers. Front Mol Biosci 9:1–11. https://doi.org/10.3389/fmolb.2022.890768

Guo J, Rahme K, He Y, Li LL, Holmes JD, O'Driscoll CM (2017) Gold nanoparticles enlighten the future of cancer theranostics. Int J Nanomed 12:6131–6152. https://doi.org/10.2147/IJN.S140772

Hapuarachchige S, Artemov D (2020) Theranostic pretargeting drug delivery and imaging platforms in cancer precision medicine. Front Oncol 10:1–8. https://doi.org/10.3389/fonc.2020.01131

Herranz F, Almarza E, Rodríguez I, Salinas B, Rosell Y, Desco M, Bulte JW, Ruiz-Cabello J (2011) The application of nanoparticles in gene therapy and magnetic resonance imaging. Microsc Res Tech 74:577–591. https://doi.org/10.1002/jemt.20992

Herrmann K, Schwaiger M, Lewis JS, Solomon SB, McNeil BJ, Baumann M, Gambhir SS, Hricak H, Weissleder R (2020) Radiotheranostics: a roadmap for future development. Lancet Oncol 21:e146–e156. https://doi.org/10.1016/S1470-2045(19)30821-6

Hu F, Huang Y, Zhang G, Zhao R, Yang H, Zhang D (2014) Targeted bioimaging and photodynamic therapy of cancer cells with an activatable red fluorescent bioprobe. Anal Chem 86:7987–7995. https://doi.org/10.1021/ac502103t

Idée J-M, Louguet S, Ballet S, Corot C (2013) Theranostics and contrast-agents for medical imaging: a pharmaceutical company viewpoint. Quant Imaging Med Surg 3:292–297. https://doi.org/10.3978/j.issn.2223-4292.2013.12.06

Jeelani S, Jagat Reddy RC, Maheswaran T, Asokan GS, Dany A, Anand B (2014) Theranostics: a treasured tailor for tomorrow. J Pharm Bioallied Sci 6:6–8. https://doi.org/10.4103/0975-7406.137249

Kamei K, Mukai Y, Kojima H, Yoshikawa T, Yoshikawa M, Kiyohara G, Yamamoto TA, Yoshioka Y, Okada N, Seino S, Nakagawa S (2009) Direct cell entry of gold/iron-oxide magnetic nanoparticles in adenovirus mediated gene delivery. Biomaterials 30:1809–1814. https://doi.org/10.1016/j.biomaterials.2008.12.015

Ke MR, Yeung SL, Ng DKP, Fong WP, Lo PC (2013) Preparation and in vitro photodynamic activities of folate-conjugated distyryl boron dipyrromethene based photosensitizers. J Med Chem 56:8475–8483. https://doi.org/10.1021/jm4009168

Kelkar SS, Reineke TM (2011) Theranostics: combining imaging and therapy. Bioconjug Chem 22:1879–1903. https://doi.org/10.1021/bc200151q

Kesharwani P, Jain K, Jain NK (2014) Dendrimer as nanocarrier for drug delivery. Prog Polym Sci 39:268–307. https://doi.org/10.1016/j.progpolymsci.2013.07.005

Kesharwani P, Ma R, Sang L, Fatima M, Sheikh A, Abourehab MAS, Gupta N, Chen ZS, Zhou Y (2023) Gold nanoparticles and gold nanorods in the landscape of cancer therapy. Mol Cancer 22:1–31. https://doi.org/10.1186/s12943-023-01798-8

Krenning EP, Kooij PPM, Bakker WH, Breeman WAP, Postema PTE, Kwekkeboom DJ, Oei HY, de Jong M, Visser TJ, Reijs AEM, Lamberts SWJ (1994) Radiotherapy with a radiolabeled somatostatin analogue, [111In-DTPA-D-Phe1]-octreotide: a case history. Ann N Y Acad Sci 733:496–506. https://doi.org/10.1111/j.1749-6632.1994.tb17300.x

Kunikowska J (2021) Theranostics—present and future. Bio-Algorithms Med-Syst 17:213–220

Langbein T, Weber WA, Eiber M (2019) Future of theranostics: an outlook on precision oncology in nuclear medicine. J Nucl Med 60:13S–19S. https://doi.org/10.2967/jnumed.118.220566

Lee W, Im HJ (2019) Theranostics based on liposome: looking back and forward. Nucl Med Mol Imaging 53:242–246. https://doi.org/10.1007/s13139-019-00603-z

Lee DY, Li KCP (2011) Molecular theranostics: a primer for the imaging professional. Am J Roentgenol 197:318–324. https://doi.org/10.2214/AJR.11.6797

Lee MH, Sharma A, Chang MJ, Lee J, Son S, Sessler JL, Kang C, Kim JS (2018) Fluorogenic reaction-based prodrug conjugates as targeted cancer theranostics. Chem Soc Rev 47:28–52. https://doi.org/10.1039/c7cs00557a

Levine R, Krenning EP (2017) Clinical history of the theranostic radionuclide approach to neuroendocrine tumors and other types of cancer: historical review based on an interview of Eric P. Krenning by Rachel Levine. J Nucl Med 58:3S–9S. https://doi.org/10.2967/jnumed.116.186502

Li S, Su W, Wu H, Yuan T, Yuan C, Liu J, Deng G, Gao X, Chen Z, Bao Y, Yuan F, Zhou S, Tan H, Li Y, Li X, Fan L, Zhu J, Chen AT, Liu F, Zhou Y, Li M, Zhai X, Zhou J (2020) Targeted tumour theranostics in mice via carbon quantum dots structurally mimicking large amino acids. Nat Biomed Eng 4:704–716. https://doi.org/10.1038/s41551-020-0540-y

Lim EK, Kim T, Paik S, Haam S, Huh YM, Lee K (2015) Nanomaterials for theranostics: recent advances and future challenges. Chem Rev 115:327–394. https://doi.org/10.1021/cr300213b

Lodhi MS, Khan MT, Aftab S, Samra ZQ, Wang H, Wei DQ (2021) A novel formulation of theranostic nanomedicine for targeting drug delivery to gastrointestinal tract cancer. Cancer Nanotechnol 12:4. https://doi.org/10.1186/s12645-021-00098-4

Luo L, Liu C, He T, Zeng L, Xing J, Xia Y, Pan Y, Gong C, Wu A (2018) Engineered fluorescent carbon dots as promising immune adjuvants to efficiently enhance cancer immunotherapy. Nanoscale 10:22035–22043. https://doi.org/10.1039/c8nr07252c

Luo N, Li J, Dong R, Lu J (2022) Exosome-based theranostics for liver diseases. Dis Mark 2022:7888906. https://doi.org/10.1155/2022/7888906

Luster M, Clarke SE, Dietlein M, Lassmann M, Lind P, Oyen WJG, Tennvall J, Bombardieri E (2008) Guidelines for radioiodine therapy of differentiated thyroid cancer. Eur J Nucl Med Mol Imaging 35:1941–1959. https://doi.org/10.1007/s00259-008-0883-1

Mostafavi M, Ghanavi J, Shaabani M, Asadi H, Alizadeh M (2021) Metallic nanomaterials in cancer theranostics: a review of iron oxide and gold-based nanomaterials. J Nanomed Nanotechnol 12:564

Nasser TA, Adel R, Badr A, Teleb M, Bekhit AA, Elkhodairy KA, Abdelhamid AS, Elzoghby AO (2023) Combined cancer immunotheranostic nanomedicines: delivery technologies and therapeutic outcomes. ACS Omega 8:4491–4507. https://doi.org/10.1021/acsomega.2c05986

Niknam S, Dehdast SA, Pourdakan O, Shabani M, Koohi MK (2022) Tungsten disulfide nanomaterials (WS2 NM) application in biosensors and nanomedicine: a review. Nanomed Res J 7:214–226. https://doi.org/10.22034/nmrj.2022.03.001

Notni J, Hermann P, Havlíčková J, Kotek J, Kubíček V, Plutnar J, Loktionova N, Riss PJ, Rösch F, Lukeš I (2010) A triazacyclononane-based bifunctional phosphinate ligand for the preparation of multimeric 68GA tracers for positron emission tomography. Chem A Eur J 16:7174–7185. https://doi.org/10.1002/chem.200903281

Orecchioni M, Cabizza R, Bianco A, Delogu LG (2015) Graphene as cancer theranostic tool: progress and future challenges. Theranostics 5:710–723. https://doi.org/10.7150/thno.11387

Patel KD, Singh RK, Kim HW (2019) Carbon-based nanomaterials as an emerging platform for theranostics. Mater Horizons 6:434–469. https://doi.org/10.1039/c8mh00966j

Pene F, Courtine E, Cariou A, Mira JP (2009) Toward theragnostics. Crit Care Med 37:S50. https://doi.org/10.1097/CCM.0b013e3181921349

Penet MF, Chen Z, Kakkad S, Pomper MG, Bhujwalla ZM (2012) Theranostic imaging of cancer. Eur J Radiol 81:S124–S126. https://doi.org/10.1016/S0720-048X(12)70051-7

Räty JK, Liimatainen T, Wirth T, Airenne KJ, Ihalainen TO, Huhtala T, Hamerlynck E, Vihinen-Ranta M, Närvänen A, Ylä-Herttuala S, Hakumäki JM (2006) Magnetic resonance imaging of viral particle biodistribution in vivo. Gene Ther 13:1440–1446. https://doi.org/10.1038/sj.gt.3302828

Ray S, Li Z, Hsu CH, Hwang LP, Lin YC, Chou PT, Lin YY (2018) Dendrimer-and copolymer-based nanoparticles for magnetic resonance cancer theranostics. Theranostics 8:6322–6349. https://doi.org/10.7150/thno.27828

Ren Q, Yang K, Zou R, Wan Z, Shen Z, Wu G, Zhou Z, Ni Q, Fan W, Hu J, Liu Y (2019) Biodegradable hollow manganese/cobalt oxide nanoparticles for tumor theranostics. Nanoscale 11(47):23021–23026

Roesch F, Martin M (2023) Radiometal-theranostics: the first 20 years. J Radioanal Nucl Chem 332:1557–1576. https://doi.org/10.1007/s10967-022-08624-3

Rong L, Lei Q, Zhang XZ (2020) Recent advances on peptide-based theranostic nanomaterials. Viewpoints 1:1–7. https://doi.org/10.1002/VIW.20200050

Rong X, Liu C, Li X, Zhu H, Wang K, Zhu B (2022) Recent advances in chemotherapy-based organic small molecule theranostic reagents. Coord Chem Rev 473:214808. https://doi.org/10.1016/j.ccr.2022.214808

Roy S, Jaiswal A (2017) Graphene-based nanomaterials for theranostic applications. Rep Adv Phys Sci 1(04):1750011

Sakthi Devi R, Girigoswami A, Siddharth M, Girigoswami K (2022) Applications of gold and silver nanoparticles in theranostics, vol 194. Springer US, p 4187

Schaefer N, Prior JO, Schottelius M (2020) From theranostics to immunotheranostics: the concept. Nucl Med Mol Imaging 54:81–85. https://doi.org/10.1007/s13139-020-00639-6

Shao W, Arghya P, Yiyong M, Rodes L, Prakash S (2013) Carbon nanotubes for use in medicine: potentials and limitations. Synth Appl Carbon Nanotubes Their Compos 13:285–311. https://doi.org/10.5772/51785

Siafaka PI, Okur NÜ, Karantas ID, Okur ME, Gündoğdu EA (2021) Current update on nanoplatforms as therapeutic and diagnostic tools: a review for the materials used as nanotheranostics and imaging modalities. Asian J Pharm Sci 16:24–46. https://doi.org/10.1016/j.ajps.2020.03.003

Sk UH, Kojima C (2015) Dendrimers for theranostic applications. Biomol Concepts 6:205–217. https://doi.org/10.1515/bmc-2015-0012

Sonali VMK, Singh RP, Agrawal P, Mehata AK, Pawde DM, Narendra SR, Muthu MS (2018) Nanotheranostics: emerging strategies for early diagnosis and therapy of brain cancer. Nano 2:70–86. https://doi.org/10.7150/ntno.21638

Strosberg J, El-Haddad G, Wolin E, Hendifar A, Yao J, Chasen B, Mittra E, Kunz PL, Kulke MH, Jacene H, Bushnell D, O'Dorisio TM, Baum RP, Kulkarni HR, Caplin M, Lebtahi R, Hobday T, Delpassand E, Van Cutsem E, Benson A, Srirajaskanthan R, Pavel M, Mora J, Berlin J, Grande E, Reed N, Seregni E, Öberg K, Lopera Sierra M, Santoro P, Thevenet T, Erion JL, Ruszniewski P, Kwekkeboom D, Krenning E (2017) Phase 3 trial of 177 Lu-Dotatate for midgut neuroendocrine tumors. N Engl J Med 376:125–135. https://doi.org/10.1056/nejmoa1607427

Sukumar UK, Bose RJC, Malhotra M, Babikir HA, Afjei R, Robinson E, Zeng Y, Chang E, Habte F, Sinclair R, Gambhir SS, Massoud TF, Paulmurugan R (2019) Intranasal delivery of targeted polyfunctional gold–iron oxide nanoparticles loaded with therapeutic microRNAs for combined theranostic multimodality imaging and presensitization of glioblastoma to temozolomide. Biomaterials 218:119342. https://doi.org/10.1016/j.biomaterials.2019.119342

Sun C, Fang C, Stephen Z, Veiseh O, Hansen S, Lee D, Ellenbogen RG, Olson J, Zhang M (2008) Tumor-targeted drug delivery and MRI contrast enhancement by chlorotoxin-conjugated iron oxide nanoparticles. Nanomedicine 3:495–505. https://doi.org/10.2217/17435889.3.4.495

Takáč P, Michalková R, Čižmáriková M, Bedlovičová Z, Balážová Ľ, Takáčová G (2023) The role of silver nanoparticles in the diagnosis and treatment of cancer: are there any perspectives for the future? Life 13:466. https://doi.org/10.3390/life13020466

Tang L, Xiao Q, Mei Y, He S, Zhang Z, Wang R, Wang W (2021) Insights on functionalized carbon nanotubes for cancer theranostics. J Nanobiotechnol 19:1–28. https://doi.org/10.1186/s12951-021-01174-y

Tangthong T, Piroonpan T, Thipe VC, Khoobchandani M, Katti K, Katti KV, Pasanphan W (2021) Water-soluble chitosan conjugated DOTA-bombesin peptide capped gold nanoparticles as a targeted therapeutic agent for prostate cancer. Nanotechnol Sci Appl 14:69–89. https://doi.org/10.2147/NSA.S301942

Tsunoi Y, Sato S, Kawauchi S, Akutsu Y, Miyagawa Y, Araki K, Shiotani A, Terakawa M (2015) Theranostic system for drug delivery and pharmacokinetic imaging based on nanosecond pulsed light-induced photomechanical and photoacoustic effects. Jpn J Appl Phys 54:116601. https://doi.org/10.7567/JJAP.54.116601

Vats S, Singh M, Siraj S, Singh H, Tandon S (2017) Role of nanotechnology in theranostics and personalized medicines. J Health Res Rev 4:1. https://doi.org/10.4103/2394-2010.199328

Wang C, Ravi S, Garapati US, Das M, Howell M, Mallela J, Alwarappan S, Mohapatra SS, Mohapatra S (2013) Multifunctional chitosan magnetic-graphene (CMG) nanoparticles: a theranostic platform for tumor-targeted co-delivery of drugs, genes and MRI contrast agents. J Mater Chem B 1:4396–4405. https://doi.org/10.1039/c3tb20452a

Wang BS, Wang J, Chen JY (2014) Conjugates of folic acids with zinc aminophthalocyanine for cancer cell targeting and photodynamic therapy by one-photon and two-photon excitations. J Mater Chem B 2:1594–1602. https://doi.org/10.1039/c3tb21620a

Wang Y, Wang Y, Chen G, Li Y, Xu W, Gong S (2017) Quantum-dot-based theranostic micelles conjugated with an anti-EGFR nanobody for triple-negative breast cancer therapy. ACS Appl Mater Interfaces 9:30297–30305. https://doi.org/10.1021/acsami.7b05654

Xie J, Lee S, Chen X (2010) Nanoparticle-based theranostic agents. Adv Drug Deliv Rev 62:1064–1079. https://doi.org/10.1016/j.addr.2010.07.009

Xing H, Hwang K, Lu Y (2016) Recent developments of liposomes as nanocarriers for theranostic applications. Theranostics 6:1336–1352. https://doi.org/10.7150/thno.15464

Xu B, Li S, Shi R, Liu H (2023) Multifunctional mesoporous silica nanoparticles for biomedical applications. Signal Transduct Target Ther 8:435. https://doi.org/10.1038/s41392-023-01654-7

Xue Y, Gao Y, Meng F, Luo L (2021) Recent progress of nanotechnology-based theranostic systems in cancer treatments. Cancer Biol Med 18:336–351. https://doi.org/10.20892/j.issn.2095-3941.2020.0510

Yakati V, Vangala S, Madamsetty VS, Banerjee R, Moku G (2022) Enhancing the anticancer effect of paclitaxel by using polymeric nanoparticles decorated with colorectal cancer targeting CPKSNNGVC-peptide. J Drug Deliv Sci Technol 68:103125. https://doi.org/10.1016/j.jddst.2022.103125

Yan F, Wu H, Liu H, Deng Z, Liu H, Duan W, Liu X, Zheng H (2016) Molecular imaging-guided photothermal/photodynamic therapy against tumor by iRGD-modified indocyanine green nanoparticles. J Control Release 224:217–228. https://doi.org/10.1016/j.jconrel.2015.12.050

Youssef Z, Yesmurzayeva N, Larue L, Jouan-hureaux V, Colombeau L, Arnoux P, Acherar S, Vanderesse R, Frochot C (2019) New targeted gold nanorods for the treatment of glioblastoma by photodynamic therapy. J Clin Med 8:1–23. https://doi.org/10.3390/jcm8122205

Yu BO, Tai HC, Xue W, Lee LJ, Lee RJ (2010) Receptor-targeted nanocarriers for therapeutic delivery to cancer. Mol Membr Biol 27:286–298. https://doi.org/10.3109/09687688.2010.521200.Receptor-targeted

Zhang Z, Zhou FL, Davies GL, Williams GR (2022) Theranostics for MRI-guided therapy: recent developments. Viewpoints 3:1–15. https://doi.org/10.1002/VIW.20200134

Zhao Y, Houston ZH, Simpson JD, Chen L, Fletcher NL, Fuchs AV, Blakey I, Thurecht KJ (2017) Using peptide aptamer targeted polymers as a model nanomedicine for investigating drug distribution in cancer nanotheranostics. Mol Pharm 14:3539–3549. https://doi.org/10.1021/acs.molpharmaceut.7b00560

Zhong X, Dai X, Wang Y, Wang H, Qian H, Wang X (2022) Copper-based nanomaterials for cancer theranostics. Wiley Interdiscip Rev Nanomed Nanobiotechnol 14:1–43. https://doi.org/10.1002/wnan.1797

Zhuang X, Xiang X, Grizzle W, Sun D, Zhang S, Axtell RC, Ju S, Mu J, Zhang L, Steinman L, Miller D, Zhang HG (2011) Treatment of brain inflammatory diseases by delivering exosome encapsulated anti-inflammatory drugs from the nasal region to the brain. Mol Ther 19:1769–1779. https://doi.org/10.1038/mt.2011.164

Nanotheranostics: A Potential Diagnostic Tool in the Cancer Management: Befriending the Barriers

4

Pallavi M. Chaudhari

Abstract

Nanotechnology has been an area of interest in medical science. Cancer nanotherapeutic is now progressing towards solving the limitations of conventional systems and the latest advancement in nanotechnology, in form of nanotheranostic that combine diagnosis and therapy, has been utilized for cancer management. In this approach, the radioactive drugs are combined to diagnose a disease as well as deliver it to the target, thus indicating the purpose of theranostic to provide one system at the same time. This strategic-based approach implies use of different nanoparticles such as polymeric nanoparticles, polymeric micelles, polymer conjugates, liposomes, gold nanoparticles, and iron oxide nanoparticles in effectively killing the cancer cells. The significance of nanotheranostic is to facilitate high bioavailability, target the site and release drug in controlled manner. This precision treatment, assures targeting only the tumor cells without harming the healthy cells and tissues. Nanotheranostics will provide a significant breakthrough that aim to utilize nanoscale materials for clinical applications and has become a reality, from "bench to bedside."

Keywords

Nanotheranostics · Polymeric nanoparticles · Polymer conjugates · Liposomes · Gold nanoparticles · Iron oxide nanoparticles

P. M. Chaudhari (✉)
Dr. D. Y. Patil College of Pharmacy, Akurdi, Pune, India
e-mail: pallavichaudhari@dyppharmaakurdi.ac.in

M. A. Barkat et al. (eds.), *Nanotheranostics for Diagnosis and Therapy*,
https://doi.org/10.1007/978-981-97-3115-2_4

4.1 Introduction

Worldwide, cancer remains a colossal threat to human health, as being the major contributor for death and increase disease burden (Sung et al. 2021). As per World Health Organization (WHO), mortalities due to cancer are estimated to be 11 million by 2030 (Kundu et al. 2020). The intensive research, knowledge of researchers, and the study conducted has helped to know the causes of cancer, which has further led to development of improved techniques for treating this deadly disease. But the selection of proper therapy remains a complex question to be answered, as the treatment includes radiation therapy, surgery, chemotherapy, hormonal therapy, or even combination of these (Hisada and Mackman 2019). These therapies possess limitations unable to cross the biological barriers, targeting of non-specifically rapidly dividing cells, poor pharmacokinetic characteristics of anticancer drugs, drug resistance, limited bio-distribution, and adverse effects of chemotherapies (Navya et al. 2019; Schirrmacher 2019). Thus, there is a need for paradigm shift to find new therapy that will improve the existing therapies and diagnostics in innovative way.

The concept of nano-based modalities is the right solution for cancer treatment that has promising role that can befriend the barriers, enable selective targeting of tumor cells and offer sustained drug release (Sharma and Bhargava 2013). "Theranostic," a portmanteau of therapeutics and diagnostics, combines therapy and diagnostics (Fig. 4.1) (Pene et al. 2009). This term was coined by John Funkhouser, in terms of union of therapy and diagnosis, to lead potential for treatment (Idée et al. 2013).

Thus, nanotheranostics uses nanotechnology for combined imaging and treatment of cancer by the use of single nanomedicine and integrated by nanoparticles (Xie et al. 2010). Nanoparticles are the essential component of nanotheranostic agent (Madanayake et al. 2019). As a consequence, theranostic nanomedicines—nanoparticles (NPs) embrace therapeutic and diagnostic agent, for cancer therapy and overcome the limitations of the existing therapies. Nanotheranostic provides means as diagnostic tools, targeted therapy, tumor specific nano delivery system, and nano imaging therapeutic integrated medicine (Fig. 4.2) (Viswanadh et al. 2018; Zhang et al. 2019; Mendes et al. 2018; Sivasubramanian et al. 2019).

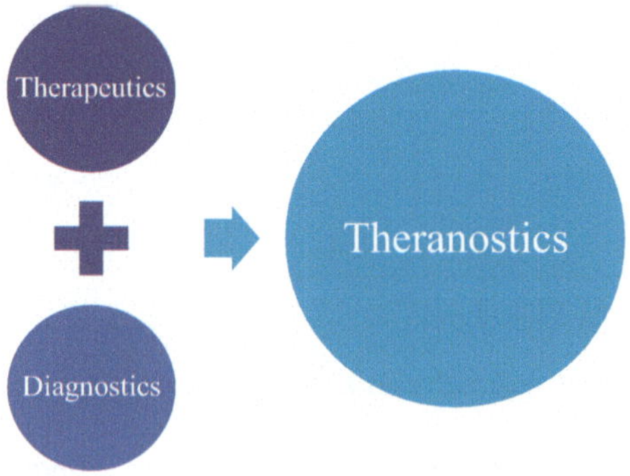

Fig. 4.1 Theranostics—amalgamation of therapeutics and diagnostics

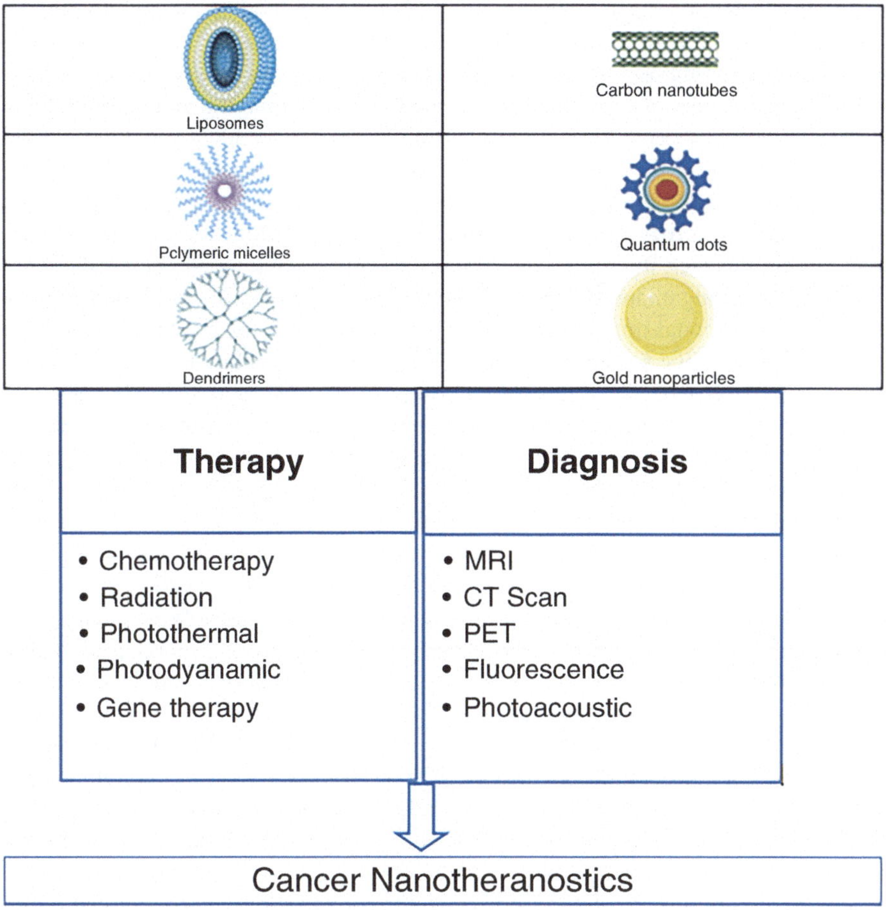

Fig. 4.2 Therapy and diagnosis in cancer theranostic

4.2 Nanotheranostics in Diagnosis and Treatment of Cancer

The conventional method of chemotherapy, though widely used for cancer treatment is based on the principle of killing indiscriminately the tumor and the normal cells, but this caused various side effects of hair loss and bone marrow suppression (Zitvogel et al. 2008). Hence, there is need for development of such treatment that will target only the tumor cells with the least adverse effects. Certainly, there are number of innovative treatments for cancer that has improved the therapeutic efficacy. The advancement in nanotechnology and imaging nano-agents has motivated as novel application in diagnostic and therapy of cancer. Different imaging agents, that can detect cancer in early stages, will indeed be an additional advantage to the medical field (Pelaz et al. 2017).

Diagnosis at early stage of cancer can contribute for good therapeutic treatment. But the detection of molecular markers that assist the diagnosis of the disease is a major challenge (Parveen et al. 2012; Parhi et al. 2012). As discussed earlier, theranostic is innovative platform that combines loading of imaging and therapeutic in nanoformulation, so as to further assist simultaneous in vivo diagnostic imaging and delivery of the drug for cancer treatment. Thus, nanotheranostics (i.e., theranostic nanomedicine) imbibe imaging and therapeutic function in single entity (i.e., nanoparticles) to certify the efficiency of cancer therapy (Mura and Couvreur 2012). In the field of cancer, the role of nanotheranostics is expanding and hence these are developed to amalgamate disease, diagnosis, controlled release, and monitoring of drug. The multimodal imaging capability blends with the advanced techniques that can be utilized for the nanoparticles for precise diagnosis and monitoring, with fabrication of these particles with desired composition, size, and shape (Cleary and Peters 2010). There exists a correlation between the composition and size of nanoparticles, as NPs of size <00 nm can be attached to the cells (Foroozandeh and Aziz 2018).

There are numerous imaging systems like computed tomography (CT), magnetic resonance imaging (MRI), that are used when the cancer cells become visible, but during that time, the growth has increased and 1 billion cancer cells are found at 1 cm^3. But at this stage, the detection is late as there are phenotypic changes, so nowadays nuclear medicine imaging is one of the non-invasive imaging tools that grant information at both molecular and cellular level that utilize radiolabeled compounds also known as radiopharmaceuticals, for diagnosis and treatment of disease (National Research Council (US) and Institute of Medicine (US), Committee on State of the Science of Nuclear Medicine 2007). The recent advancement in twenty-first century, radiopharmaceuticals basically in positron emission tomography (PET) have led to significant increase in diagnostic and therapeutic applications. PET measures the tumor metabolism metabolic rate, as well the biochemical processed traced are 2-[^{18}F] fluoro-2-deoxy-D-glucose (FDG), as cancers exhibit elevated rate of aerobic metabolism (Som 1980; Eary 1999). In recent times, nanotheranostics offer promising advantages over the conventional methods in terms of their preparation and achieve targeting, with characteristics of modification of their surfaces with different moieties, coating, or conjugation (Hou et al. 2018). The types of NPs used for imaging are illustrated in Fig. 4.3.

The magnetic NPs are used as nanotheranostics for cancer, due to its distinctive characteristics of being less toxic, good biodegradability, and easy surface modifications (Sun et al. 2008; Veiseh et al. 2010). These act as MRI agents, due to the magnetic field applied and relaxation abilities of T2/T2*, it has the privilege to be used for cancer (Koenig and Kellar 1995). The ability of radiolabeled NPs has also been used for cancer imaging along with theranostic (Bluemel et al. 2015; Imlimthan et al. 2021). Radiation therapy eradicates cancer cells due to necrosis or DNA damage, by deposit of energy in cells, not harming the normal cells (Liu et al. 2018). The two forms of radiations that are used are α particles and β particles for therapeutic application, while for diagnostic purpose, γ emission is utilized. The theranostic first utilized for cancer was applied 90 years ago by use of ^{131}I (Jadvar 2017).

Fig. 4.3 Types of NPs used for imaging

Table 4.1 Examples of theranostic pairs used for cancer

Therapeutic agent(s)	Diagnostic agent(s)	Target	Type of cancer
223Ra	18F-sodium fluoride, 99mTc-medronate	Bone hydroxy-apatite	Prostate cancer
177Lu-PSMA, 225Ac-PSMA	68Ga-PSMA, 18F-PSMA	PSMA	Prostate cancer
[131I]I-MIBG	[123I]MIBG, [124I] MIBG, [18F]MFBG	NET (SLC6A2)	Neuroendocrine tumors medullary thyroid carcinoma
131I (NaI)	123I (NaI)	NIS	Thyroid cancer

223Ra Radium-223, *99mTc* Technetium-99 m, *PSMA* prostate-specific membrane antigen, *177Lu-PSMA* Lutetium-177 vipivotidetetraxetan PSMA therapy, *225Ac-PSMA* actinium labelled PSMA therapy, *[131I]I-MIBG* radioactive iodine meta-iodo-benzyl-guanidine, *NET* (*SLC6A2*) norepinephrine transporter Solute Carrier Family 6 Member 2, *131I* (*NaI*) sodium iodide I-131, *[18F]MFBG* 18F-meta-fluorobenzylguanidine, *NIS* sodium iodide symporter

Diagnostic and therapeutic radiopharmaceuticals that access alike biological process and cellular structure (share same target) are called theranostic pairs that are given in Table 4.1.

The classic example of theranostic system is radioiodine-based used for thyroid cancer, and with hybrid techniques, have led to development of nanotheranostics for cancer to endow with personalized treatment. To obtain this approach, certain factors are to be considered such as combining imaging and therapeutic potential, targeting tumors, reduction in toxicity with exceptional biocompatibility and pharmacokinetic stability (Zhou et al. 2018). The next nanotheranostic system is MRI-based nanoparticles for cancer which has been also explored for cancer management. This approach to have enhanced chemotherapeutic efficacy has to overcome the tumor barriers that assist tumor penetration, overcome active targeting by dual receptor approach, enhance antitumor activity, and overcome its own MRI limitations (Anani et al. 2021). As mentioned earlier, the MRI-tumor targeted delivery system has to overcome barriers and target the urokinase plasminogen activator receptor (uPAR) for treatment of pancreatic cancer (Juhl et al. 2019). A study by Lee et al. developed theranostic ATF-IONP-Gem nanoparticles that possessed potential of targeting therapeutic as well as imaging approach for pancreatic cancers

(Lee et al. 2013). In another study, Li et al. provided panoramic outlook for fluorine magnetic resonance imaging (19F MRI) as optimistic imaging theranostic technique for cancer (Li et al. 2022). The barriers like blood brain barrier that encloses brain against pathogens are key barrier to be overcome for cancer treatment. Chang et al. documented different nanomaterials—polymeric nanoparticles, liposomes, and inorganic materials, and as nanotheranostic for cancer treatment (Chang et al. 2021).

The various nanotheranostics such as polymeric NPs (e.g., polymeric micelles, polymeric conjugate complexes, dendritic polymers, etc.), liposomes, and inorganic materials (e.g., iron oxide NPs, gold NPs, carbon nanotubes, quantum dots, etc.) are used for cancer management.

4.3 Polymeric Nanoparticles

As cancer being the killing disease, developments on chemotherapy that utilize the NPs can be of prime importance. Researchers are utilizing chemotherapeutic agents by use of NPS for cancer therapy (Yan et al. 2020). These colloidal particles (NPs) are of size between 10 and 1000 nm, the size much smaller than normal human cells, that offer good therapeutic strategies in terms of bioavailability, targeting, and stability (Gaumet et al. 2008). They are easy to prepare, have high drug loading capability, and are biodegradable and biocompatible (Avnesh et al. 2013). These nanoparticles can be crafted or modified, a feature ideal for diagnosis and treatment of cancer (Ulbrich et al. 2016). As per the studies reported, formulation of these nanotheranostic agents include gold ions, silica carbon, and silver. They have shown promising results for early detection of cancer but possess some of disadvantages of toxicity, poor excretion rate, and immunogenicity. To overcome these limitations, natural polymers are explored, but they too suffer from limitations of limited application, short half-life, so synthetic biodegradable polymers can be opted. Hence, biodegradable polymers like poly(*N*-vinyl pyrrolidone) (PVP), poly(ethylene glycol) (PEG), poly(D,L-lactic acid-*co*-glycolic acid), poly(*ε*-caprolactone), poly(ethylene glycol), poly(*N*-isopropyl acrylamide) (PNIPAM), polydioxanone (PDO), polyethyleneimine (PEI), and polyvinyl alcohol (PVA), etc. are explored (Ahmed et al. 2012; Luk and Zhang 2014; Rizi et al. 2022).

The biodegradable polymeric NPs possess targeting moiety that demonstrate therapeutic drug in combination with an imaging agent to acquire a theranostic approach that can transport imaging agent to lesion, administered through oral, inhalation, or intravenous route. The commonly used identities for cancer are epidermal growth factor receptor (EGFR), interleukins (ILs), α-fetoprotein (AFP), cancer antigen 125 (CA125), cancer antigen 153 (CA15–3), cancer antigen 19–9 (CA19–9), and tumor necrosis factor-alpha (TNF-α) which are identified by nanotechnology and used for transport imaging and therapeutic to site of cancer tissues (Ali et al. 2020). Grewal et al. in their review, highlighted the application of polymeric NPs for breast cancer and their optimistic role in molecular targeting in form of NPs (Grewal et al. 2021). Boyes et al. documented review stating the formulation of gadolinium NPs with use of molecular probes and its role in imaging and therapy (Boyes et al. 2010). A study by Mi et al. developed gadolinium (Gd) chelates-loaded nanocarriers, which had MRI, guided capture therapy for tumors. The results depicted micelles of calcium phosphate that was clubbed with PEG-polyanion block copolymers and MRI, which led to MRI improvement of localization of tumor (Mi et al. 2015). Janib et al. stated the theranostic NPs for drugs and contrast agents (Janib et al. 2010).

4.4 Polymeric Micelles

The polymeric micelles are self-assembly of amphiphilic polymers that possess hydrophobic tail and hydrophilic head (Webber 1998). The type of hydrophobic and hydrophilic part, the morphology of these micelles, may take different shapes either in form of tubules, spheres, inverse micelles, etc. (Torchilin 2001). These are easy to be tailor made due to the presence of number of monomers. The components of hydrophobic polymers are poly(caprolactone) (PCL), polyesters, poly(amino acids), poly(lactide) (PLA), and poly(lactide-*co*-glycolide) (PLGA), lipids, while the hydrophilic polymers can be poly(oxazolines), poly(ethylene glycol), dextran, and chitosan hyaluronic acids (Ghosh and Biswas 2021). These are nanosized between 5–200 nm that can be functionalized with targeting ligands for enhancement in tumor accumulation, which can be an optimistic approach for cancer treatment. These small sizes assist in penetration to cancer tissues. The literature documents on the review by Wu et al. on the promising approach of polymeric micelle for encapsulation of hydrophobic drugs—doxorubicin, for cancer therapy (Wu et al. 2010; Soltantabar et al. 2020).

Gregoriou et al. developed resveratrol and coumarin 6 polymeric micelles by use of Pluronic F127 and vitamin E TPGS for theranostic targeting of breast cancer cells. The formulation had good reduction of breast cancer cell viability and no toxicity. The results explored good model for encapsulation of hydrophobic drugs that delivered natural compound against breast cancer with potential togetherness of diagnosis and treatment (Gregoriou et al. 2021). Valenzuela-Oses et al. formulated miltefosine-loaded polymeric micelles by use of Pluronic-F127. The formulation, when tested against HeLa carcinoma cells, showed cytoxicity and proved to be a promising platform for cancer treatment (Valenzuela-Oses et al. 2017).

4.5 Polymeric Conjugate Complexes (PDC)

Polymeric conjugate complexes are composed of polymeric backbones that are covalently linked to drug with aid of linker or targeting moiety. These are also known as polymer therapeutics or polymeric prodrugs. They impart drug stability, high loading capacity, and premature release of drugs. The polymeric conjugates when integrated with imaging agent, and therapeutic drugs act as theranostic in single system. With the imaging agent, it helps to monitor biodistribution and localization of drugs (Qian et al. 2017; Manandhar et al. 2021). Polymers utilized for preparation of PDC can be natural and synthetic. Natural polymers that can be utilized are alginate, chitosan, gelatin, albumin, polyglutamic acid, hyaluronic acid, and hydroxyethyl starch, while the synthetic polymers comprise of polyphosphazene, polyphosphoesters (PPEs), polyglycerol, polyethylene glycol (PEG), poly *N*-(2-hydroxypropyl)-methacrylamide-(poly-HPMA), poly(lactic-*co*-glycolicacid) (PLGA), and polylactic acid (PLA). The study conducted by Aryal et al. developed cisplatin conjugate nanoparticles that showed high drug yield, enhanced cytotoxicity, and excellent drug release kinetics as compared to free cisplatin. Thus, this PDC holds a promising approach to suppress the cancer cells and improve the therapeutic efficacy (Aryal et al. 2010). Another study by Sakhi et al. decorated Trastuzumab polymeric conjugated nanoformulation for treatment of breast cancer. The nanoformulation showed increase in cytotixicity of the drug when it was surface conjugated as compared to plain nanoformulation of the available commercial brand, and possessed the desired characteristics and enhanced targeting (Sakhi et al. 2022).

4.5.1 Dendritic Polymers

These are highly branched polymers that possess unique architecture and act as new scaffolds for drug delivery. They are of six subclasses, namely, dendrimers, multiarm star polymers, hyperbranched polymers, dendronized or dendrigrafts polymers, hypergrafts or hypergrafted polymers, and dendritic-linear block polymers (Ma et al. 2016). Dendrimers have five parts—core, repeating units, branches arising from the core, one branch junction, and terminal functional groups. Li et al. documented the recent advancement of dendritic polymer-based nanomedicines for cancer and how their features endow great potential for cancer diagnosis by use of diagnostic agents and imaging strategies for tumors (Li et al. 2021a).

Ouyang et al. reviewed dendrimer-based nanohybrids for cancer. The use of dendrimers with the nano hybrids showed good biocompatibility, stability, imaging for cancer treatment, and for combination of imaging-related cancer therapy for precision cancer applications as mentioned in Table 4.2 (Ouyang et al. 2021).

4.5.2 Liposomes

Liposomes are drug delivery systems that possess vital characteristics of biocompatibility, biodegradability, less toxicity, nanosize, and surface modification. These are the vesicles that may contain single or multiple concentric lipid bilayers known as lamellae (Muthu and Feng 2010). The size range for these is 200 nm or less than that, and hence termed as nanoliposomes. These nanoliposomes are composed of phospholipids/cholesterol, natural lipids that assist to target the cancer cells and overcome the adverse effects. The development of stealth liposomes was a hallmark in liposomes development. The conjugation of polyethylene glycol (PEG) with bioactive molecule of polymer prevents the elimination of liposomes, and these can

Table 4.2 Examples of dendrimer hybrids

Dendrimer-based hybrids	Functional components	Applications	Tumor types	References
Carbon nanohybrids	G4 PPI dendrimer	Fluorescence imaging, PTT/PDT	A780/AD cells in vitro	Taratula et al. (2015)
Metal and metal sulfide nanocomposites	G5 PAMAM dendrimer, Au NFs and Fe_3O_4	MR/PA/CT imaging, PTT/RT	4T1 tumors	Li et al. (2021b)
	G2–G4 PAMAM dendrimer, Au NRs	PTT	Colon carcinoma tumors	Ouyang et al. (2021)
	G5 PAMAM dendrimer, Au DSNRs	PTT	PC-9 tumors	

G4 PPI fourth generation poly(propylene imine), *PTT* photo thermal therapy, *PDT* photodynamic therapy, *PAMAM* poly(amidoamine), *MR* magnetic resonance, *PA* photoacoustic, *CT* computed tomography, *Au NFs* Au (gold) nano flowers, *Au DSNRs* dendrimer-stabilized gold nanorods

mount up in solid tumors (Muthu et al. 2011). Prasad et al. developed a liposomal nanotheranostic system with gold nanoparticles and grapheme quantum dots with photo triggered strategy, for enhancement in tumor accumulation of nanohybrids. Doxorubicin hydrochloride was encapsulated, and further functionalized with folic acid targeting ligands. These encapsulated agents depicted diagnosis of tumors and the near IR light (NIR 750 nm) showed reduction in tumor due to heat generation and reactive oxygen species. Thus, the study showed the antitumor activity and opened new direction for targeted imaging and cancer treatment (Prasad et al. 2020). Another review by Muthu et al. highlighted the role of stealth liposomes for cancer therapy. The consideration of receptor-targeted liposomes is also mentioned, and the advantage of vitamin E TPGS-coated liposomes as nanotheranostic has been explored (Muthu and Feng 2013).

4.6 Inorganic Materials

4.6.1 Iron Oxide Nanoparticles

The field of nanomedicine is budding research in medicine. Till date, the cancer nanomedicines that are accepted are the liposomes and magnetic iron oxide nanoparticles. Since the twentieth century, these iron oxide nanoparticles have been utilized for cancer therapy, with number of innovations and improvements (Lebrun et al. 2017). Nabavinia et al. reviewed the progress of superparamagnetic iron oxide nanoparticles (SPIONs) for cancer treatment. Due to the unique properties, he superparamagnetic enhances the chemotherapy efficiency (Nabavinia and Beltran-Huarac 2020). Shakil et al. documented the iron oxide nanoparticles for breast cancer cells. These have been used for targeting hyperthermia and imaging, and they can be conjugated with drugs, antibodies, enzymes, and proteins to target the tissues, organs, and tumors. Thus, iron oxide nanoparticles can be successfully used as theranostic agents for breast cancer (Shakil et al. 2019). Soetaert et al. reported the application of iron oxide nanoparticles for cancer therapy, wherein this magnetic system, though harnesses the success of injectable iron for cancer treatment, is still striving to establish a clinical acceptance (Soetaert et al. 2020).

4.6.2 Gold Nanoparticles

The unsatisfactory answers of poor performance and side effects of the conventional treatments for cancer have motivated researchers to focus on such systems that interpret the past questions and exhibit excellent antitumor effects. These are used due to their optochemical properties, high absorption of X-rays, and biological efficiency (Wang et al. 2016). Most of the theranostic probes comprise imaging and therapeutic modalities which delay the monitoring. So to reduce the time for this, Wang and his team demonstrated one approach called as self-therapeutic feedback image guided therapy. For this therapy, they designed fluorescence-based switch on gold nanostars that traced the cell damage. This study opened doors to cancer therapy (Wang et al. 2018). Vickers et al. reported how hollow gold nanospheres depicted better yield than gold nanorods, stating comparison between the two and showed promising nanotheranostic application (Vickers et al. 2018). These gold nanostructures with noble metals enhance the photo thermal therapy (PTT) efficiency and maximize cell death. To mention, gold nanostructures-based theranostic probes are designed and developed for cancer, some reached clinical trial phase, but still studies on improvement of the properties are yet to be explored. As mentioned by Chavda et al. nanotheranostics assist as a tool for diagnosis to monitor tumors, deliver chemotherapeutic agents, and endow with synergistic effect for patient care (Chavda et al. 2023).

4.6.3 Carbon Nanotubes

Among the nanocarriers, carbon nanotubes (CNTs) have interested many researchers due to their potentials in cancer diagnosis and treatment, in terms of their thermal conductivity and electrical properties. These are carbon allotrope; structure consists of single and rolled up grapheme sheet. There are three types of CNT structures, namely, chiral, zigzag, and armchair. The structure of CNT is determined by the value of 2 integer coefficients of chiral vector (n, m). The value if $m = 0$, then it is zigzag, if it is equal then it is armchair, and if the coefficients are different, then it is chiral (Mahar et al. 2007).

CNTs are studied as cancer theranostic that enables in PTT which transforms laser energy to acoustic signals and exhibit Raman scattering, and this property is utilized for cancer imaging (Deshmukh et al. 2020). One study as documented by Chen et al. developed nanotubes that are cross linked with poly(glycidyl methacrylate) and poly(*N*-isopropyl acrylamide) wherein doxorubicin drug was encapsulated. The results depicted good biocompatibility against KB-3-1 cancer cells and efficient anticancer activity (Chen et al. 2014). Another study by Chen et al. developed CNTs for effective delivery of 5-FU for treatment of gastric cancer (Chen et al. 2020). Another approach that has been utilized for treatment of liver cancer is by use of Sorafenib (SOR) and epidermal growth factor receptor (EGFR) through CNT-based drug delivery system. This system increased release of Sorafenib and improved stability of siRNA, thus protecting it from degradation. The results indicated inhibition of the replication of liver cancer cells, thus indicating decrease in tumor volume and assisting in cancer treatment (Wen et al. 2021). Thus, CNTs are smart carriers that deliver biomolecule along with small molecule anticancer drugs.

4.6.4 Quantum Dots (QDs)

Quantum dots are being studied as probe for biomedical imaging due to their unique electronic and optical characteristics. These are nanocrystals that on excitation emit fluorescence due to a light source. The surface modifications of these QDs make them principle in cancer imaging (Zhang et al. 2008). These QDs, when conjugated with biomolecular agents such as small molecules, peptides, and antibodies that can impart sensitivity and specificity for targeting cancer molecules, have been shown by number of studies in literature (Tholouli et al. 2008). The application of QD-multifunctional probes has also led to the development of anticancer drug and siRNA delivery (Medarova et al. 2007). Literature also reveals QDs can be used to monitor the dynamic changes of tumor environment, which will obviously be a turning point in cancer research and guide for personalized therapy. Thus, molecular QD targeted nano approach offers numerous advantages, in terms of specificity, affinity, and bypass the biological barriers, for enhanced targeting of drugs (Fang et al. 2012). As per the reports of Wang et al. QDs are utilized for detection of ovarian carcinoma marker CA125, in tissues, fixed cells. The QD signals were brighter and specific than the conventional organic dye (Wang et al. 2004).

One study reported the use of non-cadmium-based QDs, for pancreatic cancer, as non-toxic and effective probe for imaging. The study also highlighted that when they were conjugated with monoclonal antibodies as anti-claudin-4, it allowed specific targeting of pancreatic cancer (Yang et al. 2009). Ayoush et al. targeted drug delivery system based on Ag–In–Zn–S quantum dots. The modification of QD was utilized for doxorubicin targeting to folate receptors. The nanoconjugates induced significant DNA break, with good cytotoxicity and effective in the treatment of lung cancer (Ruzycka-Ayoush et al. 2021). Thus, QDs are marvels that can bring revolution in cancer diagnosis and treatment.

4.7 Conclusion

The development in nanotechnology can work wonders in cancer management. These nanomedicines carriers have attracted scientists due to their vital properties, and hence nanotheranostics can assist in both diagnosis and treatment of cancer. Nanotheranostic is a powerful tool to offer basic research and clinical applications. The different types of tumors can thus be addressed through the use of nanotheranostics, but different factors in terms of their toxicity, pharmacokinetic properties, biodegradation, and elimination need to be focused. In near future, the research done by various researchers shall indeed attract pharmaceutical companies for development of victorious theranostic nanoplatforms with prologue of novel nanotheranostics into the market, with personalized medicine for the patients.

References

Ahmed N, Fessi H, Elaissari A (2012) Theranostic applications of nanoparticles in cancer. Drug Discov Today 17:929–934

Ali I, Alsehli M, Scotti L, Tullius Scotti M, Tsai ST, Yu RS, Hsieh MF, Chen JC (2020) Progress in polymeric nano-medicines for theranostic cancer treatment. Polymers (Basel) 12(3):598

Anani T, Rahmati S, Sultana N, David AE (2021) MRI-traceable theranostic nanoparticles for targeted cancer treatment. Theranostics 11(2):579–601

Aryal S, Hu CM, Zhang L (2010) Polymer–cisplatin conjugate nanoparticles for acid-responsive drug delivery. ACS Nano 4(1):251–258

Avnesh S, Thakor S, Gambhir S (2013) Nanooncology: the future of cancer diagnosis and therapy. CA Cancer J Clin 63:395–418

Bluemel C, Herrmann K, Giammarile F, Nieweg OE, Dubreuil J, Testori A et al (2015) EANM practice guidelines for lymphoscintigraphy and sentinel lymph node biopsy in melanoma. Eur J Nucl Med Mol Imaging 42(11):1750–1766

Boyes SG, Rowe MD, Serkova NJ, Kim FJ, Lambert JR, Werahera PN (2010) Polymer-modified gadolinium nanoparticles for targeted magnetic resonance imaging and therapy. Nano Life 1:263–275

Chang D, Ma Y, Xu X, Xie J, Ju S (2021) Stimuli-responsive polymeric nanoplatforms for cancer therapy. Front Bioeng Biotechnol 9:707319

Chavda VP, Khadela A, Shah Y, Postwala H, Balar P, Vora L (2023) Current status of cancer nanotheranostics: emerging strategies for cancer management. Nanotheranostics 7(4):368–379

Chen G, Chen R, Zou C, Yang D, Chen Z-S (2014) Fragmented polymer nanotubes from sonication induced scission with a thermo-responsive gating system for anti-cancer drug delivery. J Mater Chem B 2:1327–1334

Chen W, Yang S, Wei X, Yang Z, Liu D, Pu X et al (2020) Construction of aptamer-siRNA chimera/PEI/5-FU/carbon nanotube/collagen membranes for the treatment of peritoneal dissemination of drug-resistant gastric cancer. Adv Healthc Mater 9(21):e2001153

Cleary K, Peters TM (2010) Image-guided interventions: technology review and clinical applications. Annu Rev Biomed Eng 12:119–142

Deshmukh MA, Jeon JY, Ha TJ (2020) Carbon nanotubes: an effective platform for biomedical electronics. Biosens Bioelectron 150:111919

Eary JF (1999) Nuclear medicine in cancer diagnosis. Lancet 354(9181):853–857

Fang M, Peng CW, Pang DW, Li Y (2012) Quantum dots for cancer research: current status, remaining issues, and future perspectives. Cancer Biol Med 9(3):151–163

Foroozandeh P, Aziz AA (2018) Insight into cellular uptake and intracellular trafficking of nanoparticles. Nanoscale Res Lett 13:339

Gaumet M, Vargas A, Gurny R, Delie F (2008) Nanoparticles for drug delivery: the need for precision in reporting particle size parameters. Eur J Pharm Biopharm 69:1–9

Ghosh B, Biswas S (2021) Polymeric micelles in cancer therapy: state of the art. J Control Release 332:127–147

Gregoriou Y, Gregoriou G, Yilmaz V, Kapnisis K, Prokopi M, Anayiotos A, Strati K, Dietis N, Constantinou AI, Andreou C (2021) Resveratrol loaded polymeric micelles for theranostic targeting of breast cancer cells. Nanotheranostics 5(1):113–124

Grewal I, Singh S, Arora S, Sharma N (2021) Polymeric nanoparticles for breast cancer therapy: a comprehensive review. Biointerface Res Appl Chem 11(4):11151–11171

Hisada Y, Mackman N (2019) Tissue factor and cancer: regulation, tumour growth, and metastasis. Semin Thromb Hemost 45(4):385–395

Hou M, Yan C, Chen Z, Zhao Q, Yuan M, Xu Y et al (2018) Multifunctional NIR-responsive poly(vinylpyrrolidone)-Cu–Sb–S nanotheranostic agent for photoacoustic imaging and photothermal/photodynamic therapy. Acta Biomater 74:334–343

Idée JM, Louguet S, Ballet S, Corot C (2013) Theranostics and contrast-agents for medical imaging: a pharmaceutical company viewpoint. Quant Imaging Med Surg 3:292–297

Imlimthan S, Khng Y, Keinänen O, Zhang W, Airaksinen A, Kostiainen M et al (2021) A theranostic cellulose nanocrystal-based drug delivery system with enhanced retention in pulmonary metastasis of melanoma. Small 17(18):2007705

Jadvar H (2017) Targeted radionuclide therapy: an evolution toward precision cancer treatment. AJR Am J Roentgenol 209(2):277–288

Janib SM, Moses AS, MacKay JA (2010) Imaging and drug delivery using theranostic nanoparticles. Adv Drug Deliv Rev 62:1052–1063

Juhl K, Christensen A, Rubek N, Karnov KKS, Von Buchwald C, Kjaer A (2019) Improved surgical resection of metastatic pancreatic cancer using uPAR targeted in vivo fluorescent guidance: comparison with traditional white light surgery. Oncotarget 10:6308–6316

Koenig SH, Kellar KE (1995) Theory of 1/T1 and 1/T2 NMRD profiles of solutions of magnetic nanoparticles. Magn Reson Med 34:227–233

Kundu P, Singh D, Singh A, Sahoo SK (2020) Cancer nanotheranostics: a nanomedicinal approach for cancer therapy and diagnosis. Anticancer Agents Med Chem 20(11):1288–1299

Lebrun F, Klastersky J, Levacq D, Wissam Y, Paesmans M (2017) Intravenous iron therapy for anemic cancer patients: a review of recently published clinical studies. Support Care Cancer 25:2313–2319

Lee GY, Qian WP, Wang L, Wang YA, Staley CA, Satpathy M, Nie S, Mao H, Yang L (2013) Theranostic nanoparticles with controlled release of gemcitabine for targeted therapy and MRI of pancreatic cancer. ACS Nano 7(3):2078–2089

Li H, Sun J, Zhu H, Wu H, Zhang H, Gu Z, Luo K (2021a) Recent advances in development of dendritic polymer-based nanomedicines for cancer diagnosis. Wiley Interdiscip Rev Nanomed Nanobiotechnol 13(2):e1670

Li X, Ouyang Z, Li H, Hu C, Saha P, Xing L, Shi X, Pich A (2021b) Dendrimer-decorated nanogels: efficient nanocarriers for biodistribution in vivo and chemotherapy of ovarian carcinoma. Bioact Mater 6(10):3244–3253

Li Y, Cui J, Li C, Zhou H, Chang J, Aras O, An F (2022) 19F MRI nanotheranostics for cancer management: progress and prospects. ChemMedChem 17(4):e202100701

Liu Y, Zhang P, Li F, Jin X, Li J, Chen W, Li Q (2018) Metal-based nanoenhancers for future radiotherapy: radiosensitizing and synergistic effects on tumour cells. Theranostics 8(7):1824–1849

Luk BT, Zhang L (2014) Current advances in polymer-based nanotheranostics for cancer treatment and diagnosis. ACS Appl Mater Interfaces 6:21859–21873

Ma Y, Mou Q, Wang D, Zhu X, Yan D (2016) Dendritic polymers for theranostics. Theranostics 6(7):930–947

Madanayake NH, Rienzie R, Adassooriya NM (2019) Nanoparticles in nanotheranostics applications. In: Rai M, Jamil B (eds) Nanotheranostics: applications and limitations. Springer International Publishing, Cham, pp 19–40

Mahar B, Laslau C, Yip R, Sun Y (2007) Development of carbon nanotube-based sensors—a review. IEEE Sens J 7:266–284

Manandhar S, Sjöholm E, Bobacka J, Rosenholm JM, Bansal KK (2021) Polymer-drug conjugates as nanotheranostic agents. J Nanotheranostics 2:63–81

Medarova Z, Pham W, Farrar C et al (2007) In vivo imaging of siRNA delivery and silencing in tumours. Nat Med 13:372–377

Mendes LP, Lima EM, Torchilin VP (2018) Chapter 9—Targeted nanotheranostics for selective drug delivery in cancer. In: Conde J (ed) Handbook of nanomaterials for cancer theranostics. Elsevier, pp 245–277

Mi P, Dewi N, Yanagie H, Kokuryo D, Suzuki M, Sakurai Y, Li Y, Aoki I, Ono K, Takahashi H et al (2015) Hybrid calcium phosphate-polymeric micelles incorporating gadolinium chelates for imaging-guided gadolinium neutron capture tumor therapy. ACS Nano 9:5913–5921

Mura S, Couvreur P (2012) Nanotheranostics for personalized medicine. Adv Drug Deliv Rev 64(13):1394–1416

Muthu MS, Feng SS (2010) Nanopharmacology of liposomes developed for cancer therapy. Nanomedicine (Lond) 5:1017–1019

Muthu MS, Feng SS (2013) Theranostic liposomes for cancer diagnosis and treatment: current development and pre-clinical success. Expert Opin Drug Deliv 10(2):151–155

Muthu MS, Kulkarni SA, Xiong J (2011) Vitamin E TPGS coated liposomes enhanced cellular uptake and cytotoxicity of docetaxel in brain cancer cells. Int J Pharm 42:332–340

Nabavinia M, Beltran-Huarac J (2020) Recent progress in iron oxide nanoparticles as therapeutic magnetic agents for cancer treatment and tissue engineering. ACS Appl Biol Mater 3(12):8172–8187

National Research Council (US) and Institute of Medicine (US), Committee on State of the Science of Nuclear Medicine (2007) Advancing nuclear medicine through innovation. Nuclear medicine imaging in diagnosis and treatment. National Academies Press (US), Washington (DC)

Navya PN, Kaphle A, Srinivas SP et al (2019) Current trends and challenges in cancer management and therapy using designer nanomaterials. Nano Converg 6:23

Ouyang Z, Gao Y, Shen M, Shi X (2021) Dendrimer-based nanohybrids in cancer photomedicine. Mater Today Biol 10:100111

Parhi P, Mohanty C, Sahoo SK (2012) Nanotechnology-based combinational drug delivery: an emerging approach for cancer therapy. Drug Discov Today 17(17–18):1044–1052

Parveen S, Misra R, Sahoo SK (2012) Nanoparticles: a boon to drug delivery, therapeutics, diagnostics and imaging. Nanomedicine (Lond) 8(2):147–166

Pelaz B, Alexiou C, Alvarez-Puebla RA, Alves F, Andrews AM, Ashraf S et al (2017) Diverse applications of nanomedicine. ACS Nano 11:2313–2381

Pene F, Courtine E, Cariou A, Mira JP (2009) Toward theranostics. Crit Care Med 37:S50–S58

Prasad R, Jain NK, Yadav AS, Chauhan DS, Devrukhkar J, Kumawat MK, Shinde S, Gorain M, Thakor AS, Kundu GC, Conde J, Srivastava R (2020) Liposomal nanotheranostics for multimode targeted in vivo bioimaging and near-infrared light mediated cancer therapy. Commun Biol 3(1):284

Qian CG, Chen YL, Feng PJ, Xiao XZ, Dong M, Yu JC, Hu QY, Shen QD, Gu Z (2017) Conjugated polymer nanomaterials for theranostics. Acta Pharmacol Sin 38:764–781

Rizi HAY, Shin DH, Rizi SY (2022) Polymeric nanoparticles in cancer chemotherapy: a narrative review. Iran J Public Health 51(2):226–239

Ruzycka-Ayoush M, Kowalik P, Kowalczyk A et al (2021) Quantum dots as targeted doxorubicin drug delivery nanosystems in human lung cancer cells. Cancer Nano 12:8

Sakhi M, Khan A, Iqbal Z, Khan I, Raza A, Ullah A, Nasir F, Khan SA (2022) Design and characterization of paclitaxel-loaded polymeric nanoparticles decorated with trastuzumab for the effective treatment of breast cancer. Front Pharmacol 13:855294

Schirrmacher V (2019) From chemotherapy to biological therapy: a review of novel concepts to reduce the side effects of systemic cancer treatment (review). Int J Oncol 54(2):407–419

Shakil MS, Hasan MA, Sarker SR (2019) Iron oxide nanoparticles for breast cancer theranostics. Curr Drug Metab 20(6):446–456

Sharma P, Bhargava M (2013) Applications and characteristics of nanomaterials in industrial environment. Res Dev (IJCSEIERD) 3(4):63–72

Sivasubramanian M, Chuang YC, Chen NT, Lo LW (2019) Seeing better and going deeper in cancer nanotheranostics. Int J Mol Sci 20:3490

Soetaert F, Korangath P, Serantes D, Fiering S, Ivkov R (2020) Cancer therapy with iron oxide nanoparticles: agents of thermal and immune therapies. Adv Drug Deliv Rev 163:65–83

Soltantabar P, Calubaquib EL, Mostafavi E, Biewer MC, Stefan MC (2020) Enhancement of loading efficiency by coloading of doxorubicin and quercetin in thermoresponsive polymeric micelles. Biomacromolecules 21:1427–1436

Som P (1980) A fluorinated glucose analog, 2-fluoro-2-deoxy-D-glucose (F18): nontoxic tracer for rapid tumour detection. J Nucl Med 21:670–675

Sun C, Lee JS, Zhang M (2008) Magnetic nanoparticles in MR imaging and drug delivery. Adv Drug Deliv Rev 60:1252–1265

Sung H, Ferlay J, Siegel RL, Laversanne M, Soerjomataram I, Jemal A, Bray F (2021) Global cancer statistics 2020: GLOBOCAN estimates of incidence and mortality worldwide for 36 cancers in 185 countries. CA Cancer J Clin 71(3):209–249

Taratula O, Patel M, Schumann C, Naleway MA, Pang AJ, He H, Taratula O (2015) Phthalocyanine-loaded graphene nanoplatform for imaging-guided combinatorial phototherapy. Int J Nanomed 10:2347–2362

Tholouli E, Sweeney E, Barrow E et al (2008) Quantum dots light up pathology. J Pathol 216:275–285

Torchilin VP (2001) Structure and design of polymeric surfactant-based drug delivery systems. J Control Release 73:137–172

Ulbrich K, Hola K, Subr V, Bakandritsos A, Tucek J, Zboril R (2016) Targeted drug delivery with polymers and magnetic nanoparticles: covalent and noncovalent approaches, release control, and clinical studies. Chem Rev 116:5338–5431

Valenzuela-Oses JK, García MC, Feitosa VA, Pachioni-Vasconcelos JA, Gomes-Filho SM, Lourenço FR, Cerize NNP, Bassères DS, Rangel-Yagui CO (2017) Development and characterization of miltefosine-loaded polymeric micelles for cancer treatment. Mater Sci Eng C 81:327–333

Veiseh O, Gunn JW, Zhang M (2010) Design and fabrication of magnetic nanoparticles for targeted drug delivery and imaging. Adv Drug Deliv Rev 62:284–304

Vickers ET, Garai M, Naghadeh SB et al (2018) Two-photon photoluminescence and photothermal properties of hollow gold nanospheres for efficient theranostic applications. J Phys Chem C 122(25):13304–13313

Viswanadh MK, Singh RP, Agrawal P, Mehata AK, Pawde DM, Sonkar R, Muthu MS (2018) Nanotheranostics: emerging strategies for early diagnosis and therapy of brain cancer. Nanotheranostics 2:70–86

Wang HZ, Wang HY, Liang RQ et al (2004) Detection of tumour marker CA125 in ovarian carcinoma using quantum dots. Acta Biochim Biophys Sin (Shanghai) 36:681–686

Wang H et al (2016) High sensitivity of gold nanoparticles co-doped with Gd_2O_3 mesoporous silica nanocomposite to nasopharyngeal carcinoma cells. Sci Rep 6:34367

Wang J, Zhou Z, Zhang F et al (2018) A novel nanocomposite based on fluorescent turn-on gold nanostars for near-infrared photothermal therapy and self-theranostic caspase-3 imaging of glioblastoma tumour cell. Colloids Surf B Biointerfaces 170:303–311

Webber SE (1998) Polymer micelles: an example of self-assembling polymers. J Phys Chem B 102:2618–2626

Wen Z, Feng Y, Hu Y, Lian L, Huang H, Guo L, Chen S, Yang Q, Zhang M, Wan L, Xu K, Degejirifu, Yan X (2021) Multiwalled carbon nanotubes co-delivering sorafenib and epidermal growth factor receptor siRNA enhanced tumor-suppressing effect on liver cancer. Aging (Albany NY) 13(2):1872–1882

Wu Y, Sefah K, Liu H, Wang R, Tan W (2010) DNA aptamer-micelle as an efficient detection/delivery vehicle toward cancer cells. Proc Natl Acad Sci USA 107:5–10

Xie J, Lee S, Chen X (2010) Nanoparticle-based theranostic agents. Adv Drug Deliv Rev 62(11):1064–1079

Yan L, Shen J, Wang J, Yang X, Dong S, Lu S (2020) Nanoparticle-based drug delivery system: a patient-friendly chemotherapy for oncology. Dose Response 18(3):1559325820936161

Yang L, Mao H, Cao Z et al (2009) Molecular imaging of pancreatic cancer in an animal model using targeted multifunctional nanoparticles. Gastroenterology 136:1514–1525

Zhang H, Yee D, Wang C (2008) Quantum dots for cancer diagnosis and therapy: biological and clinical perspectives. Nanomedicine (Lond) 3(1):83–91

Zhang P, Chen H, Liu J, Liu G (2019) Genetically engineered plasma membrane nanovesicles for cancer-targeted nanotheranostics. Methods Mol Biol 2054:283–294

Zhou J, Li T, Zhang C, Xiao J, Cui D, Cheng Y (2018) Charge-switchable nanocapsules with multistage pH-responsive behaviours for enhanced tumour-targeted chemo/photodynamic therapy guided by NIR/MR imaging. Nanoscale 10:9707–9719

Zitvogel L, Apetoh L, Ghiringhelli F, Kroemer G (2008) Immunological aspects of cancer chemotherapy. Nat Rev Immunol 8:59–73

Bio-inspired Nanomaterials in Cancer Theranostics

5

Monika Kaushik, Sanjesh Kumar, Mansi Singh, Himanshu Sharma, Mithun Bhowmick, Pratibha Bhowmick, Sumel Ashique, Heya Khatoon, Radheshyam Pal, and Mohammad Azam Ansari

Abstract

Cancer causes many fatalities worldwide. Non-traditional cancer treatments use nanotechnology to target and control drug release. Nanomedicine, a multidisciplinary field, may lead to new cancer diagnosis, therapy, and detection approaches. Nanoparticles (NPs) outperform small-molecule treatments due to their high surface-to-volume ratio, which allows them to enter live cells. Cancer treatments including radiation therapy, chemotherapy, targeted therapy, and immunotherapy are hindered by cytotoxicity, lack of selectivity, and multiple drug resistance. Numerous methods use plants to produce nanoparticles. Because medicinal plants can treat cancer, plant-based nanoparticle studies must be thoroughly evaluated. Plant biomolecules, secondary metabolites, and

M. Kaushik
Department of Pharmacology, Amity Institute of Pharmacy, Amity University Gwalior, Gwalior, Madhya Pradesh, India

S. Kumar · M. Singh
Institute of Pharmaceutical Research, GLA University, Mathura, UP, India

H. Sharma
Teerthanker Mahaveer College of Pharmacy, Teerthanker Mahaveer University, Moradabad, UP, India

M. Bhowmick · P. Bhowmick · S. Ashique (✉) · R. Pal
Department of Pharmaceutical Sciences, Bengal College of Pharmaceutical Sciences and Research, Durgapur, West Bengal, India

H. Khatoon
Department of Pharmacy, Pandaveswar School of Pharmacy, Pandaveswar, West Bengal, India

M. A. Ansari
Department of Epidemic Disease Research, Institute for Research and Medical Consultations, Imam Abdulrahman Bin Faisal University, Dammam, Saudi Arabia

M. A. Barkat et al. (eds.), *Nanotheranostics for Diagnosis and Therapy*, https://doi.org/10.1007/978-981-97-3115-2_5

coenzymes nanoparticle size metal ions. These nanoparticles may contain antioxidants and cancer-fighters. However, nanomaterials provide innovative cancer treatments beyond typical pharmaceutical delivery. Cancer treatment with nanoparticles (1–100 nm) has many benefits, including enhanced permeability and retention, lower toxicity, superior stability, accurate targeting, and biocompatibility. This chapter reviews numerous bio-inspired nanoparticles for cancer theranostics.

Keywords

Nanomaterials · Cancer therapy · Conventional approaches · Advanced approaches · Bio-inspired nanoparticle · Clinical trials · Future window

5.1 Introduction

Nanotechnology and biological research have provided breakthrough cancer diagnosis and treatment methods in recent decades. Cancer theranostics with bioinspired smart nanoparticles are intriguing (Ferraiuolo et al. 2020). Nanomaterials with therapeutic and diagnostic properties satisfy the growing need for more accurate and effective cancer therapies. Cancer theranostics aims to treat and diagnose cancer (Ferlay et al. 2021). This method addresses the poor specificity and unexpected adverse effects of traditional cancer drugs. Biological process-inspired nanocarriers and smart nanoparticles have several applications (Mbemi et al. 2020; Ansari et al. 2023). Nanoparticles increase targeting, toxicity, and therapeutic efficacy in biological systems (Ansari et al. 2021, 2024; Tripathi et al. 2023). Nanoparticles with stimuli-responsive components can respond to cancer microenvironment signals (Xu et al. 2022). This decreased tissue damage because the therapeutic payload targeted cancer cells (Niculescu and Grumezescu 2022). Controlled medicine may be released in response to pH, temperature, or enzyme activity (Abudurexiti et al. 2023). Imaging agents can improve nanoparticle diagnostics (Dutta et al. 2022), enabling real-time sickness tracking and medicine efficacy (Jain et al. 2023). This nanoparticle mimics biological architecture and has bioinspired features (Akhtar et al. 2022). Biomimetic approaches are offered by Lim et al. (2023). Ligands that imitate cell–cell interactions may help these nanoparticles stick to cancer cells. Drug efficacy and diagnosis accuracy improved (Govindan et al. 2023). Different kinds of bio-inspired nanoparticles are showed in Fig. 5.1 (Briolay et al. 2021).

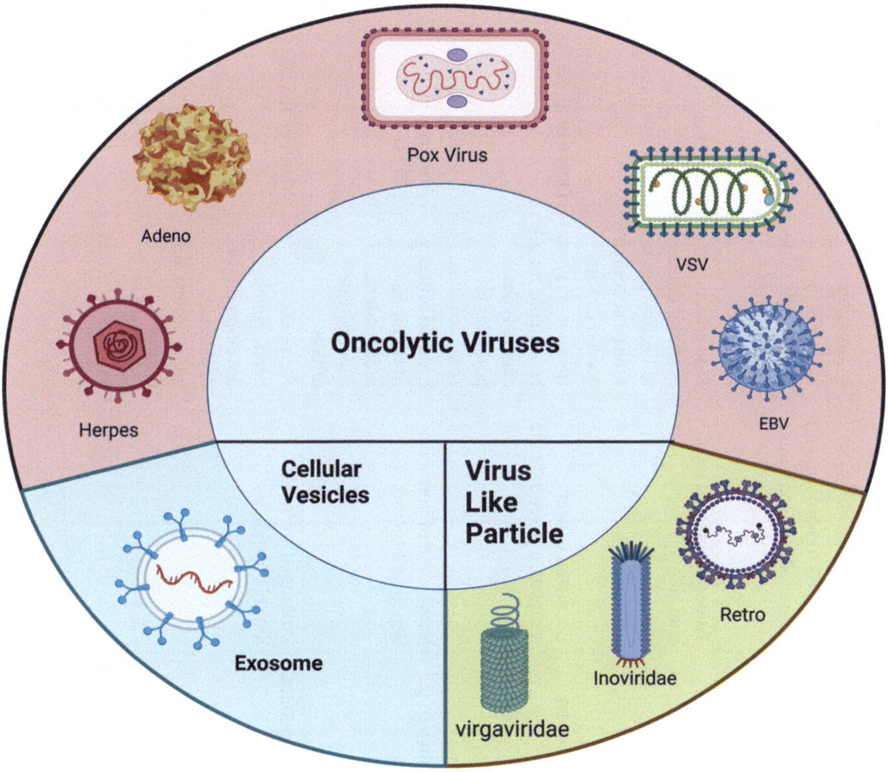

Fig. 5.1 Biological and bio-inspired nano vectors for cancer therapy. These nano vectors have been derived from different types of organisms and exhibit high biocompatibility and extensive engineering possibilities. [(http://creativecommons.org/licenses/by/4.0/) (Briolay et al. 2021)]

5.2 Traditional Methods for Treating Cancer

The modern cancer treatment plan employs a wide range of techniques, including immunotherapy, chemotherapy, radiation therapy, surgery, and targeted treatments (Table 5.1) (Baskar et al. 2012). Traditional cancer treatment treatments, which encompass a range of techniques aimed at destroying or controlling malignant cells, have long been the cornerstone of oncology (Raja et al. 2023). These conventional methods primarily include surgery, chemotherapy, and radiation treatment (Dobre et al. 2023). They are often used alone or in combination to treat various aspects of cancer genesis and spread (Padmanabhan et al. 2022). Surgery is a crucial part of cancer treatment, including the actual removal of tumors or cancerous tissues (Marei et al. 2023). Preventing the tumors from becoming bigger and maybe spreading to other parts of the body is the aim of targeted tumor excision (Abraham et al. 2023). Surgery may not be the best course of action for cancers that have spread extensively, even while it is effective for solid tumors that are readily removed. Chemotherapy is a

Table 5.1 Conventional treatment approaches for cancer treatment and their limitations

Treatment strategy	Mechanism	Type of cancers targeted	Therapeutic outcomes	Limitations	References
Chemotherapy	Targets rapidly dividing cells, disrupting DNA synthesis and cell division	Various (e.g., breast, ovarian, leukemia)	Reduced tumor size, and improved survival rates	Non-specific cytotoxicity, side effects, drug resistance	Baskar et al. (2012)
Radiation therapy	Uses ionizing radiation to damage DNA in targeted cancer cells	Localized (e.g., breast, lung, prostate)	Tumor shrinkage, local control	Side effects, potential damage to healthy tissues, limited access in certain cancers	Raja et al. (2023)
Surgery	Physically removes tumors or cancerous tissue	Localized (e.g., breast, colon, lung)	Localized cancer removal, a potential cure	Invasive, may not be suitable for all cancer types, potential for complications	Dobre et al. (2023)
Hormone therapy	Alters hormone levels to inhibit cancer growth	Hormone-sensitive (e.g., breast, prostate)	Regression of hormone-dependent tumors	Limited efficacy in hormone-independent cancers, side effects	Abraham et al. (2023)
Immunotherapy	Stimulates the immune system to recognize and attack cancer cells	Various (e.g., melanoma, lung, renal)	Prolonged survival, durable responses	Response heterogeneity, development of resistance, autoimmune side effects	Marei et al. (2023)

Targeted therapy (e.g., Trastuzumab)	Blocks specific molecules involved in cancer growth	HER2-positive breast cancer	Improved survival, and reduced recurrence	Resistance development, limited efficacy in non-HER2-amplified tumors	Kroemer et al. (2023)
Angiogenesis inhibitors	Blocks the formation of new blood vessels that tumors need to grow	Various (e.g., colorectal, kidney)	Slows tumor growth, and metastasis inhibition	Limited survival benefits, potential side effects	Elebiyo et al. (2022)
Stem cell transplantation	Replaces damaged bone marrow with healthy stem cells	Blood cancers (e.g., leukemia, lymphoma)	Restoration of blood cell production	High risk of complications, graft-versus-host disease, limited applicability	Sheykhhasan et al. (2022)
Bone marrow transplantation	Replaces damaged bone marrow with healthy donor marrow	Various (e.g., leukemia, lymphoma)	Restoration of blood cell production	Risk of complications, graft rejection, and limited donor availability	Abdallah et al. (2023) and Krishnaswami et al. (2023)
Photodynamic therapy	Uses light-sensitive compounds to kill cancer cells	Skin, esophageal, lung cancers	Localized tumor treatment, minimal damage to surrounding tissues	Limited depth of light penetration, photosensitivity, and potential for side effects	Yi et al. (2022)
Hyperthermia	Raises the temperature of the affected body part to enhance treatment efficacy	Various (e.g., breast, cervical, prostate)	Increased sensitivity to radiation, and improved blood flow	Risk of burns, difficulty in precise temperature control, limited evidence	Yi et al. (2022)

(continued)

Table 5.1 (continued)

Treatment strategy	Mechanism	Type of cancers targeted	Therapeutic outcomes	Limitations	References
Cryotherapy	Freezes and destroys cancer cells	Prostate, liver, and skin cancers	Localized treatment, minimal invasiveness	Risk of damage to adjacent healthy tissues, limited depth of treatment	Mokbel et al. (2023)
Electrochemotherapy	Combines chemotherapy with electric pulses to enhance drug uptake by cells	Skin cancers, metastatic nodules	Increased drug delivery to tumor cells	Limited evidence, potential for side effects	Helderman et al. (2023)
Topical chemotherapy	Applies chemotherapy directly to the skin surface	Skin cancers	Targeted treatment for skin malignancies	Limited depth of penetration, potential skin irritation	Xie et al. (2023)
Intraperitoneal chemotherapy	Administers chemotherapy directly into the abdominal cavity	Ovarian, and colorectal cancers	Increased drug concentration at the tumor site	Limited applicability to specific cancers, the potential for complications	Mohammed et al. (2023)
Adjuvant therapy	Supplementary treatment following primary therapy to prevent recurrence	Various (e.g., breast, colon)	Reduced risk of recurrence, improved overall survival	Side effects, may not benefit all patients	Xie et al. (2023)

systemic treatment that targets and kills rapidly proliferating cancer cells in the body using drugs. These drugs may be given orally or by injection (Kroemer et al. 2023). They enter the bloodstream and make their way to cancer cells in several organs. Chemotherapy is unfortunately associated with side effects since it is non-specific and affects both malignant and healthy cells with high rates of growth (Elebiyo et al. 2022). In radiation therapy, high doses of ionizing radiation are used to damage the DNA of cancer cells, stopping them from growing and multiplying (Sheykhhasan et al. 2022). External beam radiation or brachytherapy may be used to provide this tailored treatment externally (Krishnaswami et al. 2023). While radiation therapy might be useful in focusing on certain areas, it can potentially negatively impact nearby healthy tissues (Abdallah et al. 2023). In recent years, immunotherapy has emerged as a revolutionary adjuvant to traditional cancer treatments (Yi et al. 2022). By using the body's immune system, this technique locates and eliminates cancer cells (Helderman et al. 2023). Checkpoint inhibitors, adoptive cell therapies (Xie et al. 2023), and monoclonal antibodies are a few types of immunotherapies that provide a more specialized and targeted approach to treating cancer (Mohammed et al. 2023).

5.3 Limitations of Conventional Approaches for Cancer Treatment

Despite their historical importance, conventional cancer treatments have significant limits that highlight the need for oncological innovation (Table 5.1). Vinca alkaloids, taxanes, and anthracyclines have been used in chemotherapy to treat breast, ovarian, and leukemia (Yang et al. 2023). Despite their efficacy, their non-specific cytotoxicity damages healthy tissues, emphasizing the need for tailored therapies (McNamara et al. 2023). These inhibitors include nivolumab and pembrolizumab, for instance. Nonetheless, a subset of patients remains refractory, accentuating the complexity of the tumor microenvironment and the need for refinements in immunotherapeutic approaches (Healey Bird et al. 2022). Technological innovations like intensity-modulated radiation therapy (IMRT) and stereotactic body radiation therapy (SBRT) confer heightened precision, yet challenges persist, particularly in malignancies like glioblastoma where the blood–brain barrier poses a formidable impediment to therapeutic access. The limitations of conventional cancer treatments underscore the imperative for ongoing research and the development of more targeted, personalized, and less toxic therapeutic modalities to address the diverse array of challenges posed by different cancer types and patient populations (Tree et al. 2022).

5.4 Nanomedicine in Cancer Theranostics

Nanotechnology's advances in cancer imaging and diagnostics encourage researchers worldwide to develop novel nanoformulation uses. This idea inspires multimodal theranostic nanoparticles for cancer diagnosis and treatment. Due to their molecular property that assists diagnosis and their ability to contain medicinal and imaging compounds, these theranostics nanoparticles may outperform standard nanoformulations. Theranostic nanoparticles are used in personalized medicine due to their dual function (Ashique et al. 2022a; Ansari et al. 2022).

5.4.1 Need for Nanomedicine in Cancer Therapy

Due to its intricate biology, cancer is a leading killer. Immunotherapy, targeted therapy, radiation, and chemotherapy are conventional cancer treatments. Multidrug resistance, cytotoxicity, and lack of selectivity limit cancer treatment. Nanotechnology transformed cancer detection and treatment. Due to their improved permeability, retention, stability, toxicity, and precision targeting, nanoparticles (1–100 nm) can treat cancer. Many categories of nanoparticles exist. Nanoparticle drug delivery uses tumor and environment features, making it unique. Nanoparticles solve multidrug resistance and cancer treatment issues (Ashique et al. 2022b). Because they biodegrade, nanoparticles treat cancer better than microparticles (Palazzolo et al. 2018). Their extracellular matrix renders nanoparticles too big for blood arteries. Immature vasculatures block lymphatic outflow as tumors grow (Gavas et al. 2021). Inhibiting lymphatic outflow lets nanoparticles reach cells. The "enhanced permeability and retention effect" (EPR) drives passive nanoparticle targeting. Superparamagnetic nanomaterials facilitate cancer diagnosis and treatment. Nanoparticles of superparamagnetic iron oxide provide this. They may treat cancer due to their size, targeted selectivity, and immune evasion (Gavas et al. 2021). Active and passive nanomaterial-based drug delivery has improved. Passive targeting improves permeability and retention, while active targeting uses antibodies or small molecule-conjugated nanoparticles. Active targeting increases retention and efficiency, replacing passive targeting and improving tumor localization (Ashique et al. 2023a). Nanomedicines have a longer half-life, more loading capacity, reduced cytotoxicity, stronger selectivity, and better absorption than chemicals. Advances in nanoscience, technology, and cancer pathology have led to many cancer therapeutic nanomaterials. Not many nanomaterial-based medications have been clinically tested. Cancer treatment requires drug delivery, and non-conjugated polymers with luminescence Excel (Ashique et al. 2023b). Drug size matters. The dense extracellular matrix makes medication entry harder. Tumor-induced angiogenesis impeded lymphatic outflow and created immature vasculatures (Lim et al. 2018). Due to their "leaky," nanoparticles can penetrate targets. Targeting peptide attachment is easy with PNPs' high surface-to-volume ratio. These surfactants help solubilize and fluidize. Ultrasound-responsive PNPS delivers anticancer medicines better than chemicals. Theranostics, a new diagnosis-treatment approach, needs fluorescent PNPs. Complex fluorescent PNPs have been exploited as theragnostic materials recently.

5.4.2 Mechanism of Nanomedicine in Cancer Therapy

Effective nanomedicines require tumor interstitium penetration even with big tumors (Gavas et al. 2021; Handa et al. 2021). Nanocarrier physicochemical properties and tumor microenvironment affect penetration. High IFP and extensive collagen, hyaluronan, and proteoglycan networks characterize tumor microenvironments.

These features biologically hinder tumor invasion (Ashique et al. 2023c). Tumor penetration rises with nanocarrier shrinking (Gavas et al. 2021). Tumor site nanomedicine accumulation, retention, and drug release are optimized to improve therapy efficacy. Figure 5.2 describes the mechanism of bio-inspired nanoparticles against cancer cell (Tiwari et al. 2023).

5.4.2.1 Passive Targeting

Hypoxia creates or destroys blood vessels; making them permeable, and accelerating tumor growth is known as neovascularization. Large leaky tumor blood channel pores diminish perm-selectivity (Torchilin 2011). These massive pores, or fenestrations, can be 200–2000 nm depending on malignancy, location, and tumor microenvironment (Jain 1998). NPs can form and accumulate inside cancer cells as a result of rapid and defective angiogenesis, which occurs through blood routes. ECF from healthy tissues moves into lymphatic arteries at 0.1–2.0 μm/s, supporting drainage and regeneration (Gavas et al. 2021). Cancer reduces lymphatic function and ISF uptake (Hobbs 1998). Without elimination, NPs accumulate in the tumor interstitium. EPR affects retention more. Non-unique short-circulating molecules wash out of cancer cells. To improve pharmacokinetics, tumor selectivity, and adverse effects, nanosized drug carriers contain these tiny compounds (Gavas et al. 2021). For passive targeting over EPR, TME is necessary. Through glycolysis, tumors grow swiftly. Low pH in TME releases pH-sensitive NPs (Pelicano et al. 2006). They call this tumor-targeting “passive.” Different tumor biology (vascularity, leakiness) and carrier features (size and circulation time) drive passive targeting. This tumor-targeting method lacks a tumor cell-specific ligand. The extent of angiogenesis, lymphangiogenesis, perivascular tumor invasion, and intratumor pressure considerably affect the EPR impact. These parameters, along with NP physicochemical properties, determine NP drug delivery system efficiency. Taxanes and Paclitaxel are effective against several malignancies. Taxanes are most commonly used to treat breast, advanced or metastatic breast cancer (MBC), lung, and ovarian cancer. In pancreatic cancer xenograft mouse models, Abraxane alone or with gemcitabine reduces pancreatic stroma (Lim et al. 2018; Barkat et al. 2020, 2021; Harshita et al. 2019). Genexol PM is approved for MBC, and pancreatic cancer (Gavas et al. 2021). Cancer cell proliferation is reduced with DaunoXome. Daunorubicin is the active ingredient that is utilized in the treatment of Kaposi’s sarcoma.

5.4.2.2 Active Targeting

Transferrin and folate are necessary for active targeting because they attach to receptors or substances on sick organs, tissues, cells, or subcellular domains known as ligand-mediated targeting (Gavas et al. 2021). Target affinity requires ligand-NP absorption and tight retention. This technique improves cancer cell NP binding and drug penetration. NP-target crosstalk should increase without affecting biodistribution (Kamaly et al. 2012). Active substrate receptor targeting needs ligands, for example, proteins, peptides, antibodies, nucleic acids, carbohydrates, and vitamins (Byrne et al. 2008). Most research focuses on transferrin, folate, glycoproteins, and EGFR receptors. After ligand-target interaction infolds membranes, receptors endocytose NPs. Many systems allow active targeting. Most cancer targeting is done by NPs targeting tumor cells.

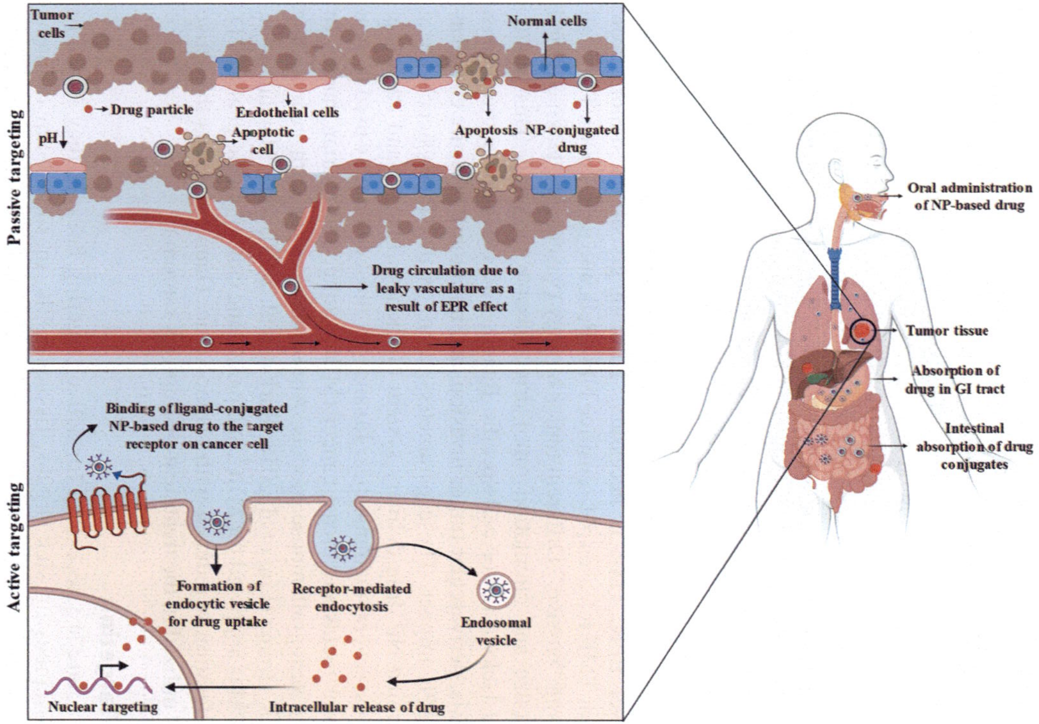

Fig. 5.2 Mechanism of passive and active targeting of NP-drug conjugates. During passive targeting, the drug-loaded NPs are transported via leaky blood vasculature through EPR effect due to poor lymphatic drainage and release of drug from the nanoparticles takes place due to acidic microenvironment of tumor. On the other hand, active targeting involves binding of ligand-conjugated drug-loaded nanoparticles to the antigenic receptor on cancer cells followed by receptor-mediated endocytic uptake (Tiwari et al. 2023) [(© 2023 by the authors. Licensee MDPI, Basel, Switzerland. This article is an open access article distributed under the terms and conditions of the Creative Commons Attribution (CC BY) license (https://creativecommons.org/licenses/by/4.0/)]

This improves cell penetration. One of the most studied receptors is transferrin. Cells get iron from serum glycoprotein. Most tumor cells, especially solid tumors, overexpress these receptors while healthy cells under-express them. NPs vary with transferrin-targeting ligands (Gavas et al. 2021). A2780 ovarian cancer cells overexpress transferrin, therefore Tf-Mpeg-pe NPs target them (Amreddy et al. 2015). Targeting angiogenic endothelial cells near cancer cells is another option. Cancer cell blood flow reduction causes hypoxia and necrosis. Normal tissues are less acidic than tumors (Gavas et al. 2021). The Warburg effect explains it well. Cancer cell glycolysis and lactic acid generation ruin cells. To compensate, cells overexpress proton pumps that acidify extracellular space with lactic acid. Thus, liposome-based pH-sensitive drug delivery is studied. Multiple-ligand NPs interact better with cancer cells. NP architecture and ligand-target chemistry affect method efficacy, making NP design difficult. The system benefits from ligand density (Gavas et al. 2021), NP size, and administration route. Cancer EGFR overexpression and squamous cell histology are targets. SCC is targeted by anti-EGFR and anti-IgG gold nanoparticles. Breast cancer cells overexpress Herceptin®. Anthracycline cardiotoxicity was reduced by HER2-targeted PEGylated liposomal doxorubicin. Targeting VCAM-1 with NPs in breast cancer models suggests function. Nucleotide synthesis needs folic acid. Folate receptors take in folic acid. FR-α is overexpressed in tumor cells and FR-β in liquid cancer cells. Selective cancer treatments involve folate receptor-targeted NPs (Ashique et al. 2021).

5.5 Advantages of Nanomaterials in Cancer Treatment

Cancer detection, treatment, and management are entering a new era using nanotechnology. Active or passive targeting of NPs increases intracellular medicine concentration to prevent toxicity in healthy tissue. Medication release can be controlled using temperature- or pH-sensitive NPs. A pH-sensitive medication delivery system can reach the acidic TME. Drugs are released by temperature-sensitive nanoparticles in response to magnetic fields and ultrasonic noises. The "physicochemical characteristics" of NPs-shape, size, molecular mass, and surface chemistry-also affect targeted drug delivery. Modified NPs can target a moiety (Gavas et al. 2021). Cytotoxicity and unequal distribution limit conventional chemotherapy and radiation therapy. Thus, cancer cell killing requires cautious dosing without harm. Under physiological conditions, spot therapy must cross renal infiltration, BBB, TME, and RES to target BMC ("blood monocytes, macrophages, and other immune cells") (Rezvantalab et al. 2018). MPS triggers "macrophages or leukocytes" to swiftly remove medicines from the liver, spleen, or lungs. This reduces medication half-life (Yona and Gordon 2015). Renal infiltration eliminates NP toxicity, essential to human function. But NPs can cross BBB. Only intraventricular or intracerebral infusions can treat brain cancer because the blood–brain barrier prevents toxic substances from entering (Liang et al. 2021; Ashique et al. 2022c). NPs are transported by transcytosis, peptide-modified endocytosis, focused ultrasound, and

EPR. Methotrexate absorption improved in glutathione PEGylated liposome-fed rats (Tran et al. 2017; Singhai et al. 2024).

5.6 Types of Different Kinds of Nanoparticles for Cancer Therapy

Three types of NPs are commonly utilized in drug delivery systems: hybrid, inorganic, and organic NPs. Figure 5.3 depicts the different kinds of bio-inspired nanoparticles (Shih et al. 2022). Table 5.2 describes numerous nanocarriers and their application in cancer treatment.

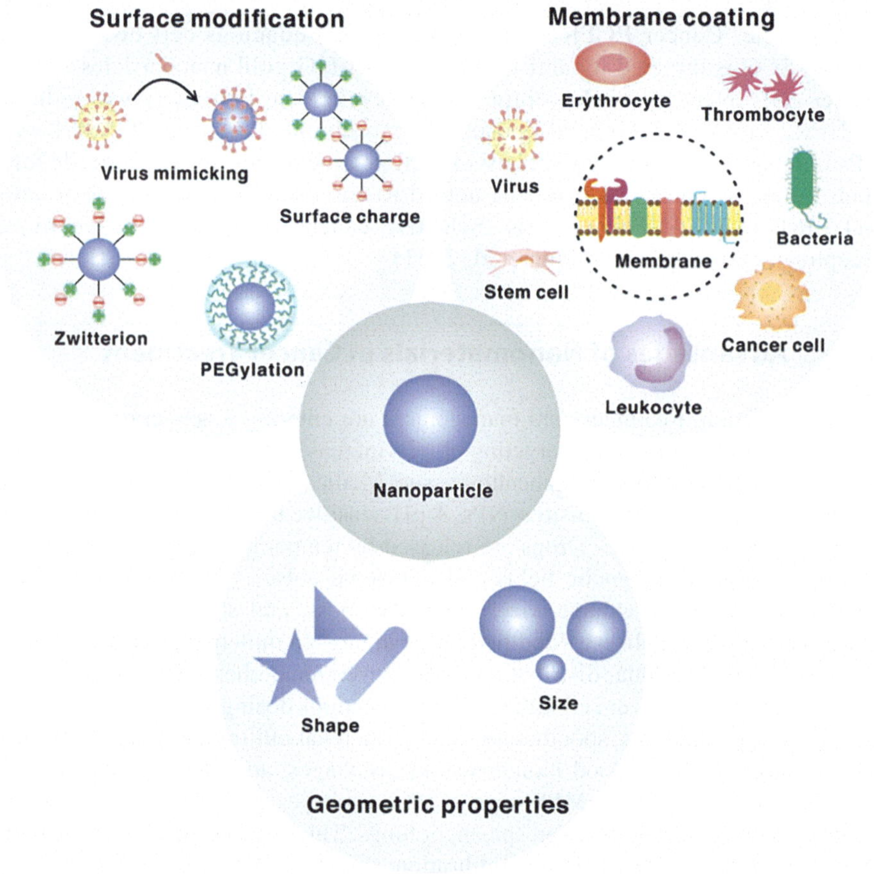

Fig. 5.3 Design principles for bioinspired interfaces of nanoparticle categorized into three different sections. Membrane-coated nanoparticles derived from mammalian cell, cancer cell, bacteria or virus (Shih et al. 2022) [(© 2022 Shih, Tang, Kuo, Chueh and Chen. This is an open-access article distributed under the terms of the Creative Commons Attribution License (CC BY)]

Table 5.2 Different nanoparticles for cancer treatment

Nanoparticles	Cancer type	Therapeutic ingredients	Treatment approach	References
1. Hybrid nanoparticles				
Lipid-polymer hybrid NPs	Melanoma and Lewis lung cancer	Combretastatin and doxorubicin	Anti-angiogenesis together with chemotherapy	Gulia et al. (2022)
	Prostate and breast cancer	Cisplatin and siRNAs	A combination of chemotherapy and RNAi treatment targeting REV1 and REV3L	
	Triple-negative breast cancer	Doxorubicin and siRNA	RNAi and chemotherapy for MRP1 treatment	
	Breast cancer	Doxorubicin and TRAIL	The use of chemotherapy and cytokines to cause apoptosis	
	Melanoma	TGFβ inhibitor SB505124 and IL-2	Promoting tumor immunotherapy	
	Lung metastasis	siRNAs and miRNAs	Apoptosis was triggered by combining RNAi treatment (targeting MYC, MDM2, and VEGFA) with miR-34a. Photodynamic treatment	
Polysiloxane gold-NPs	Advanced cervical cancer	Doxorubicin	To assess the MTD of AGuIX-NP along with cisplatin and radiation in cervical cancer patients	Gulia et al. (2022)
	MCF7 human breast cancer cells	Doxorubicin	Passive targeting by EPR	
	PANC-1 and BxPC3 human pancreatic cancer cells	Camptothecin or paclitaxel	Folic acid	
2. Inorganic nanoparticles				
Carbon nanotube	Prostate cancer	siRNA (Small interfering RNA)	CNTs prevent tumor growth by activating immune cells, utilized for DNA transfer and thermal ablation	Gulia et al. (2022)
Graphene	Non-small-cell lung cancer	TRAIL and Doxorubicin	GO and rGOs are popular in TME for treating hypoxia and aberrant angiogenesis. In cell models, GO-doxorubicin attacks breast cancer better. Cytokines and chemotherapy cause apoptosis	Gulia et al. (2022)

(continued)

Table 5.2 (continued)

Nanoparticles	Cancer type	Therapeutic ingredients	Treatment approach	References
Metallic NPs • Iron oxide NPs • Gold NPs	Prostate and testicular cancer nodal, metastases, glioma	Curcumin and doxorubicin	Iron-NP drug treatment: curcumin–doxorubicin. Late-stage glioma showed nodal metastases. Ferumoxytol and iron oxide NP treat Feraheme® iron-deficiency anemia. Treatment for prostate and testicular nodal metastases	Gulia et al. (2022)
	Cervical cancer	Doxorubicin	Heat and chemo. They are good intracellular targeted drug carriers because they can change size and surface characteristics. Cell NP trajectories may be tracked because they block light. Gold-on-silica nanoshells with anti-HER2 properties target HER2-positive breast cancer cells	Gulia et al. (2022)
Molybdenum sulfide NS	Breast cancer	Doxorubicin	Infusion of chemotherapy and hyperthermia	Gulia et al. (2022)
Fullerenes	Tumor cells	Metallofullerene	Light-absorbing fullerenes are used in photodynamic treatment for their triple yield, prolonged π-conjugation, and oxygen generation (Torchilin 2011). PEG-modified fullerenes may photodynamically damage cancer cells	Gulia et al. (2022)
Quantum dots	Prostate cancer cells	Doxorubicin-quantum dots aptamer	Prostate cancer cells are targeted by quantum dots aptamer—doxorubicin	Gulia et al. (2022)
3. Organic nanoparticles				
Dendrimers (PAMAM, PEG, PPI, and TEA)	Liver cancer Prostate cancer	DNA-Doxorubicin Doxorubicin with CpG oligonucleotides	TRAIL-encoded plasmid chemotherapy and gene therapy (Lim et al. 2023) Combined immunotherapy/chemo Nucleic acid collision	Gulia et al. (2022)

Table 5.2 (continued)

Nanoparticles	Cancer type	Therapeutic ingredients	Treatment approach	References
Liposomes or lipid-based NPs	Small cell lung cancer	Irinotecan and cisplatin	Multiple chemotherapies	Gulia et al. (2022)
	Melanoma	Combretastatin, doxorubicin	Anti-angiogenesis together with chemotherapy	
	Non-small-cell lung cancer	Doxorubicin with antisense Oligonucleotides	Chemotherapy and antisense treatment targeting MRP1 and BCL-2	
	Cervical cancer	SiRNA-Vorinostat	Targeting MCL1 with chemotherapy and RNAi treatment	
	Hepatocellular carcinoma	DNA-Docetaxel	Chemotherapy and gene therapy employing survivin suppressor	
	Non-small-cell lung cancer	siRNAs	RNAi therapy using siRNAs against MDM2, MYC, and VEGFA	Gulia et al. (2022)
	Melanoma	Oligonucleotide G3139 and d-(KLAKLAK)2 peptide	Integrating BCL-2 antisense treatment with peptide-enhancing apoptosis	
Polymeric micelles (NPs)	Non-small-cell lung cancer	Doxorubicin/ paclitaxel	Multiple chemotherapies	Gulia et al. (2022)
	Breast cancer with drug resistance	Disulfiram, doxorubicin	Anti-drug resistance and chemotherapy combination	
	Breast cancer	Paclitaxel with siRNAs	Chemotherapy and RNAi therapy targeting SNAIL and TWIST	
	Colon carcinoma	Camptothecin and DNA	Cancer treatment combined with gene therapy employing TRAIL-encoded plasmid	
	Breast cancer	Paclitaxel with DNA	Combining chemotherapy and gene therapy with IL-12-encoded plasmid	
	Prostate cancer	siRNAs	RNAi treatments using siRNAs targeting VEGFA and BCL-2	
	Triple-negative breast cancer	Antisense nucleotides	Antisense cancer treatments targeting miR-10b and miR-21 in	
Solid lipid nanoparticles (SLN)	Leukemia	Doxorubicin and idarubicin	Positive findings were seen with "P388/ADR leukemia cells" and the "murine leukemia mouse model"	Gulia et al. (2022)

5.7 Bio-Inspired Nanomaterials in Cancer Theranostics

Recently, bio-inspired nanoparticles that mimic natural body components have garnered attention as biocompatible alternative drug delivery strategies in cancer theranostic (Neubi et al. 2018). The key benefit of these nanoparticles over non-native drug delivery systems is systemic bio-distribution changes. This review provides an overview of bio-inspired nanomaterials like liposomes, lipid nanoparticles, biosynthesized metal nanoparticles, viral nanoparticles, protein nanoparticles, and others for cancer therapeutics (Madamsetty et al. 2019). Table 5.3 describes several cancer theranostics applying bio-inspired nanomaterials.

Table 5.3 Cancer theranostics using bioinspired nanomaterials

Sr. no.	Bio-inspired nanomaterials	Type of cancer	Application	References
1.	**INPs (inorganic nanoparticles in cancer)** • **AuNPs (Gold nanoparticles)** • **AgNPs (Silver nanoparticles)** • **SPIONs (iron oxide nanoparticles)**	In breast cancer, epidermoid carcinoma, leukemia	Their unique physicochemical properties make them antibacterial, anticancer, biocompatible, bio-imaging, biosensing, cancer theranostics, cell imaging medicine, and nucleic acid delivery	Madamsetty et al. (2019) and Mukherjee et al. (2019)
		Glioblastoma	Nucleic acids are synthesized on spherical gold nanoparticles to calculate NU-0129's efficiency	Madamsetty et al. (2019) and Mukherjee et al. (2019)
		Cancer	Anti-cancer medication administration may be a possible use for biosynthesized AgNPs	Madamsetty et al. (2019) and Mukherjee et al. (2019)
		Cancers of the neck and head	Determining whether SPIONs—ferumoxytol—can be used for analysis utilizing magnetic resonance imaging	Madamsetty et al. (2019) and Mukherjee et al. (2019)
2.	**LNPs (lipid nanoparticles in cancer theragnostic)** • **LBMs (lipid-based micelles)** • **LNCs (lipid nanocapsules)** • **NSLs (nano-structured lipids)** • **SLNPs (solid lipid nanoparticles)**	Hepatic tumors	The study advises activating the long-term support ligaments (LTSL) using focused ultrasound (FUS) to release cytotoxic drugs	Madamsetty et al. (2019) and Feng et al. (2017)
		Head and neck squamous cell carcinoma	To assess docetaxel microgel's safety and efficacy in recurrent or metastatic HNSCC	Madamsetty et al. (2019) and Ma et al. (2012)
		Glioblastoma	Test NFL-TBS.40-63 peptide-containing DiD co-encapsulated LNCs as glioblastoma targeting ligand Determine if DNA-LNCs or plasmid DNA can deliver genes as well as LNCs	Madamsetty et al. (2019) and Balzeau et al. (2013)
		Hepatocellular carcinoma	To investigate murine hepatocellular carcinoma imaging and antitumor activity of dual-loaded NLCs with paclitaxel and quantum dots (QDs), used a multifunctional nanosystem with NIR dye-loaded, CXCR4-targeted NLCs to study and suppress cancer progression	Madamsetty et al. (2019) and Li et al. (2017)
		Glioblastoma	Imaging glioblastoma-targeted SLNs with IR-780 iodide-loaded tumor vasculature to follow PTT Research SLN's many cancer applications	Madamsetty et al. (2019)

(continued)

Table 5.3 (continued)

Sr. no.	Bio-inspired nanomaterials	Type of cancer	Application	References
3.	**LCTs (liposomes in cancer theranostics)**	Tumors, solid tumors	To study 188Re-BMEDAliposome dispersion and radiation dosimetry, and asses the role, mechanism, and safety of EphA2 siRNA-encapsulated liposomes in the management of tumor	Madamsetty et al. (2019) and Silva et al. (2019)
4.	**PBNPs (protein-based nanoparticles for cancer theranostics)**	Solid tumors bladder cancer cutaneous T-cell Lymphoma	To understand EphB4-HSA fusion protein MTD and adverse-tumor-directed drug response (MTD) in chemotherapy patients To evaluate ABI-009's MTD, anti-tumor effectiveness, and safety Assess ONTAK's effectiveness in CTCL	Madamsetty et al. (2019)
5.	**VNPs (viral nanoparticles in cancer theranostics**	Melanoma	VLPs and VNPs prevent, diagnose, monitor, and treat sickness Non-invasive imaging technologies like MRI and PET use VNPs to image tumor angiogenesis	Madamsetty et al. (2019)

5.8 Plant-Based Nanomaterials in Cancer Theranostics

Due to their phytocomponents and medicinal qualities, plants treat many ailments, including cancer (Gulia et al. 2022). Several therapeutic plants' secondary messengers block or activate cell signal transduction pathways, preventing and combating cancer. Biogenic nanoparticle synthesis controls particle size and form better than physical and chemical methods for biomedical applications. Fast synthesis, stability, affordability, cultivability, and variety of shapes and sizes are benefits of green nanoparticles (Uzair et al. 2020). Metallic nanoparticles can carry plant medicines, siRNA, peptides, antibiotics, chemotherapeutic agents, hydrophilic and hydrophobic compounds, and tiny molecules to tumors without affecting healthy cells (Gulia et al. 2022). Nanoparticle-encapsulated drugs resist systemic enzyme breakdown. Natural extracts, essential oils, and their bioactive ingredients may help treat cancer due to their multiple targeted modes of action and low side effects. Later-diagnosed cancer may be tougher to treat. Cancer therapy and prevention require early identification. CT, MRI, ultrasound, and PET are popular imaging modalities (Gulia et al. 2022). Novel imaging technologies are needed due to traditional imaging methods' declining pharmacological efficacy, quick clearance, continuous spread, and poor results (Manikandan et al. 2015). Many nanoscale biosensor materials improve imaging. In optical imaging, greenly synthesized iron, gold, titanium, silver, and copper nanoparticles spread the light a million times more than molecules due to plasmon excitation (Gulia et al. 2022).

Biologically synthesized silver nanoparticles are cheaper and easier to use for cancer theranostics (Patil et al. 2017). The therapy of cancer commonly involves silver nanoparticles (Gulia et al. 2022). Green chemistry for silver nanoparticle production is superior to standard methods because: It is simple, one-stage, rapid, affordable, and highly dependable; utilizes less toxic compounds; can be used with water, a flexible solvent; and can use natural resources including plants, fungi, algae, and microbes as stabilizing bases and reduction agents. Non-invasive in vivo imaging of silver nanoparticles using *Zinnia elegans* leaf extract works in C57BL6/J mice. Silver nanoparticles from *Olax scandens* extract treated with A549 and B16 cells may glow red for cancer diagnosis. Gold nanoparticles from plants are stable, permeable, and nontoxic (Gulia et al. 2022). Making iron oxide/gold nanoparticles required only ferric chloride and grape seed proanthocyanidin. Superparamagnetism and CT contrast reveal that our green, biocompatible, and nonpolluting nanoparticle production outperforms conventional imaging approaches (Hall 2002). Bifunctional imaging with CT, RMI, and optoacoustic signals uses *Hypericum lanceolatum* and *Hubertia ambavilla* gold nanoparticles. Many research studies have studied plant-based nanoparticles' medicinal benefits, including cancer imaging. Evaluation of readily available cancer thermometry biosensors requires extensive investigation. Safe, affordable, and physiologically compatible green nanoparticles that can replace chemically amalgamated nanoparticles in cancer diagnostics need further study. Table 5.4 describes several clinical trials related to cancer nanotherapy.

Table 5.4 Recent clinical trials for cancer theranostics

Clinical trial ID/ phase	Nanocarrier systems	Description	Mechanisms	Place of origin	Cancer type	References
NCT02106598 phase 1&2	Silica nanoparticles	Real-time imaging of metastases from lymph nodes	Not specified	Memorial Sloan Kettering Cancer Center	Colon, head and neck, breast cancer	Shah et al. (2022)
NCT03308604 Phase 1	Polysiloxane Gd-nanoparticles liposomes	Evaluating the MTD of AGUIX-NP in patients with cervical cancer in conjunction with cisplatin and radiation	Not specified	Gustave Roussy, NCI, France	Advanced cervical cancer	Maury et al. (2023)
NCT03409198 Phase 2B	Carbon nanoparticles	Assessing the effects of immunogenic chemotherapy in patients with metastatic luminal B breast cancer combined with Ipilimumab and Nivolumab	Immunogenic chemotherapy	Oslo University Hospital	Breast cancer	Ji et al. (2023)
NCT03350945 Not Applicable	Carbon nanoparticles	Utilizing carbon nanoparticles during laparoscopic surgery to identify and isolate primary tumors as well as lymph nodes	Laparoscopic surgery	Not specified	Not specified	Ali (2022)
NCT02495896 Phase 1	Protein-based nanoparticles	Recognizing the adverse effects and MTD of EphB4-HSA fusion protein in patients with solid tumors receiving chemotherapy	EphB4-HSA fusion protein with chemotherapy	Aiguo, Lu	Solid tumors	Hoffmann et al. (2021)
NCT02805894 Phase 1	Hafnium oxide (HfO2) nanoparticle	Determining MTD of NBTXR3 activated by brachytherapy and radiation	NBTXR3 activated by radiation, brachytherapy	Vasgene Therapeutics, Inc.	Solid tumors	Patil et al. (2023)
NCT03217838 Phase 1	Polymeric nanoparticles	Analysing AZD2811's MTD, safety, PK, and PD in patients with recurrent AML both with and without azacitidine	AZD2811 with/without azacitidine	Not specified	Acute myeloid leukemia	Huang et al. (2019)

(continued)

Table 5.4 (continued)

Clinical trial ID/ phase	Nanocarrier systems	Description	Mechanisms	Place of origin	Cancer type	References
NCT03020017 Early Phase 0	Gold nanoparticles	An assessment of NU-0129's effectiveness (nucleic acids produced on the outside of a spherical gold nanoparticle)	NU-0129 (nucleic acids)	Vasgene Therapeutics, Inc.	Glioblastoma	Huang et al. (2023)
NCT03774680 Phase 1	Polymeric nanoparticles	Assessing CRLX101 with enzalutamide's antitumor efficacy in patients with prostate cancer who have previously had enzalutamide treatment	CRLX101 plus enzalutamide	Not specified	Prostate cancer	Madej et al. (2022)
NCT03712423 Phase 1	Polymeric nanoparticles	Assessing docetaxel micelle's effectiveness for recurrent or metastatic HNSCC	Docetaxel micelle	Not specified	Head and neck squamous cell carcinoma	Miedema et al. (2022)
NCT03531827 Phase 2	Liposomes	Examining 188 Re-BMEDAliposomes' radiation dosimetry and dispersion profile	188 Re-BMEDAliposomes	Not specified	Solid tumors	Schmidt et al. (2022)
NCT02639858 Phase 2	Liposomes	Examining the MTD of EphA2 siRNA-encapsulated liposomes and assessing their effectiveness in tumor cells that are incurable	EphA2 siRNA-encapsulated liposomes	Not specified	Solid tumors	Dragovich et al. (2006)
NCT02271516 Phase 1	Lipid-based nanoparticles	Calculating ABI-009's MTD and assessing its anti-tumor and safety	ABI-009	Not specified	Various Lung Cancer	Naziris and Demetzos (2022)

NCT02191878 Phase 3	Protein-based nanoparticles	Assessing ONTAK's efficacy in cutaneous T-cell lymphoma (CTCL)	ONTAK	Clarity Pharmaceuticals Ltd	CTCL	Baboci et al. (2020)
NCT02009332, Phase 1	Protein-based nanoparticles	Examining BIND-014's effectiveness in different lung cancer patients	BIND-014	M.D. Anderson Cancer Center	NSCLC	Anselmo and Mitragotri (2016)
NCT00211198 Phase 4	PSMA-targeted polymeric nanoparticles	Determining MTD for solid tumors using conventional chemotherapy and rapamycin-loaded albumin	Rapamycin-loaded albumin with standard chemotherapy	National Cancer Institute	Solid tumors	Zhang et al. (2022)

MTD maximum tolerated dose, *PK* pharmacokinetics, *PD* pharmacodynamics, *NSCLC* non-small cell lung cancer

5.9 Future Prospective of Nanotheranostics in Cancer Treatment

Bio-inspired smart nanoparticles have an opportunity to transform cancer care and usher in a new era of precision medicine due to their enhanced ability for cancer theranostics. Research into the development of these nanotechnological advances may lead to significant breakthroughs in several vital sectors (Chehelgerdi et al. 2023). Bioinspired smart nanoparticles provide a platform for personalized treatment and enable targeted medication delivery. By adding convinced ligands and altering their surface, these nanoparticles might be engineered to detect different biomarkers on cancer cells, enabling more accurate targeting. With its targeted approach, complications are reduced and overall patient care is improved by limiting damage to healthy tissues (Goel et al. 2017; Ashique et al. 2023d). By adding imaging compounds to bioinspired nanoparticles, medication platforms improve diagnostics. In the future, more advanced imaging modalities including positron emission tomography (PET), magnetic resonance imaging (MRI), and photoacoustic imaging could be created (Kim et al. 2023). These technologies have the potential to diagnose cancer earlier and more with precision, providing medical practitioners with the critical information they need to act swiftly (Madamsetty et al. 2019). Further investigation into the stimuli-responsive elements of nanoparticles will progress the development of more sophisticated drug delivery systems (Barkat et al. 2020).

5.10 Clinical Trials for Theranostics in Cancer

Many nanosystems have shown promise in recent research for treating a variety of malignancies, including liposomes, carbon nanoparticles, protein-based nanoparticles, polymeric nanoparticles, silica nanoparticles, polysiloxane Gd-nanoparticles, and gold nanoparticles (Smith et al. 2022). Interestingly, the focus of clinical research conducted at Memorial Sloan Kettering Cancer Centre (NCT02106598) is on using silica nanoparticles for real-time imaging of lymph node metastases (Shah et al. 2022). Alternatively, the Gustave Roussy research NCT03308604 investigates the potential use of polysiloxane Gd-nanoparticles liposomes in conjunction with cisplatin and radiation therapy to identify the maximum tolerated dose (MTD) for patients with cervical cancer (Maury et al. 2023). The newest big thing in the cutting-edge field of cancer theranostics is carbon nanoparticles. For patients with metastatic luminal B breast cancer, immunogenic chemotherapy in conjunction with nivolumab and ipilimumab is being investigated. BIO-inspired smart nanoparticles in increased cancer theranostics clinical research link lab outcomes to cancer patient applications (Zhang et al. 2023). Initial phase I and II trials identify appropriate doses, safety, and feasibility (Ji et al. 2023). Studying how these nanoparticles interact with the body is laying the groundwork for future discoveries (Ali 2022). For theranostics patient selection, cancer-specific biomarkers must be identified (Hoffmann et al. 2021). These indicators assess bio-inspired nanoparticles' ability to precisely target and interact with cancer cells, enabling molecular profile-specific treatment (Madamsetty et al. 2019). Increasing multicenter trials improve patient variety, medical organization collaboration, and study generalizability (Khalaf et al. 2019). Data, information, and resources are collected collaboratively to speed research (Patil et al. 2023). To study combination medicines, bioinspired smart nanoparticles are tested with standard pharmaceuticals. Comparison of efficacy research shows synergies and inspires new treatment methods. A long-term monitoring technique maintains the treatment's longevity and checks any compounding or delayed effects of these particular nanoparticles. Safety monitoring during the trial quickly identifies and addresses adverse events. To receive FDA or EMA approval, investigations must be effective (Olsen et al. 2001). Before being used in cancer treatments, bioinspired smart nanoparticles must prove their safety and efficacy. Assessing quality of life and patient outcomes completes the picture. Understanding bio-inspired smart nanoparticles' therapeutic advantages and hazards requires understanding how these therapies affect people's daily lives. Overall, these clinical studies are crucial to developing bioinspired smart nanoparticles into personalized cancer treatments. Because these trials could improve cancer treatment and personalize it, researchers, patients, and doctors are eagerly awaiting their results. Even though these studies are still in their early stages, the volume of data they produce could alter cancer therapy by opening new avenues for precise diagnosis and customized treatments in the ever-changing field of cancer research.

5.11 Conclusion

In the realm of cancer theranostics, the burgeoning field of bio-inspired nanoparticles has emerged as a beacon of innovation and promise. The review explores a myriad of bio-inspired nanomaterials, encompassing liposomes, lipid nanoparticles, bio-synthesized metal nanoparticles, viral nanoparticles, and protein nanoparticles, all poised to revolutionize cancer diagnosis and treatment. Liposomes, with their unilamellar lipid bilayers and aqueous cores, epitomize versatility, offering biocompatibility, biodegradability, and the ability to incorporate both hydrophilic and hydrophobic therapeutic compounds. The prospect of liposomes extends to targeted cancer therapy, with ongoing clinical trials and approvals underscoring their potential clinical translation. Lipid nanoparticles (LNPs), characterized by biocompatibility and scalability, present a lucrative platform for cancer theranostics. From co-loading siRNA and near-infrared (NIR) dyes on targeted LNPs to utilizing cyanine fluorescent dyes for imaging and thermal ablation, LNPs show adaptability in delivering both therapeutic and imaging payloads. Tailored solutions are offered by subcategories including lipid-based micelles, nano-structured lipids, and solid lipid nanoparticles (SLNs). For instance, IR-780-loaded SLNs are helpful for photothermal therapy, whereas paclitaxel-loaded micelles have shown greater therapeutic efficacy in breast cancer models. One of the best examples of protein-based nanoparticles that expand the use of bio-inspired theranostics is human serum albumin (HSA). Abraxane, an albumin-bound paclitaxel formulation, is one example of a protein-based nanoparticle formulation that has been effectively employed in medicine. Another illustration of albumin's versatility is the production of multifunctional nanoparticles, such as HSA/dc-IR825/GA complexes, which demonstrate the efficacy of photothermal therapy and diagnostics. Moreover, other protein-based nanoparticles, such as apoferritin, which may build and deconstruct in response to pH variations, open up new possibilities for controlled drug delivery. The study of protein-based nanoparticles is not restricted to albumin because other proteins, including ferritin, gelatin, elastin, gliadin, legumin, mucin, soy proteins, silk, zein, and milk proteins, exhibit promise in preclinical settings. While Apoferritin, a self-assembling protein nanocage, demonstrates pH-responsive regulated delivery of therapeutic medicines, H-ferritin (HFn) serves as a biocompatible nanocarrier for targeted drug administration and effectively suppresses the growth of tumors. The groundbreaking potential of bioinspired nanoparticles in the realm of cancer theranostics is highlighted in the study's conclusion. These state-of-the-art platforms, which include adaptable liposomes, LNPs, and multifunctional protein-based nanoparticles, hold the key to targeted and customized cancer treatments. Clinical studies and preclinical research are underway to support the efficacy of these bioinspired nanomaterials, suggesting an exciting future where cutting-edge technology and nature-inspired design will revolutionize cancer diagnostics and treatment.

Conflict of Interest None

Funding None

References

Abdallah R, Atallah D, Bitar N, Chahine G, Ghanem H, Ghosn M, Kattan J, Nasr F, Makdessi J, Shamseddine A (2023) Consensus on the management of platinum-sensitive high-grade serous epithelial ovarian cancer in Lebanon. Gynecol Oncol Rep 47:101186

Abraham J, Ocen J, Stafford J (2023) Hormonal therapy for cancer. Medicine 51(1):28–31

Abudurexiti M, Zhao Y, Wang X, Han L, Liu T, Wang C, Yuan Z (2023) Bio-inspired nanocarriers derived from stem cells and their extracellular vesicles for targeted drug delivery. Pharmaceutics 15(7):2011

Akhtar N, Mohammed HA, Yusuf M, Al-Subaiyel A, Sulaiman GM, Khan RA (2022) SPIONs conjugate supported anticancer drug doxorubicin's delivery: current status, challenges, and prospects. Nanomaterials 12(20):3686

Ali F (2022) Regulatory perspectives of nanomaterials for theranostic application. In: Nanotheranostics for treatment and diagnosis of infectious diseases. Academic Press, pp 373–384

Amreddy N, Muralidharan R, Babu A, Mehta M, Johnson EV, Zhao YD, Munshi A, Ramesh R (2015) Tumor-targeted and pH-controlled delivery of doxorubicin using gold nanorods for lung cancer therapy. Int J Nanomed 10:6773–6788

Ansari MA, Thiruvengadam M, Farooqui Z, Rajakumar G, Jamal QM, Alzohairy MA, Almatroudi A, Alomary MN, Chung IM, Al-Suhaimi EA (2021) Nanotechnology, in silico and endocrine-based strategy for delivering paclitaxel and miRNA: prospects for the therapeutic management of breast cancer. Semin Cancer Biol 69:109–128

Ansari MA, Thiruvengadam M, Venkidasamy B, Alomary MN, Salawi A, Chung IM, Shariati MA, Rebezov M (2022) Exosome-based nanomedicine for cancer treatment by targeting inflammatory pathways: current status and future perspectives. Semin Cancer Biol 86:678–696

Ansari MA, Govindasamy R, Begum MY, Ghazwani M, Alqahtani A, Alomary MN, Jamous YF, Alyahya SA, Asiri S, Khan FA, Almessiere MA (2023) Bioinspired ferromagnetic $CoFe_2O_4$ nanoparticles: potential pharmaceutical and medical applications. Nanotechnol Rev 12(1):20230575

Ansari MA, Shoaib S, Chauhan W, Gahtani RM, Hani U, Alomary MN, Alasiri G, Ahmad N, Jahan R, Yusuf N, Islam N (2024) Nanozymes and carbon-dots based nanoplatforms for cancer imaging, diagnosis and therapeutics: current trends and challenges. Environ Res 241:117522

Anselmo AC, Mitragotri S (2016) Nanoparticles in the clinic. Bioeng Transl Med 1(1):10–29

Ashique S, Sandhu NK, Chawla V, Chawla PA (2021) Targeted drug delivery: trends and perspectives. Curr Drug Deliv 18(10):1435–1455

Ashique S, Almohaywi B, Haider N, Yasmin S, Hussain A, Mishra N, Garg A (2022a) siRNA-based nanocarriers for targeted drug delivery to control breast cancer. Adv Cancer Biol Metastas 4:100047

Ashique S, Upadhyay A, Hussain A, Bag S, Chaterjee D, Rihan M, Mishra N, Bhatt S, Puri V, Sharma A, Prasher P (2022b) Green biogenic silver nanoparticles, therapeutic uses, recent advances, risk assessment, challenges, and future perspectives. J Drug Deliv Sci Technol 77:103876

Ashique S, Upadhyay A, Kumar N, Chauhan S, Mishra N (2022c) Metabolic syndromes responsible for cervical cancer and advancement of nanocarriers for efficient targeted drug delivery—a review. Adv Cancer Biol Metastas 4:100041

Ashique S, Afzal O, Hussain A, Zeyaullah M, Altamimi MA, Mishra N, Ahmad MF, Dua K, Altamimi AS, Anand K (2023a) It's all about plant-derived natural phytoconstituents and phytonanomedicine to control skin cancer. J Drug Deliv Sci Technol 84:104495

Ashique S, Garg A, Hussain A, Farid A, Kumar P, Taghizadeh-Hesary F (2023b) Nanodelivery systems: an efficient and target-specific approach for drug-resistant cancers. Cancer Med 12(18):18797–18825

Ashique S, Garg A, Mishra N, Raina N, Ming LC, Tulli HS, Behl T, Rani R, Gupta M (2023c) Nano-mediated strategy for targeting and treatment of non-small cell lung cancer (NSCLC). Naunyn Schmiedebergs Arch Pharmacol 396:2769–2792

Ashique S, Upadhyay A, Gulati M, Singh D, Chawla PA, Chawla V (2023d) One-dimensional polymeric nanocomposites in drug delivery systems. Curr Nanosci 19(6):825–839

Baboci L, Capolla S, Di Cintio F, Colombo F, Mauro P, Dal Bo M, Argenziano M, Cavalli R, Toffoli G, Macor P (2020) The dual role of the liver in nanomedicine as an actor in the elimination of nanostructures or a therapeutic target. J Oncol 24:2020

Balzeau J, Pinier M, Berges R, Saulnier P, Benoit JP, Eyer J (2013) The effect of functionalizing lipid nanocapsules with NFL-TBS.40-63 peptide on their uptake by glioblastoma cells. Biomaterials 34(13):3381–3389

Barkat HA, Das SS, Barkat MA, Beg S, Hadi HA (2020) Selective targeting of cancer signaling pathways with nanomedicines: challenges and progress. Future Oncol 16(35):2959–2979. https://doi.org/10.2217/fon-2020-0198

Barkat A, Beg S, Panda SK, Alharbi SK, Rahman M, Ahmed FJ (2021) Functionalized mesoporous silica nanoparticles in anticancer therapeutics. Semin Cancer Biol 69:365–375. https://doi.org/10.1016/j.semcancer.2019.08.022

Baskar R, Lee KA, Yeo R, Yeoh KW (2012) Cancer and radiation therapy: current advances and future directions. Int J Med Sci 9(3):193–199. https://doi.org/10.7150/ijms.3635

Briolay T, Petithomme T, Fouet M, Nguyen-Pham N, Blanquart C, Boisgerault N (2021) Delivery of cancer therapies by synthetic and bio-inspired nanovectors. Mol Cancer 20:1–24

Byrne JD, Betancourt T, Brannon-Peppas L (2008) Active targeting schemes for nanoparticle systems in cancer therapeutics. Adv Drug Deliv Rev 60(15):1615–1626

Chehelgerdi M, Chehelgerdi M, Allela OQ, Pecho RD, Jayasankar N, Rao DP, Thamaraikani T, Vasanthan M, Viktor P, Lakshmaiya N, Saadh MJ (2023) Progressing nanotechnology to improve targeted cancer treatment: overcoming hurdles in its clinical implementation. Mol Cancer 22(1):169

Dobre EG, Surcel M, Constantin C, Ilie MA, Caruntu A, Caruntu C, Neagu M (2023) Skin cancer pathobiology at a glance: a focus on imaging techniques and their potential for improved diagnosis and surveillance in clinical cohorts. Int J Mol Sci 24(2):1079

Dragovich T, Mendelson D, Kurtin S, Richardson K, Von Hoff D, Hoos A (2006) A phase 2 trial of the liposomal DACH platinum L-NDDP in patients with therapy-refractory advanced colorectal cancer. Cancer Chemother Pharmacol 58:759–764

Dutta V, Verma R, Gopalakrishnan C, Yuan MH, Batoo KM, Jayavel R, Chauhan A, Lin KY, Balasubramani R, Ghotekar S (2022) Bio-inspired synthesis of carbon-based nanomaterials and their potential environmental applications: a state-of-the-art review. Inorganics 10(10):169

Elebiyo TC, Rotimi D, Evbuomwan IO, Maimako RF, Iyobhebhe M, Ojo OA, Oluba OM, Adeyemi OS (2022) Reassessing vascular endothelial growth factor (VEGF) in anti-angiogenic cancer therapy. Cancer Treat Res Commun 32:100620

Feng L, Cheng L, Dong Z, Tao D, Barnhart TE, Cai W, Chen M, Liu Z (2017) Theranostic liposomes with hypoxia-activated prodrug to effectively destruct hypoxic tumors post-photodynamic therapy. ACS Nano 11(1):927–937

Ferlay J, Colombet M, Soerjomataram I, Parkin DM, Piñeros M, Znaor A, Bray F (2021) Cancer statistics for the year 2020: an overview. Int J Cancer 149(4):778–789

Ferraiuolo RM, Manthey KC, Stanton MJ, Triplett AA, Wagner KU (2020) The multifaceted roles of the tumor susceptibility gene 101 (TSG101) in normal development and disease. Cancers 12(2):450

Gavas S, Quazi S, Karpiński TM (2021) Nanoparticles for cancer therapy: current progress and challenges. Nanoscale Res Lett 16(1):173

Goel S, England CG, Chen F, Cai W (2017) Positron emission tomography and nanotechnology: a dynamic duo for cancer theranostics. Adv Drug Deliv Rev 113(113):157–176

Govindan B, Sabri MA, Hai A, Banat F, Haija MA (2023) A review of advanced multifunctional magnetic nanostructures for cancer diagnosis and therapy integrated into an artificial intelligence approach. Pharmaceutics 15(3):868

Gulia K, James A, Pandey S, Dev K, Kumar D, Sourirajan A (2022) Bio-inspired smart nanoparticles in enhanced cancer theranostics and targeted drug delivery. J Funct Biomater 13(4):207

Hall JÁ (2002) Cellular mechanisms for heavy metal detoxification and tolerance. J Exp Bot 53(366):1

Handa M, Beg S, Shukla R, Barkat MA, Choudhry H, Singh KK (2021) Recent advances in lipid-engineered multifunctional nanophytomedicines for cancer targeting. J Control Release 340:48–59. https://doi.org/10.1016/j.jconrel.2021.10.025

Harshita, Barkat MA, Beg S, Pottoo FH, Ahmad FJ (2019) Nanopaclitaxel therapy: an evidence based review on the battle for next-generation formulation challenges. Nanomedicine (Lond) 14(10):1323–1341. https://doi.org/10.2217/nnm-2018-0313

Healey Bird B, Nally K, Ronan K, Clarke G, Amu S, Almeida AS, Flavin R, Finn S (2022) Cancer immunotherapy with immune checkpoint inhibitors-biomarkers of response and toxicity; current limitations and future promise. Diagnostics 12(1):124

Helderman RF, Bokan B, van Bochove GG, Rodermond HM, Thijssen E, Marchal W, Torang A, Löke DR, Franken NA, Kok HP, Tanis PJ (2023) Elevated temperatures and longer durations improve the efficacy of oxaliplatin and mitomycin C-based hyperthermic intraperitoneal chemotherapy in a confirmed rat model for peritoneal metastasis of colorectal cancer origin. Front Oncol 13(13):1122755

Hobbs SK (1998) Regulation of transport pathways in tumor vessels: role of tumor type and microenvironment. Proc Natl Acad Sci USA 95:4607–4612

Hoffmann C, Calugaru V, Borcoman E, Moreno V, Calvo E, Liem X, Salas S, Doger B, Jouffroy T, Mirabel X, Rodriguez J (2021) Phase I dose-escalation study of NBTXR3 activated by intensity-modulated radiation therapy in elderly patients with locally advanced squamous cell carcinoma of the oral cavity or oropharynx. Eur J Cancer 146:135–144

Huang X, Lin H, Huang F, Xie Y, Wong KH, Chen X, Wu D, Lu A, Yang Z (2019) Targeting approaches of nanomedicines in acute myeloid leukemia. Dose Response 17(4):1559325819887048

Huang H, Liu R, Yang J, Dai J, Fan S, Pi J, Wei Y, Guo X (2023) Gold nanoparticles: construction for drug delivery and application in cancer immunotherapy. Pharmaceutics 15(7):1868

Jain RK (1998) The next frontier of molecular medicine: delivery of therapeutics. Nat Med 4(6):655–657

Jain N, Shahrukh S, Famta P, Shah S, Vambhurkar G, Khatri DK, Singh SB, Srivastava S (2023) Immune cell–camouflaged surface-engineered nanotherapeutics for cancer management. Acta Biomater 155:57–79

Ji H, Hu Y, Cheng J, Lu J, Zhu W, Li Q, Li Z, Liu X, Wang S, Wu X, Yan J (2023) Use of carbon nanoparticles to improve the efficiency of harvesting lymph nodes in rectal cancer. Surg Laparosc Endosc Percutan Tech 33:382–390

Kamaly N, Xiao Z, Valencia PM, Radovic-Moreno AF, Farokhzad OC (2012) Targeted polymeric therapeutic nanoparticles: design, development and clinical translation. Chem Soc Rev 41(7):2971–3010

Khalaf DJ, Annala M, Taavitsainen S, Finch DL, Oja C, Vergidis J, Zulfiqar M, Sunderland K, Azad AA, Kollmannsberger CK, Eigl BJ (2019) Optimal sequencing of enzalutamide and abiraterone acetate plus prednisone in metastatic castration-resistant prostate cancer: a multicentre, randomized, open-label, phase 2, crossover trial. Lancet Oncol 20(12):1730–1739

Kim KR, Lee AS, Kim SM, Heo HR, Kim CS (2023) Virus-like nanoparticles as a theranostic platform for cancer. Front Bioeng Biotechnol 10:1106767

Krishnaswami V, Natarajan B, Sethuraman V, Natesan S, RajSelvaraj B (2023) Nano-based photodynamic therapy to target the tumor microenvironment. Nano Trends 1:100003

Kroemer G, Chan TA, Eggermont AM, Galluzzi L (2023) Immunosurveillance in clinical cancer management. CA Cancer J Clin 74(2):187–202

Li H, Wang K, Yang X, Zhou Y, Ping Q, Oupicky D, Sun M (2017) Dual-function nanostructured lipid carriers to deliver IR780 for breast cancer treatment: anti-metastatic and photothermal anti-tumor therapy. Acta Biomater 53:399–413

Liang T, Zhang R, Liu X, Ding Q, Wu S, Li C, Lin Y, Ye Y, Zhong Z, Zhou M (2021) Recent advances in macrophage-mediated drug delivery systems. Int J Nanomed 16:2703–2714

Lim EK, Chung BH, Chung SJ (2018) Recent advances in pH-sensitive polymeric nanoparticles for smart drug delivery in cancer therapy. Curr Drug Targets 19(4):300–317

Lim XY, Capinpin SM, Bolem N, Foo AS, Yip WC, Kumar AP, The DB (2023) Biomimetic nanotherapeutics for targeted drug delivery to glioblastoma multiforme. Bioeng Transl Med 8:e10483

Ma M, Hao Y, Liu N, Yin Z, Wang L, Liang X, Zhang X (2012) A novel lipid-based nano micelle of docetaxel: evaluation of antitumor activity and biodistribution. Int J Nanomed 7:3389–3398

Madamsetty VS, Mukherjee A, Mukherjee S (2019) Recent trends of the bio-inspired nanoparticles in cancer theranostics. Front Pharmacol 10:1264

Madej M, Kurowska N, Strzalka-Mrozik B (2022) Polymeric nanoparticles—tools in a drug delivery system in selected cancer therapies. Appl Sci 12(19):9479

Manikandan R, Manikandan B, Raman T, Arunagirinathan K, Prabhu NM, Basu MJ, Perumal M, Palanisamy S, Munusamy A (2015) Biosynthesis of silver nanoparticles using ethanolic petals extract of *Rosa indica* and characterization of its antibacterial, anticancer and anti-inflammatory activities. Spectrochim Acta Part A Mol Biomol Spectrosc 138:120–129

Marei HE, Hasan A, Pozzoli G, Cenciarelli C (2023) Cancer immunotherapy with immune checkpoint inhibitors (ICIs): potential, mechanisms of resistance, and strategies for reinvigorating T cell responsiveness when resistance is acquired. Cancer Cell Int 23(1):64

Maury P, Mondini M, Chargari C, Darricau A, Shahin M, Ammari S, Bockel S, Genestie C, Wu TD, Lux F, Tillement O (2023) Clinical transfer of AGuIX®-based radiation treatments for locally advanced cervical cancer: MR quantification and in vitro insights in the NANOCOL clinical trial framework. Nanomed Nanotechnol Biol Med 50:102676

Mbemi A, Khanna S, Njiki S, Yedjou CG, Tchounwou PB (2020) Impact of gene–environment interactions on cancer development. Int J Environ Res Public Health 17(21):8089

McNamara B, Greenman M, Pebley N, Mutlu L, Santin AD (2023) Antibody–drug conjugates (ADC) in HER2/neu-positive gynecologic tumors. Molecules 28(21):7389

Miedema IH, Zwezerijnen GJ, Huisman MC, Doeleman E, Mathijssen RH, Lammers T, Hu Q, van Dongen GA, Rijcken CJ, Vugts DJ, van der Houven M, van Oordt CW (2022) PET-CT imaging of polymeric nanoparticle tumor accumulation in patients. Adv Mater 34(21):2201043

Mohammed HA, Khan RA, Singh V, Yusuf M, Akhtar N, Sulaiman GM, Albukhaty S, Abdellatif AA, Khan M, Mohammed SA, Al-Subaiyel AM (2023) Solid lipid nanoparticles for targeted natural and synthetic drug delivery in high-incidence cancers, and other diseases: roles of preparation methods, lipid composition, transitional stability, and release profiles in nanocarriers' development. Nanotechnol Rev 12(1):20220517

Mokbel K, Kodresko A, Ghazal H, Mokbel R, Trembley J, Jouhara H (2023) The evolving role of cryosurgery in breast cancer management: a comprehensive review. Cancers 15(17):4272

Mukherjee A, Paul M, Mukherjee S (2019) Recent progress in the theranostics application of nanomedicine in lung cancer. Cancers 11(5):597

Naziris N, Demetzos C (2022) Lipid nanoparticles as platforms for theranostic purposes: recent advances in the field. J Nanotheranostics 3(2):86–101

Neubi GM, Opoku-Damoah Y, Gu X, Han Y, Zhou J, Ding Y (2018) Bio-inspired drug delivery systems: an emerging platform for targeted cancer therapy. Biomater Sci 6(5):958–973

Niculescu AG, Grumezescu AM (2022) Novel tumor-targeting nanoparticles for cancer treatment—a review. Int J Mol Sci 23(9):5253

Olsen E, Duvic M, Frankel A, Kim Y, Martin A, Vonderheid E, Jegasothy B, Wood G, Gordon M, Heald P, Oseroff A (2001) Pivotal phase III trial of two dose levels of denileukin diftitox for the treatment of cutaneous T-cell lymphoma. J Clin Oncol 19(2):376–388

Padmanabhan R, Kheraldine H, Gupta I, Meskin N, Hamad A, Vranic S, Al Moustafa AE (2022) Quantification of the growth suppression of HER2+ breast cancer colonies under the effect of trastuzumab and PD-1/PD-L1 inhibitor. Front Oncol 21:977664

Palazzolo S, Bayda S, Hadla M, Caligiuri I, Corona G, Toffoli G, Rizzolio F (2018) The clinical translation of organic nanomaterials for cancer therapy: a focus on polymeric nanoparticles, micelles, liposomes, and exosomes. Curr Med Chem 25(34):4224–4268

Patil MP, Ngabire D, Thi HH, Kim MD, Kim GD (2017) Eco-friendly synthesis of gold nanoparticles and evaluation of their cytotoxic activity on cancer cells. J Cluster Sci 28:119–132

Patil VM, Noronha V, Menon N, Singh A, Ghosh-Laskar S, Budrukkar A, Bhattacharjee A, Swain M, Mathrudev V, Nawale K, Balaji A (2023) Results of phase III randomized trial for use of docetaxel as a radiosensitizer in patients with head and neck cancer, unsuitable for cisplatin-based chemoradiation. J Clin Oncol 41(13):2350–2361

Pelicano H, Martin DS, Xu RA, Huang P (2006) Glycolysis inhibition for anticancer treatment. Oncogene 25(34):4633–4646

Raja SM, Othman SA, Roslan RM (2023) A short review on the imaging technology in radiation therapy. e-Jurnal Penyelidikan dan Inovasi 30:108–122

Rezvantalab S, Drude NI, Moraveji MK, Guvener N, Koons EK, Shi Y, Lammers T, Kiessling F (2018) PLGA-based nanoparticles in cancer treatment. Front Pharmacol 9:1260

Schmidt KT, Karzai F, Bilusic M, Cordes LM, Chau CH, Peer CJ, Wroblewski S, Huitema AD, Schellens JH, Gulley JL, Dahut WL (2022) A single-arm phase ii study combining NLG207, a nanoparticle camptothecin, with enzalutamide in advanced metastatic castration-resistant prostate cancer post-enzalutamide. Oncologist 27(9):718–e694

Shah S, Famta P, Bagasariya D, Charankumar K, Sikder A, Kashikar R, Kotha AK, Chougule MB, Khatri DK, Asthana A, Raghuvanshi RS (2022) Tuning mesoporous silica nanoparticles in novel avenues of cancer therapy. Mol Pharm 19(12):4428–4452

Sheykhhasan M, Manoochehri H, Dama P (2022) Use of CAR T-cell for acute lymphoblastic leukemia (ALL) treatment: a review study. Cancer Gene Ther 29(8–9):1080–1096

Shih CP, Tang X, Kuo CW, Chueh DY, Chen P (2022) Design principles of bioinspired interfaces for biomedical applications in therapeutics and imaging. Front Chem 10:990171. https://doi.org/10.3389/fchem.2022.990171

Silva CO, Pinho JO, Lopes JM, Almeida AJ, Gaspar MM, Reis C (2019) Current trends in cancer nanotheranostics: metallic, polymeric, and lipid-based systems. Pharmaceutics 11(1):22

Singhai M, Gupta GD, Khurana B, Arora D, Ashique S, Mishra N (2024) Nanotechnology-based drug delivery systems for the treatment of cervical cancer: a comprehensive review. Curr Nanosci 20(2):22–47. https://doi.org/10.2174/1573413719666230413084140

Smith L, Byrne HL, Waddington D, Kuncic Z (2022) Nanoparticles for MRI-guided radiation therapy: a review. Cancer Nanotechnol 13(1):38

Tiwari H, Rai N, Singh S, Gupta P, Verma A, Singh AK, Kajal SP, Singh SK, Gautam V (2023) Recent advances in nanomaterials-based targeted drug delivery for preclinical cancer diagnosis and therapeutics. Bioengineering 10(7):760

Torchilin V (2011) Tumor delivery of macromolecular drugs based on the EPR effect. Adv Drug Deliv Rev 63(3):131–135

Tran S, DeGiovanni PJ, Piel B, Rai P (2017) Cancer nanomedicine: a review of recent success in drug delivery. Clin Transl Med 6:1–21

Tree AC, Ostler P, van der Voet H, Chu W, Loblaw A, Ford D, Tolan S, Jain S, Martin A, Stafford J, Armstrong J (2022) Intensity-modulated radiotherapy versus stereotactic body radiotherapy for prostate cancer (PACE-B): 2-year toxicity results from an open-label, randomized, phase 3, non-inferiority trial. Lancet Oncol 23(10):1308–1320

Tripathi D, Hajra K, Maity D (2023) Recent advancement of bio-inspired nanoparticles in cancer theragnostic. J Nanotheranostics 4(3):299–322

Uzair B, Liaqat A, Iqbal H, Menaa B, Razzaq A, Thiripuranathar G, Fatima Rana N, Menaa F (2020) Green and cost-effective synthesis of metallic nanoparticles by algae: safe methods for translational medicine. Bioengineering 7(4):129

Xie L, Xu L, Xu M, Xu F, Zhao Q (2023) Present and future of metal nanoparticles in tumor ablation therapy. Nanoscale 15(44):17698–17726

Xu X, Li T, Jin K (2022) Bioinspired and biomimetic nanomedicines for targeted cancer therapy. Pharmaceutics 14(5):1109

Yang CY, Shiranthika C, Wang CY, Chen KW, Sumathipala S (2023) Reinforcement learning strategies in cancer chemotherapy treatments: a review. Comput Methods Progr Biomed 229:107280

Yi GY, Kim MJ, Kim HI, Park J, Baek SH (2022) Hyperthermia treatment as a promising anticancer strategy: therapeutic targets, perspective mechanisms, and synergistic combinations in experimental approaches. Antioxidants 11(4):625

Yona S, Gordon S (2015) From the reticuloendothelial to mononuclear phagocyte system—the unaccounted years. Front Immunol 6:328

Zhang X, Qi S, Liu D, Du J, Jin J (2022) PSMA-targeted supramolecular nanoparticles prepared from cucurbit[8]uril-based ternary host–guest recognition for prostate cancer therapy. Front Chem 10:847523

Zhang H, Mi J, Xin Q, Cao W, Song C, Zhang N, Yuan C (2023) Recent research and clinical progress of CTLA-4-based immunotherapy for breast cancer. Front Oncol 13:1256360

Stratagems for Breast Cancer Management in the Era of Nanotheranostics

6

Honey Goel, Pankaj Arora, Riya, Viney Chawla, Richu Singla, and Pooja A. Chawla

Abstract

Breast cancer undoubtedly tops the charts as one of the most common forms of malignancy with 25.8% of rate incidence, amongst all the reported cancers in women globally. It is no wonder, the number of fatalities over the last 30 years have gone up as a result of changes in risk factor profiles, better cancer registration, and detection control. Considering the increasing incidence of one of its worst forms, triple negative breast cancer, relative to other kinds, the role of smart nanotheranostic approaches have become more indispensable for achieving high specificity, effective screening and tracking (in order to thwart the peril of false positives and superfluous biopsies), improved diagnosis, along with therapeutic efficacy and utmost safety potential. Such nanotheranostic systems could be designed in a manner (by combining with chemotherapies, hormonal therapies biological therapies or immunotherapies) to annihilate specific tumor sites and stubborn breast cancer stem cells via novel strategies such as local heat and reactive oxygen species to prevent tumor relapse. Therefore, owing to dearth of feasible diagnostic techniques, nanotheranostic systems could pave the way to simultaneously diagnose the disease along with delivery of anticancer compounds selectively to breast cancer cells. Several imaging dyes/radioactive compounds, polymeric nanoparticles, micellar systems, lipid vesicular systems, contrasting agents, carbon nanotubes, dendrimers and ligand based targeted

Honey Goel, Pankaj Arora and Riya contributed equally with all other contributors.

H. Goel · P. Arora · Riya · V. Chawla · P. A. Chawla (✉)
University Institute of Pharmaceutical Sciences and Research, Baba Farid University of Health Sciences, Faridkot, Punjab, India

R. Singla
Multidisciplinary Research Unit, Guru Gobind Singh Medical College and Hospital, Faridkot, Punjab, India

M. A. Barkat et al. (eds.), *Nanotheranostics for Diagnosis and Therapy*,
https://doi.org/10.1007/978-981-97-3115-2_6

nanocarriers have been implicated for advanced breast cancer theranostic applications include identifying distinct metastatic locations of malignant tumors. The present outlay of literature aims to comprehend the basic understanding of the etiology, pathophysiological mechanisms depending upon the stage and molecular subtypes (luminal A, luminal B, HER-2 negative enriched, and basal-like), existing therapeutic regimens (including mix of multiple modalities primarily surgery, radiation, chemotherapy, hormonal treatment, and biological therapies), recent nanotheranostic carriers for breast cancer therapy, as well as associated safety and hazard problems are being tested in the field of oncology.

Keywords

Breast cancer · Nanotheranostics · Nanocarriers · Nanoparticles · Carbon nanotubes · Dendrimers · Therapies · Ligands

6.1 Introduction

Breast cancer is an extremely common and major cause of cancer morbidity and mortality in women globally. According to the American Cancer Society statistics, breast cancer will be solely responsible for 32% of new cases amongst every kind of cancer recorded among women in the United States (US) in 2024 (Siegel et al. 2013). The published data from National Breast Cancer Foundation in 2023 projects that at least 1 out of 8 women in the US will be confirmed to have breast carcinoma in her lifetime with total estimates of 297,790 cases of invasive breast cancer in women only (Breastcancer.org 2023). According to WHO, statistics published in 2023, approximately 2.3 million women were diagnosed with breast cancer and 685,000 of deaths were reported globally in 2020. As of the end of 2020, there were a total of 7.8 million women alive who had been confirmed to have breast cancer in the previous 5 years, which makes it the world's most frequently diagnosed malignancy (World Health Organization n.d.).

Further, it is a matter of the fact that the 5-year global survival rate for breast cancer varies greatly from higher rate of 80% among rich countries compared to 40% respectively, in under-developed countries (Coleman et al. 2008). Limited resources and infrastructural issues in such nations make it quite challenging to accomplish the objective to enhance breast cancer outcomes through early detection, diagnosis, and treatment (Anderson et al. 2008).

As per WHO recommendations, early detection and treatment of breast cancer continues to be the cornerstone for the breast cancer management. A wide range of novel drugs have been used to treat breast cancer including prophylactic chemotherapeutic therapies for individuals with a family history of breast cancer such as anti-estrogens, e.g., raloxifene or tamoxifen, which may prevent the illness in those who are susceptible to it (Peng et al. 2009). Other treatment choices for breast cancer include specific interventions like hormone therapy, radiation therapy, and surgical procedures. Management of breast cancer patients with distant metastases is typically aimed at increasing their quality of life and rates of survival (Chaurawal et al. 2022; Reeder and Vogel 2008). However, the limitations of all these conventional treatment modalities for breast cancer comprise adverse effects associated within their dose ranges which prompt the scientific community to adopt alternative approaches such combining therapeutic agents of plant or herbal origin to improve the tolerability and safety margins (Abdull and Noor 2013).

6.2 Stages of Breast Cancer and Primary Response to Check Its Progression

Figure 6.1 depicts the various stages of breast cancer. The in-situ stage (Stage-0) indicates the presence of aberrant tissue development in the breast lobules, typically is not related to invasive carcinoma, however elevates the chance of developing it again in either breast by around 7% over a 10-year period (Chuba et al. 2005). At this stage, no systemic and local treatments are generally advised, except close clinical examination for breast cancer at every 6 months and an annual mammogram. However, chemopreventive agents with selective estrogen receptor modulators (SERMs), such as tamoxifen may also be employed to check the DCIS progresses to aggressive breast cancer in few cases. In early invasive phase (stages I and II), treatment comprises modified radical mastectomy (recommended in individuals with susceptibility to relapse) or breast conservative surgery accompanied by radiation therapy which exhibit high success rate in early stages to improve better survival (Clarke et al. 2005).

Further, the mode treatment in locally advanced (Stage-III, lymph node >5 cm) is induction chemotherapy supplemented by regional procedures, radiation therapy, or combined which improves the overall survival for 5 years. In metastatic phase (Stage IV), 23% of the patients who have relapsed complying with treatment for early stages (II and III) achieve 5-year survival. Systemic therapy is determined by hormone receptor status, illness development pace, and patient willingness to accept therapeutic side effects. Trastuzumab, with or without chemotherapy, is the mainstay as initial therapeutic possibility for metastatic disease overexpressing ERBB2. Table 6.1 summarizes the therapeutic response to different stages in breast cancer.

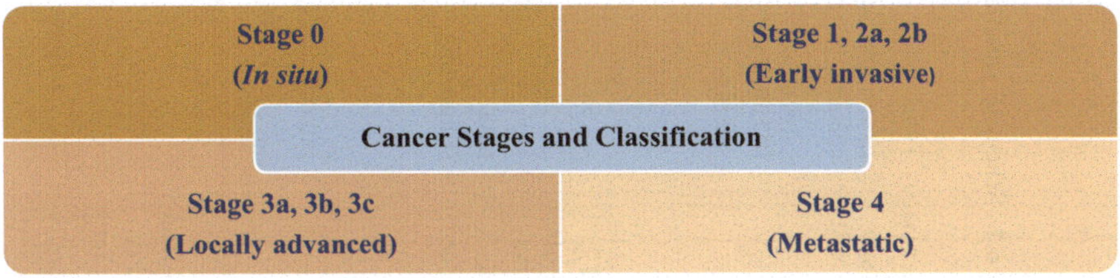

Fig. 6.1 Different stages of breast cancer

Table 6.1 Schematic summary the therapeutic response to different stages in breast cancer

Cancer stage	Type	Primary treatment	Hormone receptor positive	Hormone receptor negative	Node evaluation	ERBB2 overexpression	Reference
0: In situ	In situ lobular cancer	No therapy or use tamoxifen as a preventative measure	–	–	–	–	Rockville (2002)
	In situ ductal carcinoma	Chemotherapy and surgery to preserve the breast (may contemplate amputation if widespread or multicenter)					
1 and 2: Early-stage invasive	–	Radiotherapy and breast-saving surgery	Chemotherapy and endocrine therapy	Chemotherapy	SLN biopsy or ALN dissection	Chemotherapy and trastuzumab (Herceptin)	Clarke et al. (2005)
3: Locally advanced	Non-stimulating	Radiation treatment and breast-conserving surgery are administered after initiation chemotherapy	Chemotherapeutic initiation and hormonal therapy following surgery	Treatment with chemotherapy commencement	ALN dissection or SLN biopsy	Intraoperative radiotherapy and trastuzumab	Classe et al. (2009)
	Inflammatory	The use of radiation and a mastectomy come after initial treatment					
4: Metastatic initial or recurrent		Manage patient's medical objectives for bone pain, consider chemotherapy or bisphosphonates	Chemotherapeutic or endocrine treatment alone	Chemotherapy	–	In conjunction with or without chemotherapy, trastuzumab	Friberg and Nystrom (2015)

6.3 Etiological Causes of Breast Cancer

It is needless to argue that the highest incidence of breast carcinoma has been observed mostly in the women of countries with the highest socioeconomic levels. Therefore, this chapter aims to decipher some of the most likely etiological causes in order to understand the pathophysiology of breast cancer.

Figure 6.2 illustrates percentage wise diverse segment of risk factors linked with the incidence of breast carcinoma among the wide percentage range of female population globally. Literature studies have predicted that the average lifetime chance for a woman in the US to undergo breast cancer diagnosis is 12.3% (Siegel et al. 2021). Among the myriad multiple risk factors (which primarily include-age, family history, environmental toxins, alcoholism, cigarette smoking, early radiograph exposure, obesity, late menarche, late birth of child, early menopause, fat diet, contraceptives, and other estrogens, etc. (Alkabban and Ferguson 2022), some of the prominent factors which attributed significant influence on the etiology of breast cancer has been depicted in Fig. 6.3.

6.3.1 Early Menarche

Early menarche is the one of the most significant links to elevated risk of cancer of breast primarily because of low amount of circulating sex-hormone-binding globulin and greater levels of estradiol (Yang et al. 2022). Clinical investigations revealed that females with breast cancer had a greater mean age at natural menopause than women without the disease. Moreover, breast cancer patients have longer menarche and menopause periods (after deducting the time spent pregnant and lactating). Furthermore, if an oophorectomy is performed at age <40, there is a correlation between the surgery and lower risk of breast cancer.

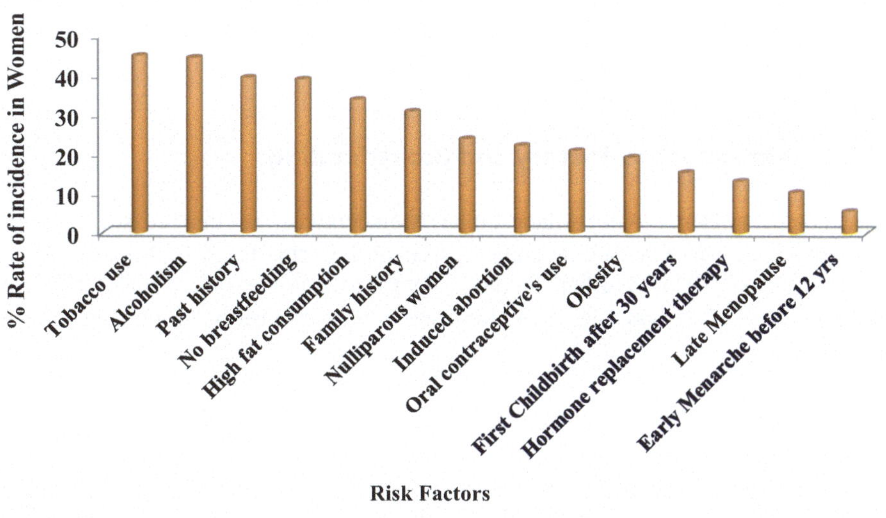

Fig. 6.2 Percentage rate of risk factors associated with the incidence of breast cancer among the women population globally

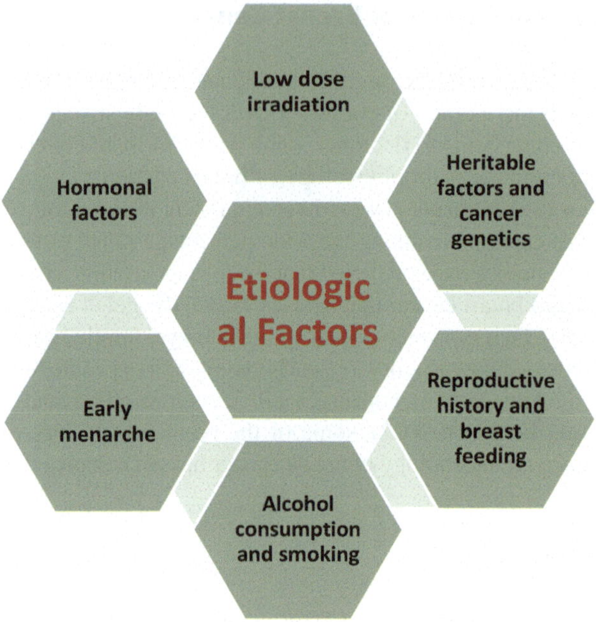

Fig. 6.3 Prominent etiological factors for the burden of breast cancer

6.3.2 Alcohol Consumption and Smoking

Alcoholism and the incidence of cancer were shown to have a substantial dose-response association in a 4-year follow-up research including approximately 90,000 US nurses between the ages of 34 and 59. Further, studies have also shown that smoking cigarettes elevates the 2-hydroxylation of estradiol in the body which may hinder the regulation of estrogen metabolism (Starek-Świechowicz et al. 2023).

6.3.3 Reproductive History and Breastfeeding

Based on the epidemiological studies, it has been demonstrated that breastfeeding for a minimum of 6 months significantly impacts on lowering the risk of breast cancer (Schwarz et al. 2024). Further, literature data also supports that women who had breast cancer often had fewer children and a shorter history of nursing (Wang et al. 2023).

6.3.4 Low Dose Irradiation

There is suspicion of an enhanced incidence of breast cancer within the young demographic exposed to low-dose radiation from surplus radiation during fluoroscopic examinations or scalp irradiation when compared to the elderly (Bollet et al. 2023).

6.3.5 Heritable Factors, Genetics, and Hormonal Significance

An initial history of families with breast cancer considerably elevates the chances of developing breast carcinoma due to gene related abnormalities (owing to pathogenic variations in either the breast tumor susceptibility gene 1, BRCA1, or the breast carcinoma vulnerability gene 2, BRCA2). Literature studies clearly suggest twofold risk of breast cancer in a woman, if she was born into a family with a first-degree relative who had breast cancer ever in her life (Infante et al. 2024). Another cohort study in more than 50 lakh women population conclude 20% elevated in risk of cancer in situ and 30% increase in risk of metastasis, who had a second-degree sibling has cancer of breast (Mukama et al. 2020). Further, studies have also revealed that the incidence of breast cancer rises till 35 years for BRCA1 gene carriers in contrast to BRCA2 gene carriers (with 3% more risk associated for BRCA1 gene carriers) (Van der Kolk et al. 2010). Further, some investigations have implicated the role of endogenous or exogenous variables such as hormone levels during the menstrual cycle. Although no discernible variations in the blood amounts of prolactin, luteinizing hormone, follicle-stimulating hormone, estrogen, and progesterone or estriol were reported, however, a steady downward trend in all other than estriol was observed in risky groups (McCann et al. 2024). Furthermore, few investigations have suggested the irregularity in estrogen conjugation during a particular phase of the ovulatory cycle as the most significant link to an elevated risk of cancer in breast. Therefore, on the basis of etiological basis, Fig. 6.4 depicts the complete cascade of the events occur during the pathophysiological mechanism which ultimately leads to progression of breast cancer.

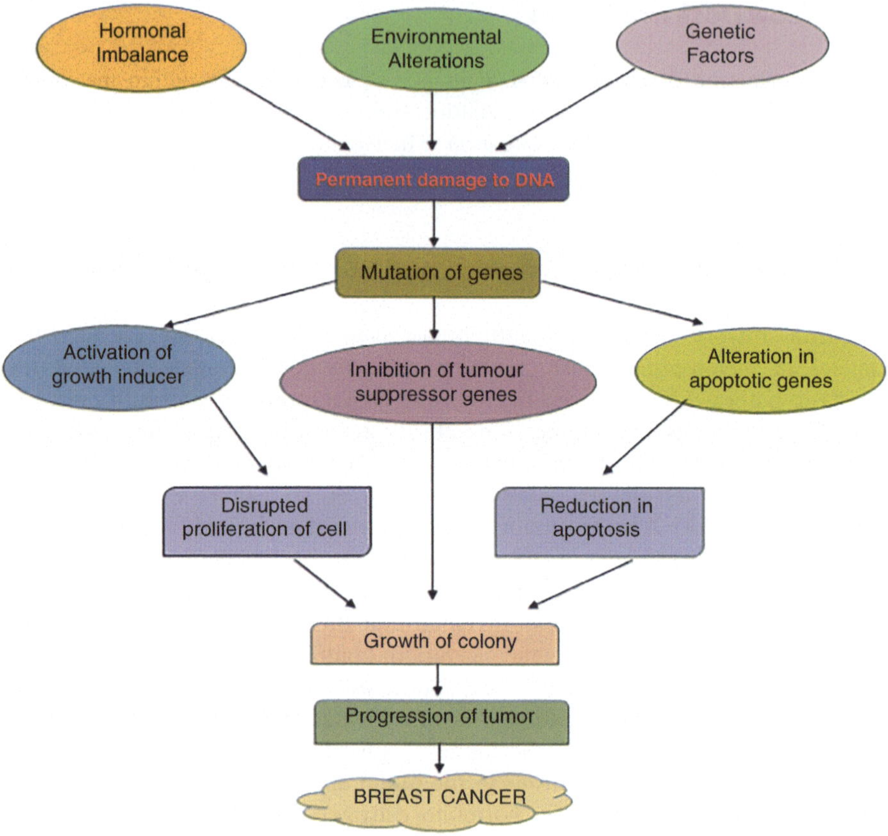

Fig. 6.4 Pathophysiological mechanisms involved in the progression of breast cancer

6.4 Prophylactic Measures for Control of Breast Cancer

The most prominent measures that contribute to lessen the progression or annihilating the development of the malignant disease/tumor include primary preventive control (comprising regulation of life style modifications such as diet, obesity control, cessation of alcohol consumption and physical inactivity) and secondary preventive control (which include diagnostic control and screening employing mammography, ultrasound imaging, MRI, and self-examination for breasts, etc.) as depicted in Fig. 6.5.

For primary preventive control, investigations have demonstrated that dietary modifications (comprising high intake of fruits/vegetables/grains/legumes with low intake of high-calorie foods/beverages and edible fats may regulate reversible epigenetic modifications that contribute to oncogenesis, such as aberrant histone modification, Hypermethylation of the promoter zone of genes that suppress tumors and hypomethylation in DNA (Rodriguez-Miguel et al. 2015). Further, obesity resulting

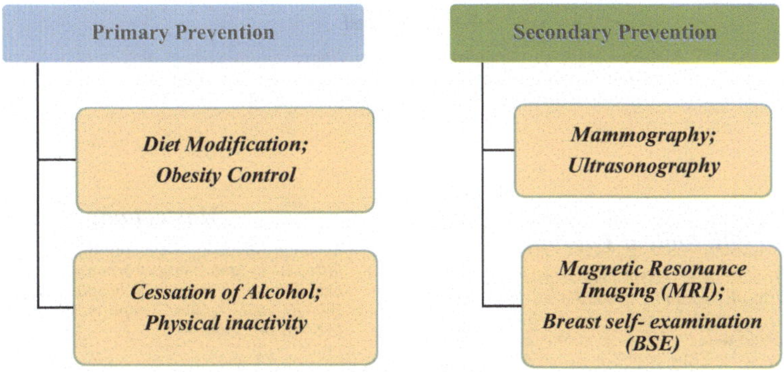

Fig. 6.5 Prophylactic measures for the control of breast cancer

from consumption of high-calorie meals, high alcohol intake (especially affects the estrogen pathway [i.e., effects on estrogen levels and estrogen receptors] by causing aromatase to become active) and physical inactivity due to sedentary life style (especially among women who have gone through the menopause) has been linked to an elevated likelihood of cancer in women (Lahart et al. 2015; Liu et al. 2015; Norat et al. 2015).

On the other hand, in case of secondary control, the main focus of the prevention is early identification by screening methods/diagnostic techniques (breast self-examination, mammography screening, ultrasonography breast examination, and MRI investigation) towards specific groups within the healthy population that are considered to be at a higher risk in order to reduce cancer-related mortality (Weigel et al. 2015). Generally, precautions must be taken while undergoing such diagnostic procedures (e.g., since ionizing radiation may trigger tumor growth, mammography is not advised for women <40 with moderate risk of developing breast cancer). However, non-invasive investigations with safety and specificity may be adopted to supplement other diagnostic procedures (Godavarty et al. 2015; Mehnati and Tirtash 2015).

6.5 Current Therapies in Oncological Settings for Breast Cancer

Figure 6.6 illustrates various diverse lines of therapies which are being adopted at various oncological settings worldwide for the therapeutic management patients of breast cancer. However, European Society for Medical Oncology (ESMO) guidelines recommend the consideration of factors such as tumor size, surgical feasibility, clinical profile, and patient willingness to preserve breast tissue all serve important roles in a patient's response to breast cancer care, particularly in patients with the initial stages breast carcinoma (Cardoso et al. 2019; Draganescu and Carmocan 2017; Heimes and Schmidt 2018; Ishii et al. 2019; Kwapisz 2017; Maximiano et al. 2016; Nguyen et al. 2021; Tarantino et al. 2020; Yang and Ho 2013).

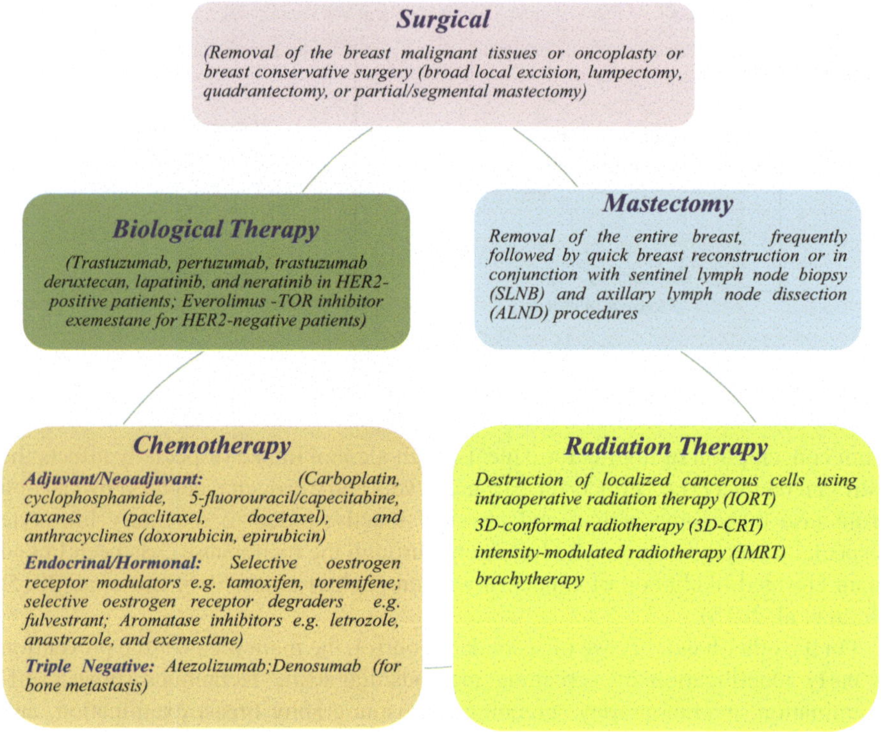

Fig. 6.6 Current lines of therapies for treatment of breast cancer

6.6 Monotherapies for Breast Cancer

Presently, chemotherapeutic regimens (monotherapies as well as combination therapies) have been designed against specific targets to obtain maximum efficacy and minimal toxicity relative to the normal chemotherapy and radiation therapies profiles (Sinha and Honey 2012a, b). Several molecular targeted therapies (such as PI3K/AKT/mTOR inhibitors and AMPK activators) are being bio-modulated to target multiple subtypes of breast cancer to enhance overall survival in people who do not react to hormonal treatments (Goel and Sinha 2016; Goel et al. 2016). Table 6.2 lists a few of the most significant and complex single therapies (including chemotherapy, hormonal, targeted, and immunotherapies) used to treat cancer of the breast.

Table 6.2 Schematic representation of diverse monotherapies used in breast cancer

Monotherapies		Drug description	Mechanism	Reference
Chemotherapy	Cisplatin	Platinum agents	Disruption of DNA synthesis, which prevents dividing cells	Zhang et al. (2022)
	Carboplatin			
	Doxorubicin	Anthracyclines	Encouraging DNA damage in cancer cells, which triggers dying	Shah and Gradishar (2018)
	Epirubicin			
	Paclitaxel	Taxanes	Preventing dividing cells	Willson et al. (2019)
	Docetaxel			
Hormone therapies	Tamoxifen	SERM	Suppression of the estrogen/ER relationship	Osborne (1998), Patel and Bihani (2018)
	Fulvestrant	SERD	Degradation of ER in breast cancer cells	Nicolini et al. (2016)
	Letrozole	Aromatase inhibitors	Inactivation of the aromatase an enzyme, which results in the synthesis of less estrogen	Sood et al. (2021)
	Exemestane			
	Anastrozole			
	Ribocilib	CDK 4/6 inhibitor	Regulation of Rb phosphorylation, control of cell cycle progression	Greenwalt et al. (2020)
	Capivasertib	AKT inhibitor	Stops the AKT mechanism from phosphorylating the building block	
	Niraparib	PARP inhibitor	PARP restriction, DNA replication interruption, and accelerated malignancy death of cells	
	Talazoparib			
	Olaparib			
	Abemacilib	CDK 4/6 inhibitor	The process of cell cycle progression and Rb phosphorylation regulated	
	Palbocilib			
	Everolimus	mTOR inhibitors	Repression of the PI3K/AKT/mTOR pathway and control of cell division	Jones et al. (2020), Martorana et al. (2021), Porta et al. (2014)
	Voxtalisib	Dual PI3K/mTOR inhibitor		
	Alpelisib	P13K inhibitor	Restraint of the PI3K/AKT/mTOR pathway and control of the division of cells	
	Pilaralisib			
	Bevacizumab	VEGF inhibitor	Monoclonal antibody against VEGF, prevention of blood vessel expansion	Chen and Hurwitz (2018), Sabatier et al. (2019), Hyseni et al. (2010)
Immuno therapies	Pertuzumab	Immunotherapy targeted HER2	HER2 overexpression-induced suppression of the HER2 transmission technique and stimulation of immune system reaction	Murthy et al. (2020), Schneble et al. (2015), Maximiano et al. (2016)
	Trastuzumab			
	Atezolizumab	PD-1/PD-L1 inhibitor	Immunological regulation loop suppression prevents T-lymphocytes from interacting with cancer cells	Liu et al. (2021)

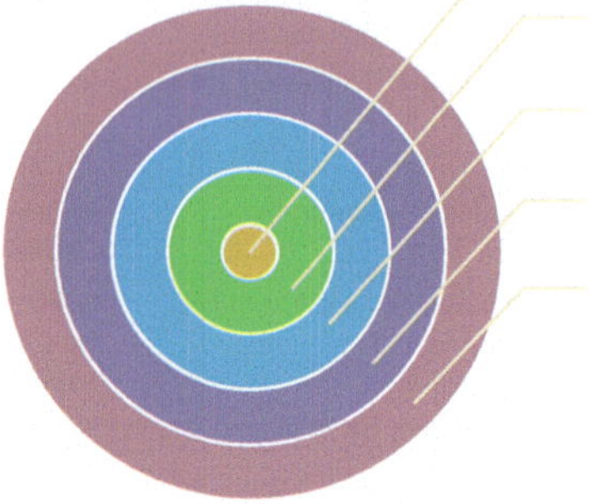

Fig. 6.7 Diverse combination strategies for achieving the desired therapeutic outcome in management of breast carcinoma

6.7 Combination Therapies for Breast Cancer

Nevertheless, monotherapies have shown tremendous results in few cases, but have also shown high failure rates in achieving the desired therapeutic outcomes. Therefore, combination therapies (comprising two or more therapeutic substances with different molecular targets) have proved to augment patient outcomes while lowering side effects and resistance building. Figure 6.7 depicts diverse combination therapies exhibiting multiple pathways and molecular targets (such as P13K, mTOR inhibitors, and dual mTOR, PI3K, CDK4/6 inhibitors) for the efficacious management of breast carcinoma (especially metastatic HR positive, HER-2 advance stage cancer of breast). Table 6.3 summarizes various clinical trials of combination therapies being investigated for the management of breast carcinoma.

It is a matter of fact that not all combination strategies have been therapeutically effective in oncological settings, e.g., Palbociclib combined with endocrine treatment (Anti-estrogen or AI) for a period of time did not enhance disease-free survival in individuals with remaining invasive disease after chemotherapy with neoadjuvant therapy (Loibl et al. 2021). Therefore, clinically viable combination approaches must be optimized for treating resistant breast tumors to monotherapies via assessment of biomarkers as well as the tumor microenvironment before deciding on a course of combination therapy. Table 6.4 summarizes the list of commercially available chemotherapeutic regimens being implemented in clinical practice for breast cancer.

Hence, these results imply that, while more work is needed to adjust the settings, boosting breast cancer outcomes in cases of treatment resistance, tumor repetition, tumor variation, resistance to multiple drugs, or metastasis can improve the patient's quality of life.

Table 6.3 Overview of clinical studies of multiple therapies under investigation for the treatment of breast carcinoma

Combination therapy	Clinical trial no.	Phase (status)	Condition of the disease	Expected therapeutic outcome
Abemaciclib (TT) + Fulvestrant, Anastrozole, Letrozole (HT)	NCT02246621	III (first and only CDK4/6 inhibitor (FDA approved in 2021)	Post-menopause HER2− ABC, HR+ with	PFS: 14.76 months of HT alone versus 28.18 months of CBT
Abemaciclib (TT) + Tamoxifen, Anastrozole, Letrozole (HT)	NCT03155997 NCT01698918	III (on-going)	HR+, node-positive, HER2−, high-risk, early breast cancer	2-year IDFS: 25% lower risk of death and readmission with 92.2% of CBT compared to 88.7% with HT alone
Everolimus (TT) + Letrozole, Exemestane (HT)	NCT01231659	II/III (completed)	Women in postmenopausal who have ER+ and HER2− MBC/locally ABC	PFS: 3.2 months of exemestane, 9.0 months of letrozole alone, and 22.0 months of CBT
Pilaralisib, Voxtalisib (TT) + Letrozole (HT)	NCT01082068	I/II (completed)	HR+, HER2−, refractory, recurrent, non-steroidal AI, MBC	PFS: 24 weeks in 22% of cases
Capivasertib, palbociclib, Alpelisib (TT) + Fulvestrant (HT)	NCT01992952 NCT01942135 NCT02422615	II/III (on-going)	Women who are menopause and have either HER2−, HR+, or PI3K/AKT/mTOR mutations, MBC/ABC	PFS and OS of CBT are noticeably better than those of HT alone
Letrozole, tamoxifen (HT) + Pembrolizumab (IT)	Cases reported	(on-going)	MBC + HR	PFS: More than 21 months
Anastrozole, Letrozole (HT) + Pertuzumab, Trastuzumab (IT)	NCT01491737	II (on-going)	HER2+ and HR+ MBC/ ABC locally	PFS: 12.45 months vs. to 21.72 months

(continued)

Table 6.3 (continued)

Combination therapy	Clinical trial no.	Phase (status)	Condition of the disease	Expected therapeutic outcome
Olaparib (TT) + Durvalumab (IT)	NCT02734004	I/II (on-going)	MBC with a BRCA mutation	PFS: In the OlympiAD study, 8.2 months of CBT compared to 7.0 months of Olaparib alone and 4.2 months of Olaparib alone
Niraparib (TT) + Pembrolizumab (IT)	NCT02657889	I/II (completed)	Cancer of the ovaries with advanced or metastatic TNBC	8.3 months is the median PFS, with a 21% subjective response percentage and a 49% control of the disease rate
Pembrolizumab (IT) + nab-paclitaxel, paclitaxel, gemcitabine plus carboplatin (CT)	NCT03036488 NCT02819518	III (first-line treatment; FDA approved in 2021)	TNBC	PFS: 9.7 months of CBT vs. 5.6 months of CT alone
Atezolizumab (IT) + albumin-bound paclitaxel (CT)	NCT02425891	III (first-line treatment; FDA approved in 2021)	Expanded TNBC	PFS: 9.3 months with CBT as opposed to 6.1 months with CT alone; OS: 28.9 months with CBT as opposed to 20.8 months with CT alone
Atezolizumab (IT) + paclitaxel (CT)	NCT03125902	III	Metastatic/ advanced TNBC	With possible security concerns, PFS or OS are not improved when compared to CT alone.
Bevacizumab, Everolimus (TT) + doxorubicin (CT)	NCT02456857	II (on-going)	Locally diagnosed TNBC that is not responsive to conventional treatment	21% was the desired rate for response
Bevacizumab (TT) + etoposide, cisplatin (CT)	NCT01281696	II (completed)	BC with malignancies to the brain	9.1 months for PFS, 10.7 months for OS

(continued)

Table 6.3 (continued)

Combination therapy	Clinical trial no.	Phase (status)	Condition of the disease	Expected therapeutic outcome
Olaparib (TT) + Capecitabine, Eribulin, Vinorelbine (CT)	NCT02000622	III (on-going)	BRCA mutation-related recurrent cancers of the breast	PFS: 7.0 months as opposed to 4.2 months with only chemotherapeutic
HY19991 (IT) + paclitaxel (CT) + Thioridazine (ACT)	–	Preclinical study	MCF-7 MBC mice given PTX/THZ/HY liposome treatment	93.45% was the tumor-inhibiting frequency
Trastuzumab (IT) + Capecitabine (CT) + Tucatinib (TT)	NCT02614794	III (first-line treatment; FDA approved in 2020)	Progressive ABC or MBC or HER2+	Improved OS, PFS, and secure

Abbreviations: *TT* targeted therapy, *HT* hormone therapy, *CT* chemotherapy, *IT* immunotherapy, *ACT* anti-CSC therapy

Table 6.4 List of FDA-approved commercial chemotherapeutic formulations used in oncological practice for breast cancer

Drug/active therapeutic agent	FDA approval year	Formulations in market	Type of forms in market	Dose Recommended	Breast cancer conditions	Type of drug delivery
Methotrexate	1953	47	IV injections	15 mg/mL	Primary metastatic breast cancer	Combined
			Tablet	2.5 mg		
			Topical gel	1% w/w (20 g)		
Fluoxymesterone	1956	2	Tablet	10–40 mg	Androgen-responsive; developed metastatic (bones) Breast carcinoma	Individually
Tepadina	1957	4	IV injection	0.3–0.4 mg/kg	First line and second line breast cancer	Combined
Cyclophosphamide	1959	25		200 mg	Retinoblastoma mammary tumor	Individually and combined
			Tablet	50 mg		
Vinblastine sulfate	1965	4	IV injection	10 mg/mL	Neo-adjuvant, additive, or first-line treatment for metastatic breast carcinoma	Combined
				10 mg/mL		
5-Flurouracil	1970	7	Capsule	10 mg	Adenocarcinoma breast cancer	Individually
			Tablet	14 mg		
			IV injection	600 mg/m^2		
Methyltestosterone	1973	2	IV injection; Capsule	1 mL; 10–50 mg	Breast cancer, postpartum breast discomfort, and bloating in women	Individually and combined
Doxorubicin	1974	58	IV injection	10 mg/5 mL	Metastatic breast cancer	Individually
				2 mg/mL		
Tamoxifen	1977	21	Tablet	10 mg/per day	Oestrogen receptor-positive malignant cancer of the breast	Individually

Goserelin acetate	1989	3	IV injection	3.6 mg	Progressive breast carcinoma in pre- and perimenopausal female	Individually and combined
				6.8 mg		
Paclitaxel	1992	7		20 mL	First-line therapy for tripled adverse breast carcinoma	
				30 mL		
Anastrozole	1995	34	Soft capsule	1 mg	First-line therapy for tripled adverse breast carcinoma	Individually
			Tablet	1 mg		
Pamidronate	1996	1	IV injection	30 mg/90 mg	Osteolytic bone malignancies of breast carcinoma	
Gemcitabine		40		200 mg	Initial treatment for aggressive breast carcinoma	Individually and combined
Docetaxel		56		20 mg	Locally progressed metastatic breast carcinoma	
Letrozole	1997	1	Tablet	2.5 mg	First-line or second-line treatment for metastatic breast carcinoma	Individually
Toremifene				60 mg	ER-positive malignant breast carcinoma	
					Cancer of breast	
Megestral acetate	1998	6		40 mg; 160 mg	Hormonal treatment for progressive breast carcinoma	Individually and combined
Transtuzumab		18	IV injection	440 mg; 600 mg	HER_2 cancer of breast	Individually
Capecitabine		33	Tablet	500 mg	Metastatic cancer	Individually and combined
Exemestane	1999	7		25 mg	Advance cancer of breast	Individually
Epirubicin		33	IV injection	2–10 mg/m^2	Primary cancer of breast	
Fulvestrant	2002	1		5 mL	HR positive cancer of breast	
Ixabepilone	2007			15 mg	Metastatic cancer of breast	Individually and combined
Lapatinib		8	Tablet	250 mg		Combined
Everolimus	2009	14		5 mg	Advance HER_2 negative cancer	Individually and combined
				0.25 mg		

(continued)

Table 6.4 (continued)

Eribulin mesylate	2010	5	IV injection	1.5 mg/m^2	Hormonal treatment for severe breast carcinoma	Individually
				0.5 mg/m^2		
Pertuzumab	2012	1		420 mg/14 mL	HER_2 positive cancer of breast	Individually and combined
Palbociclib	2015		Capsule	125 mg	HER_2 negative or breast cancer	Combined
Abemaciclib	2017		Tablet	150 mg	Initial therapy for breast cancer	Combined
Ribociclib				200 mg	HER_2 negative breast cancer	Combined
Neratinib maleate		1		240 mg	HER_2 positive metastatic cancer of breast	Individually
Olaparib	2018			150 mg	HER_2 negative, localised progressed or metastatic breast carcinoma	Individually and combined
Tolazoparib			Capsule	1 mg		Individually
Atezolizumab	2019		IV injection	840 mg/mL	Triple adverse breast carcinoma	Individually and combined
Alpelisib	2019	15		90 mg	HER_2 negative metastatic breast cancer	
			Tablet	1 mg		
Pembrolizumab	2020	1	IV injection	100 mg	Metastatic triple negative cancer of breast	
Margetuximab				15 mg/kg over 120 min	HER_2 positive or metastatic breast cancer	
Tucatinib			Tablet	300 mg		Combined

6.8 Comprehensive Assessment of Approaches for Breast Cancer Management

For the effective therapeutic management of breast cancer, there is imperative need for implementation of multidimensional strategies comprising diagnostic evaluation followed by stringent screening (supplemented by additional imaging techniques such as thermal imaging, breast-specific X-ray/MRI scans, positron emission tomography (PET) mammography, and optical scanning in order to curb ultrasound industry's high proportion of pseudo-positive results and low positive predictive value), pathological evaluation (processing of the diseased tissue for differentiation between closely similar. Pathologists have diagnostic challenges when dealing with illnesses such aberrant ductal hypertrophy and in situ illness and microinvasion, or carcinoma of the ducts and lobular cancer supplemented by ancillary immunohistochemistry (for the assessment of status of the estrogen receptor (ER), progesterone receptor (PR), and her-2/neu/HER2 receptors) and molecular testing, in situ hybridization, and (RT-PCR) reverse transcription-polymerase chain reaction in the characterization of ambiguous morphology (especially tumor size with precise location). Figure 6.8 depicts the multidisciplinary framework for the comprehensive screening, diagnosis, pathological evaluation to therapeutic response for the management of breast cancer.

The NCCN (National Comprehensive Cancer Network) has also anticipated recommendations for thorough diagnostic evaluation of staging of the disease. In early-stage breast cancer, genetic evaluation becomes necessary to assess the chance of remote recurring and to assist judgements about systemic treatment. Breast tumors may be divided into various prognostic categories using high throughput

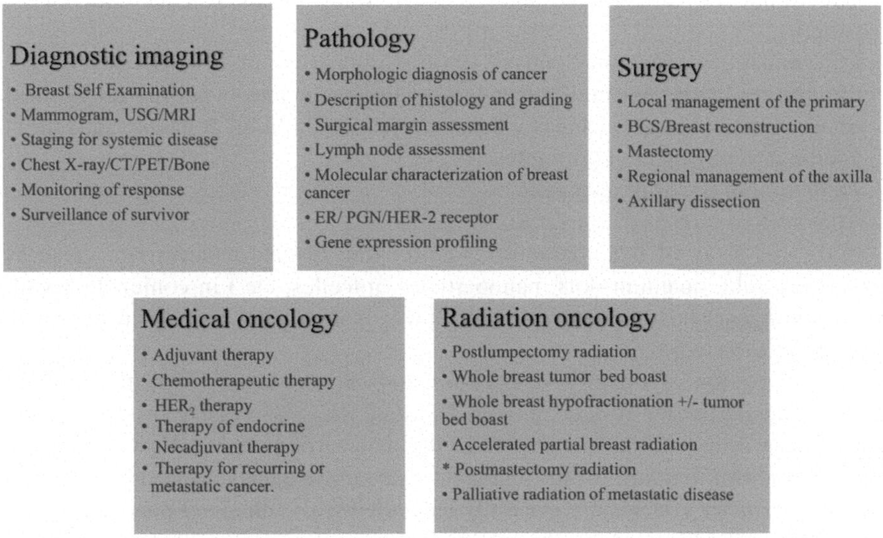

Fig. 6.8 Framework for multidisciplinary therapeutics

polymerase chain reactions and reverse transcription evaluation for many genes as well as DNA microarrays. These investigations assist in enhancing the use of other, individual risk factors in medical care. Further, the role of predictive tumor markers, protein expression tests, which are unaffected by the morphologic features of the tumor, are vital to make important treatment decisions. Targeted cancer genomes which are of significant interest may also be tested for next-generation gene sequencing-based precision treatments (Bennett and Farah 2014).

6.9 The Rise of Nanotheranostics

In recent years, nanotheranostic shave spawned a new era in the management of breast cancer therapeutics by enabling precise diagnosis, analyzing biopharmaceutical variables, enhanced tumor penetrability, optimization the medication release pattern in an appropriate manner for a desired specific therapeutic effect compared to conventional anticancer therapies (Tian et al. 2024). Nanotheranostic framework not only intends to shorten the time necessary for managing patients and monitor medication delivery (for proper medication, dose estimate, and decreasing the danger of drug abuse or side effects related to potentially dangerous metabolites) throughout the body, however anticipates to bolster the potential of emerging anticancer drugs and nanocarriers (by optimizing their pharmacokinetic and pharmacodynamic characteristics as well as supporting good vascular perfusion and absorption), guaranteeing appropriate financial returns in comparison to expenditures, improving the standard of life, and being adaptable to a range of administrative ways.

Some of the most lucrative applications of nanotheranostic framework include:

- Envisaging the drug performance through study on its kinetics and release pattern.
- Determination of real time analysis of drug biodistribution in vivo.
- Monitoring the therapeutic efficacy in continuous manner via drug release and accumulation at targeted site (by measuring probe build-up at the targeted site).
- Facilitation of triggered drug release.
- Prediction of the drug responses.

A myriad array of nanotheranostics employing diverse nanocarriers (such as liposomes, gold, quantum dots, nanoparticles, micelles, etc.) in conjunction with diagnostic agents has been pre-clinically investigated in breast cancer studies as depicted in Table 6.5.

The integration of imaging dyes/compounds along with therapeutic agent in nanotheranostic carrier systems aids to facilitate the non-invasive insights on the dynamics of drug emission in vivo from the nanocarrier system. However, it is a matter of the fact that most of the therapeutically active agents utilized in the nanotheranostic carrier systems are generally entangled within the inert nanocarrier and generally employ imaging dyes/agents such as radionuclides which provide analogous signals in both bound/entrapped form as well as unbound/free form. Therefore,

Table 6.5 Summary of preclinical studies on nanotheranostic systems in breast cancer

Nanotheranostic platform				
Nanocarrier	Drug compound	Diagnostic compound	Therapeutic outcome in breast cancer	Reference
Targeted PEG liposomes	Mitoxantrone	SPIONs	Enhanced cell toxicity to the MCF-7 female breast tumor line of cells	He et al. (2014)
	Doxorubicin	Fluorescent probe	Inhibition of breast tumor-bearing mice	Ma et al. (2016)
		Gadoteridol	Improved intratumor dose and full lesion shrinkage	Rizzitelli et al. (2015), Rizzitelli et al. (2016)
Polystyrene NPs		Cyanine dye	Increased therapeutic efficacy in vitro and reduced tumor burden in vivo	Cortes et al. (2020)
Graphene quantum dots		–	A mutually optimized anticancer method that offers therapy and evaluation	Ko et al. (2017)
Nanoclusters gold	Paclitaxel	Green indocyanine	Decreased tumor expansion in a mouse breast cancer study	Liu et al. (2019a,b)
Micelles	Docetaxel	DiR	Tumor reduction in breast cancer patients with minimal toxicity	Ma et al. (2012)
Iron oxide NPs	Curcumin	–	Substantial anticancer activities demonstrated when contrasted to free turmeric	Yallapu et al. (2012)

it becomes quite imperative to address these challenges in order to accomplish accurate quantity of drug emitted in a controlled fashion at targeted location without any glitch (Taran et al. 2023).

On the other hand, contrast compounds, such as gadolinium and manganese, are typically used in MRI scans and create a signal through liaisons with molecules of water in the vicinity. As a result, the nature of such associations varies greatly based on the placement of the contrast compound (within or exterior) of the water impervious liposomes; tracking of medications may be performed using resonance imaging probes.

Nanotheranostics integrated using contrast agents, as well as active compounds have also been instrumental in achieving the in vivo real time drug biodistribution, accumulation at the intended site, forecasting and enhancement of the efficacy apart from being ability to visualize and grasp a variety of crucial elements of the drug delivery process (Abolzadeh et al. 2023). Moreover, nanotheranostics can also be a very useful strategy to analyze target site accumulation in addition to real-time tracking (precise drug delivery to a specific location) of the biodistribution of nanocarrier formulations. The major goal of biodistribution studies at the intended site is to precisely examine the characteristics of naonotherapeutic drug delivery system as well as to gain a better understanding of crucial elements of such therapies. Hence, nanotheranostic platforms provide a methodical foundation for numerous imaging techniques that may enable functional, cellular, and anatomical accuracy for the targeted delivery of nanocarrier based drug delivery therapies at the specific sites in breast cancer. Image-guided drug delivery can also be used to validate the prompt drug release at the targeted site in a non-invasive way, as well as to continually assess its therapeutic effectiveness (Li et al. 2023).

It is also important to comprehend capability of the nanotheranostics framework to induce drug release from nanocarrier, drug accumulation and distribution at the target sites, and release pattern of drugs form the nanocarriers. However, these nanotheranostic systems are able to ease the triggering of drug discharge in a stimulus-sensitive nanotheranostic formulation with excellent temporal accuracy (i.e., by employing the trigger at the time points when the quantity of the formulation at the target site is ideal) as well as excellent spatial accuracy (i.e., those areas of the target tissue where the formulation is present in large quantities). Thus, in order to achieve this, two distinct visualization probes may be utilized-one PET tracer or magnetic resonance tracer to study whether the system is still intact by employing a contrast dye with free exposure to surrounding water molecules (e.g., in the case of liposomes, bound to the outside bilayer and not bound in the core), and a different probe that only creates a signal once the system breaks down (Thavendiranathan et al. 2023).

Furthermore, nanotheranostics are extremely important systems for anticipating therapeutic responses and making rational forecasts regarding the probable impact of specific therapeutic treatments, especially in the initial phases of clinical testing. Another significant benefit of nanotheranostics and image-guided drug delivery, particularly in preclinical research, to streamline longitudinal testing setups, allowing for not only assessing and minimally invasive biological distribution studies, but also more elegant and relevant effectiveness such as mutant mice and orthotopic tumor models for actual time exploration of disease progression as well as therapy outcome.

6.10 Limitations and Challenges of Nanotheranostics

- Nanocarriers are hazardous, immunogenic, and have delayed clearance from the body's tissues.
- Probability of mitochondrial, membrane disruption and production of oxidative stress respectively, by particles of silver, magnetic nanoparticles and single-walled nanotubes of carbon (SWNCTs).
- Concerns of multidrug resistance (MDR), autophagy by lysosomes.
- Wrapping an individual diagnostic/therapeutic drug in a nanotheranostic system might not offer ideal effectiveness, accuracy, or tolerance. Thus, it grows more crucial to leverage multipurpose nanotheranostics systems by integrating many scanning and therapies on a single system.

Nevertheless, in the light of aforementioned propositions, it could be implied that nanotheranostics exhibit tremendous potential to provide a promising alternative for breast cancer therapy in a few years, permitting them to be healed or managed at earlier stages.

6.11 Diverse Nanotechnology-based Nanocarriers for Breast Cancer Therapy

Nanotechnological driven drug carriers such as nanovesicular systems (e.g., liposomes), polymeric/albumin/magnetic nanoparticles (NPs), quantum dots, micelles, carbon nanotubes, dendrimers, nanodiamonds, etc. which have played quintessential role in the therapeutic management of various types of cancers especially breast cancer owing to their distinctive properties in protective the drugs from degradation, ensuring efficient drug concentration at the targeted tumor sites and assist in overcoming drug resistance. Tables 6.6 and 6.7 provides comprehensive summary of some of the prominent clinical (under different phases-I/II and III) and preclinical investigations comprising these nanocarriers have been carried out in the last decade using diverse nanobiomaterials especially for the breast cancer applications, respectively.

Table 6.6 Confirmed clinical trials employing nanotechnology for breast cancer purposes (www.clinicaltrials.gov)

Nanocarrier	Therapeutic payload	Clinical application	Clinical phase	Trial no.
Types of nanocarrier	Targeted drug molecule	Type of cancer	Clinical trial phase	Clinical trial no.
Liposomes	Daunorubicin	Breast cancer	I	NCT 00004207
	IVAC_W_bre1_uID and IVAC_M_uID			NCT02316457
	Trastuzumab and docetaxel		II	NCT00687440
	Doxorubicin and cyclophosphamide ± Trastuzumab			NCT01206881
	Bupivacaine			NCT03393117
	Methotrexate			NCT00992602
	Doxorubicin and carboplatin			NCT05949021
	Mitoxantrone hydrochloride		Not applicable	NCT06156761
Polymeric NPs	BP-C1	Breast cancer	I	NCT01861509
	Durvalumab and PTX along with or lacking the neoantigen vaccination			NCT 03606967
	DOX, cyclophosphamide and filgrastim followed by PTX			NCT 00407888
	Carboplatin and nab-paclitaxel with or without vorinostat			NCT 00616967
	PTX and cyclophosphamide			NCT 00629499P

(continued)

Table 6.6 (continued)

Nanocarrier	Therapeutic payload	Clinical application	Clinical phase	Trial no.
Types of nanocarrier	Targeted drug molecule	Type of cancer	Clinical trial phase	Clinical trial no.
Albumin bound NPs	Mifepristone	Breast cancer	I	NCT01493310
	Paclitaxel		II	NCT04538833
	Fluorouracil, Epirubicin, cyclophosphamide (FEC)			NCT00110695
	Anlotinib hydrochloride			NCT05244993
	PTX and cyclophosphamide			NCT 00629499
	PTX, gemcitabine and bevacizumab			NCT 00662129
Quantum dots	Veldoreotide	Breast cancer	I	NCT 04138342
Magnetic NPs	Superparamagnetic iron oxide	Breast cancer	I and II	NCT 05359783

Table 6.7 Preclinical studies based on the application of diverse nanocarriers in breast cancer

Nanocarrier platform	Drug/therapeutic payload	Composition (polymeric carrier/vehicle)	Expected outcome	Reference
Nanoemulsions	Doxorubicin	Perfluorocarbon (hydrocarbon) nanoemulsions	Outstanding anti-cancer effects were characterized by tumor shrinkage, and demonstrated long-lasting, robust ultrasonography contrast	Nicolini et al. (2016)
Microbubbles	Radiation	SE61 microbubbles recharged with oxygen or nitrogen	Improve mammary tumor aeration levels, allowing oxygen supply to vascular areas of the tumor, improving tumor development and survival in animals	Eisenbrey et al. (2018)
Polymeric micelles	Doxorubicin	Polyethylene glycol (PEG)-polylactic acid (PLA)	Higher cell death, low impact on regular cells, enhanced surface area, and bigger quantity of particles created by ultrasound	Jablonowski et al. (2018)
Magnetic nanoparticles	–	Polylactic-co-glycolic acid (PLGA)	Superior scanning quality and high biological compatibility	Nie et al. (2020)
Oxygen	Radiation	Span 60 and water-soluble vitamin E	A mouse study involving recurrent breast carcinoma, treatment-induced quick growth of tumors, volume expansion, tumor development interruption, and enhanced prognosis	Delaney et al. (2019)
Micelles	Curcumin	Pluronic P123/F127	Greater neuronal absorption and an extended circulation period	Martorana et al. (2021)

Table 6.7 (continued)

Nanocarrier platform	Drug/therapeutic payload	Composition (polymeric carrier/vehicle)	Expected outcome	Reference
Liposomes	Doxorubicin	Sonazoid and perfluoromethylcylopentane	The effectiveness of treatment is significantly increased by acoustical clustered treatment	Bush et al. (2020)
	Docetaxel, epirubicin, cyclophosphamide	Sonovue and saline	With ultrasound scans, one may precisely monitor a tumor's response to treatment	Soyemi et al. (2018), Wang et al. (2019), Sobczak et al. (2021), Amioka et al. (2016)
	Doxorubicin	Perfluoropentane	The effectiveness of treatment is significantly increased by acoustical clustered treatment	Sheng et al. (2021)
	Artemisinin	Polyethylene glycol 2000 (PEG 2000)	The cellular impact shown that the pegylated nanoliposomal composition has a greater fatal impact compared to the nanoliposomal	Dadgar et al. (2014)
	Paclitaxel	1,2-dioleoyl-sn-glycero-3-phosphocholine	LEP has been linked to decreased mortality with comparable effectiveness	Slingerland et al. (2013)
	Cisplatin	Distearoylphosphoethanolamine-polyethylene glycol and phosphatidylcholine	Enhancing medication effectiveness and lowering doses	Ghafari et al. (2020)
	Epirubicin-hydrochloride	Phosphatidylcholines with thin film hydration using egg yolk	Compared to medication nor silica-coated version of this medication, concentration of epirubicin-encapsulated liposomes that was higher in breast cancer cell types	Alexander et al. (2016)
	Thymoquinone	Thymoquinone (2-isopropyl-5-methyl-1,4-benzoquinone) and triton X-100; 1,2-dipalmitoyl-sn-glycero-3-phosphocholine (DPPC)	Keeps their cancer-fighting properties, rigidity, or better bioavailability	Odeh et al. (2012)
	Raloxifene	Methanol-ethyl acetate	The most significant amount of cancer fighting activity was detected, and raloxifene-loaded cochleates containing DM-β-CD have been shown to block the MMP-2 enzyme	Agardan et al. (2016)
	Resveratrol	Chloroform solutions of cadmium oxide and sucrose laurate	Reversible connected hyaluronic acid nanomaterials possess significant prospects in cancer treatment due to their exceptional biocompatibility CD44-targetability, and efficient restoration of medication tolerance	Zhao et al. (2019)
	Doxorubicin	Perfluoropropane	Produced a comparable anticancer response at roughly a fifth of the typical anticancer dosage utilized for DOX treatment	Ueno et al. (2011)
	Doxorubicin	Polyethylene glycol	The negative consequences that were not considered hematological generally modest to minor and easily controlled. Minimal decline in the ability of the left ventricle to expel gas	Rau et al. (2015)
	Cisplatin	Soy phosphatidyl choline (SPC-3), cholesterol, dipalmitoyl phos-phatidyl glycerol (DPPG), and methoxy-polyethylene glycol-distearoyl phosphatidylethanolamine (mPEG 2000-DSPE)	Diminished chance of kidney damage	Stathopoulos and Boulikas (2011)
	Doxorubicin	Polyethylene glycol	Effective molecular targeting: Immunoliposomes and other ligand-directed structures combine techniques for delivery with elements of biology that can recognize tumors	Batist et al. (2001)

Polymeric nanoparticles	Paclitaxel	Albumin nanoparticle	Provide the physician with a wide range of patient-friendly, efficient alternatives to therapy	Miele et al. (2009)
	Doxorubicin	Lipoic acid, hyaluronic acid, L-lysine methyl ester	Possibility for cancer treatment is greatly enhanced by a high level of biocompatibility, CD44-targetability, and efficient recovery of drug susceptibility	Zhong et al. (2015)
	Paclitaxel	Montmorillonite and poly(D, L-lactide-co-glycolide)	SK-BR-3 fibroblasts were used in an inside-out mortality study to more demonstrate this antibody decoration's targeted properties	Sun et al. (2020)
	Paclitaxel	Folic acid Polylactic-co-glycolic acid, polyethylene glycol succinate	Reveals that the route of administration stocked with PTX and ornamented with ALN/FA, created in this work, has a lot of potential for treating bone-metastasized breast carcinoma	Chen et al. (2020)
	Docetaxel	Albumin nanoparticle	An approach with considerable promise as a treatment is created by the beneficial oxidative description, proven nanonimmunogenicity, biologic compatibility, and increased cancer retention of HSA	Zheng et al. (2014)
	Tamoxifen citrate	Polylactide-co-glycolide	The unbound drug was shown to have a less harmful to cells than nanoparticles	Maji et al. (2014)
	Cisplatin	Luteinizing hormone-releasing hormone (LHRH)-modified dextra	Dex-SA-CDDP-LHRH nanoparticles show great potential for targeted chemotherapy of metastatic breast cancer	Li et al. (2015)
	Doxorubicin	Chitosan and pluronic F127	Effective delivery systems for the therapy of tumors, especially breast cancer tumors where receptors that recognize folate are overproduced, core-shell nanomaterials show promise	Manaspon et al. (2012)

(continued)

Table 6.7 (continued)

Nanocarrier platform	Drug/therapeutic payload	Composition (polymeric carrier/vehicle)	Expected outcome	Reference
Micellar nanoparticles	Paclitaxel	Polyethylene glycol (PEG)-polyacrylic acid (PAA) (PEG-PAA)	Validates the theory behind the tolerance and moderate effectiveness of a novel PTX method for delivering drugs composition	Kato et al. (2012)
	Aminoflavone	Anti-epidermal growth factor receptor	Superior tumor resection in a triple-negative cancer of breast transplanted rodent model that is orthotopic	Wang et al. (2017)
	Doxorubicin	Pluronic copolymer P123 polyethylene glycol-block-poly (di-isopropanolamino ethyl methacrylate) diblock copolymer	An efficient nanoplatform for addressing cancers multidrug may involve pH and NIR light-responsive micelles with hyperthermia-triggered tumor penetration and cytoplasm medication being released	Yu et al. (2015)
	Paclitaxel	Polyethylene glycol-b-polylactide	The preventive performance against the metastasis of murine breast cancer cells is enhanced by pH-responsive polymeric micelles	Wang et al. (2018)
	Paclitaxel	Dextran-g-indomethacin	For MDR tumors, DEX-IND/PTX tiny particles shown stronger tumor growth-inhibition activities	Ji et al. (2017)
	Doxorubicin and salinomycin	Polyacrylic acid and polyethylene glycol	Reversing the progression of the EMT and efficiently suppressing tumor growth without causing spleen cancer or additional major tissue damage, the simultaneous treatment of DOX + SL trapped in nano-micelles also effectively prevented tumor recurrence and metastasis following treatment separation	Cui et al. (2019)
	Doxorubicin and paclitaxel	Lauryl carbamate derivative of plant-based polymer inulin	A distinct method of administration for one or more types of chemotherapy medication is provided by INT-micelles	Kesharwani et al. (2019)
	Aminoflavone	Poly(amidoamine) dendrimer, polyethylene glycol derivatives	F-loaded, EGFR-targeted unimolecular micelle nanoparticles as a potent therapy for tumors express EGFR	Brinkman et al. (2016)
	Paclitaxel	Methoxy polyethylene glycol-polylactide (mPEG-PLA)	In those with non-small cell lung tumors, CrEL-free paclitaxel and gemcitabine showed good anticancer efficacy with minimal emetogenicities	Ahn et al. (2014)
	Paclitaxel	PEG-block-poly[(1,4-butanediol)-diacrylate-β-5-amino1-pentanol] polyethyleneimine-block-PDHA	In both orthotopic and aberrant 4 T1 following the operation development simulations, tumor regrowth is inhibited by macrophage-mediated liposomes encapsulated with PTX and res	Qiu et al. (2021)
	Fisetin	Pluronic127 folic acid	FS-PF-FA small particles may be employed to emphasize the administration of drugs that treat cancer	Pawar et al. (2018)

Nanodiamonds (NDs)	Immunoglobulin G (IgG)	NDs conjugated with immunoglobulin G (IgG–gFND)	IgG-gFND was shown to be specifically absorbed by immunological cells in in vitro conditions, whereupon immune system survivability was unaffected and triggering immune cells was considerable. After intratumoral via injection, IgG–gFND stayed at the tumor site while uFND moved to the kidneys themselves and liver. Breast cancer mouse model and natural killer/monocyte co-culture system	Kelly et al. (2021)
	Paclitaxel- and cetuximab	Drug conjugated nanodiamond nanocomposite	Addressing the EGFR of TNBC in tumor cell types of MDA MB-231, MCF-7, and BT474 enhanced mitotic catastrophe and death	Liao et al. (2019)
	Benzoquinone	NDs functionalized with benzoquinone	Induction of cell death in MDA MB-231 and MCF-7 varieties of cancer cells	Shirley et al. (2021)
	miR-34a	Nanodiamond-based layer-by-layer nanohybrids	MiR-34a administration significantly reduced TNBC cell movement, expansion, and mortality in culture. It also prevented cancer development in vivo by decreasing Fra-1 expression in TNBC cells and xenograft TNBC malignancies	Xia et al. (2018)
Carbon nanotubes	Doxorubicin (DOX)	Hyaluronic acid modified amino single-walled carbon nanotubes (NH2-SWCNTs	Preventing cells with cancer from proliferating and triggering their death MDA-MB-231	Liu et al. (2019a, b)
	Cisplatin	Loaded multi-walled carbon nanotubes	Remarkable reduction in caspase-3 and p53 expression following a 48-h exposure, concomitant with reduction of NF-κB in cancer cells MDA-MB-231 treated with MWCNT COOH-CDDP	Badea et al. (2018)
	Doxorubicin (DOX)	Glycopolymer decorated multiwalled carbon nanotubes	The glycopolymers enhance DOX's trafficking into MCF-7 and MDA-MB-231 tumor cells, enhancing the drug's chemotherapy actions	Ozgen et al. (2020)
	–	SWCNTs and MWCNTs	CNTs reduced the overall size of the tumors. In rodents and cancerous cells of MC4L2, the BCL2 expression decreased and BAX and caspase-3 were increased in the untreated cohorts	Kavosi et al. (2018)
	Platinum nanoparticles (Pt-NPs)	Polybenzimidazole (PBI)-functionalized MWCNT (MWCNT/ PBI/Pt-NPs)	Decline in bone marrow mesenchymal cancer stem cells' (CSCs') growth and multiplication compared to that of CSCs	Luo et al. (2021)
	Doxorubicin	MWCNTs) functionalized using hyaluronic acid (HA) and α-Tocopheryl succinate (α-TOS)	Comparing with other compositions the MDA-MB-231 cells treated with α-TOS-HA-MWCNTs/DOX showed a significant total apoptosis probability as well as development suppression impact	Singhai et al. (2020)
	Ginsenosides Rb3	Carbon nanotubes (CNT)	In MDA-MB-231 and BT-549 triple negative breast cancer lineages, Rb3 inhibits the PD-1/PD-L1 mechanism	Berber et al. (2020)

6.12 Summary and Future Perspectives

Breast cancer ranks as one of the most prevalent non-cutaneous cancers identified worldwide amongst women. With the advances in genetic assessment making treatment plans more tailored to the individual patient than ever before, nowadays, scientific community has laid their attention towards the design of leading innovative therapies/techniques or therapeutic drug regimens which generally orbits around the application of smart nanocarriers owing to their vivacious facets compared to conventional chemotherapeutic regimens employed for managing cancer-related therapies in oncological situations. Undoubtedly, as far as the diagnostic screening recommendations are concerned, most of the oncologists believe that there is no "one-size-fits-all" treatment or screening option, hence, a comprehensive framework for the screening, diagnosis, pathological evaluation to therapeutic response for the management of breast cancer becomes the need of the hour. Further, breast tumors possess biological characteristics (distinct tumor micro-environment comprising Compromised vasculature, atypical matrix outside of cells and elevated interstitial fluid pressure) which are divergent from normal cells/tissues which must be overcame by the adoption of nanocarrier driven nanotheranostic approaches.

Nanotheranostic systems require a strong congeniality within the nanocarrier core, the targeting molecule, and the imaging agent, and the therapeutic agent. These nanotherapeutic carrier systems must be optimized employing impeccable formulation strategies which should offer excellent reproducibility, diagnosis along with therapeutic efficacy. It is needless to argue that only a few unique tailored nanocarriers have been approved by US FDA as yet, despite being myriad nanocarriers are still being investigated clinically for breast cancer applications. However, nanotheranostic systems seems to be promising multimodal strategy for standard screening and treatments, for mitigation of breast cancer which enable simultaneous characterization of the disease and drugs employed, along with understanding the host-disease interaction compared to conventional cancer detection procedures and treatment regimens. Novel imaging procedures and diagnostic modalities such as light scans, scintillography, tomography (PET or SPECT), magnetic resonance (MRI), ultrasound, and others have been used for theranostic purposes to maximize the efficacy, clinical translation, and long-time stability of these tools in the nanocarrier systems for breast cancer. In the light of above said facts, it can be inferred that, nanotheranostics applications are presently at the nascent stages and still warrants substantial advancement to successfully translate the preclinical and clinical outcomes in breast cancer therapy using nanotheranostic systems.

References

Abdull R, Noor N (2013) Cruciferous vegetables: dietary phytochemicals for cancer prevention. Asian Pac J Cancer Prev 14:1565–1570

Abolzadeh V, Imanparast A, Nassirli H, Tayebi Meybodi N, Khalili Najafabad B, Sazgarnia A (2023) In vivo evaluation of Sono-chemo therapy via hollow gold nanoshells conjugated to mitoxantrone on breast cancer. Iran J Basic Med Sci 26(3):285–294

Agardan NB, Degim Z, Yilmaz S, AltintaS L, Topal T (2016) The effectiveness of raloxifene loaded liposomes and cochleates in breast cancer therapy. AAPS Pharm SciTech 17(4):968–977

Ahn HK, Jung M, Sym SJ, Shin DB, Kang SM, Kyung SY, Park JW, Jeong SH, Cho EK (2014) A phase II trial of Cremorphor EL-free paclitaxel and gemcitabine in patients with advanced non-small cell lung cancer. Cancer Chemother Pharmacol 74(2):277–282

Alexander P, Jainamboo M, Praseetha PK, Gopukumar ST (2016) Silica coated liposomes for drug delivery towards breast cancer cells. Rasayan J Chem 9:300–308

Alkabban FM, Ferguson T (2022) Breast cancer. In: Stat Pearls. StatPearls Publishing, Treasure Island, FL

Amioka A, Masumoto N, Gouda N, Kajitani K, Shigematsu H, Emi A, Kadoya T, Okada M (2016) Ability of contrast-enhanced ultrasonography to determine clinical responses of breast cancer to neoadjuvant chemotherapy. Jpn J Clin Oncol 46:303–309

Anderson BO, Yip CH, Smith RA et al (2008) Guideline implementation for breast healthcare in low-income and middle-income countries: overview of the breast health global initiative global summit 2007. Cancer 113(8):2221–2243

Badea MA, Prodana M, Dinischiotu A, Crihana C, Ionita D, Balas M (2018) Cisplatin loaded multi-walled carbon nanotubes induce resistance in triple negative breast cancer cells. Pharmaceutics 10(4):228

Batist G, Ramakrishnan G, Rao CS, Chandrasekharan A, Gutheil J, Guthrie T, Shah P, Khojasteh A, Nair MK, Hoelzer K, Tkaczuk K, Park YC, Lee LW (2001) Reduced cardiotoxicity and preserved antitumor efficacy of liposome-encapsulated doxorubicin and cyclophosphamide compared with conventional doxorubicin and cyclophosphamide in a randomized, multicenter trial of metastatic breast cancer. J Clin Oncol 19(5):1444–1454

Bennett NC, Farah CS (2014) Next-generation sequencing in clinical oncology: next steps towards clinical validation. Cancers Basel 6:2296–2312

Berber MR, Elkhenany H, Hafez IH, El-Badawy A, Essawy M, El-Badri N (2020) Efficient tailoring of platinum nanoparticles supported on multiwalled carbon nanotubes for cancer therapy. Nanomedicine 15(08):793–808

Bollet MA, Racadot S, Rivera S, Arnaud A, Bourgier C (2023) Radiotherapie des cancers du sein : questions dactualite 2023 [breast cancer radiation therapy: current questions in 2023]. Cancer Radiother 27(6–7):524–530

Breastcancer.org (2023) American Cancer Society: Breast Cancer Facts and Figures 2022–2024, Cancer Facts and Figures https://www.breastcancer.org/facts-statistics

Brinkman AM, Chen G, Wang Y, Hedman CJ, Sherer NM, Havighurst TC, Gong S, Xu W (2016) Aminoflavone-loaded EGFR-targeted unimolecular micelle nanoparticles exhibit anti-cancer effects in triple negative breast cancer. Biomaterials 101:20–31

Bush N, Healey A, Shah A, Box G, Kirkin V, Eccles S, Sontum PC, Kotopoulis S, Kvale S, van Wamel A, Davies CL, Bamber J (2020) Theranostic attributes of acoustic cluster therapy and its use for enhancing the effectiveness of liposomal doxorubicin treatment of human triple negative breast cancer in mice. Front Pharmacol 11:75

Cardoso F, Kyriakides S, Ohno S, Penault-Lorca F, Poortmans P, Rubio I, Zackrisson S, Senkus E (2019) Early breast cancer: ESMO clinical practice guidelines for diagnosis, treatment and follow-up. Ann Oncol 30:1194–1220

Chaurawal N, Misra C, Abul Barkat H, Jatyan R, Chitkara D, Barkat MA, Sharma T, Singh B, Raza K (2022) Oral sorafenib-loaded microemulsion for breast cancer: evidences from the in-vitro evaluations and pharmacokinetic studies. Sci Rep 12(1):13746. https://doi.org/10.1038/s41598-022-17333-6

Chen DS, Hurwitz H (2018) Combinations of bevacizumab with cancer immunotherapy. Cancer J 24:193–204

Chen SH, Liu TI, Chuang CL, Chen HH, Chiang WH, Chiu HC (2020) Alendronate/folic acid-decorated polymeric nanoparticles for hierarchically targetable chemotherapy against bone metastatic breast cancer. J Mater Chem B 8:3789–3800

Chuba PJ, Hamre MR, Yap J, Severson RK, Lucas D, Shamsa F, Aref A (2005) Bilateral risk for subsequent breast cancer after lobular carcinoma in situ: analysis of surveillance, epidemiology, and end results data. J Clin Oncol 23(24):5534–5541

Clarke M, Collins R, Darby S, Davies C, Elphinstone P, Evans V, Godwin J, Gray R, Hicks C, James S, MacKinnon E, McGale P, McHugh T, Peto R, Taylor C, Wang Y (2005) Effects of radiotherapy and of differences in the extent of surgery for early breast cancer on local recurrence and 15-year survival: an overview of the randomised trials. Lancet 366(9503):2087–2106

Classe JM, Bordes V, Campion L et al (2009) Sentinel lymph node biopsy after neoadjuvant chemotherapy for advanced breast cancer: results of ganglion Sentinelle et Chimiotherapie Neoadjuvante, a French prospective multicentric study. J Clin Oncol 27(5):726–732

Coleman MP, Quaresma M, Berrino F, Lutz JM, De Angelis R, Capocaccia R, Baili P, Rachet B, Gatta G, Hakulinen T, Micheli A, Sant M, Weir HK, Elwood JM, Tsukuma H, Koifman S, Silva GA, Francisci S, Santaquilani M, Verdecchia A, Storm HH, Young JL (2008) Cancer survival in five continents: a worldwide population-based study. Lancet Oncol 9(8):730–756

Cortes MV, Marchal SA, Blas MP, Mochon JJ, Marchal JA, Martin RM (2020) A versatile theranostic nanodevice based on an orthogonal bioconjugation strategy for efficient targeted treatment and monitoring of triple negative breast cancer. Nanomed Nanotechnol Biol Med 24:102120

Cui Y, Yang Y, Ma M, Xu Y, Sui J, Li H, Liang J, Sun Y, Fan Y, Zhang X (2019) Reductive responsive micelle overcoming multidrug resistance of breast cancer by co-delivery of DOX and specific antibiotic. J Mater Chem B 7:6075–6086

Dadgar N, Esfahani MMK, Torabi S, Alavi SE, Akbarzadeh A (2014) Effects of nanoliposomal and pegylated nanoliposomal artemisinin in treatment of breast cancer. Indian J Clin Biochem 29(4):501–504

Delaney LJ, Ciraku L, Oeffinger BE, Wessner CE, Liu JB, Li J, Nam K, Forsberg F, Leeper DB, O'Kane P, Wheatley MA, Reginato MJ, Eisenbrey JR (2019) Breast cancer brain metastasis response to radiation after microbubble oxygen delivery in a murine model. J Ultrasound Med 38(12):3221–3228

Draganescu M, Carmocan C (2017) Hormone therapy in breast cancer. Chirurgia 112:413–417

Eisenbrey JR, Shraim R, Liu JB, Li J, Stanczak M, Oeffinger B, Leeper DB, Keith SW, Jablonowski LJ, Forsberg F (2018) Sensitization of hypoxic tumors to radiation therapy using ultrasound-sensitive oxygen microbubbles. Int J Radiat Oncol Biol Phys 101:88–96

Friberg S, Nystrom A (2015) Cancer metastases: early dissemination and late recurrences. Cancer Growth Metastasis 8:43–49. https://doi.org/10.4137/CGM.S31244. PMID: 26640389; PMCID: PMC4664198

Ghafari M, Haghiralsadat F, Pour SKF, Zavar Reza J (2020) Development of a novel liposomal nanoparticle formulation of cisplatin to breast cancer therapy. J Cell Biochem 121(7):3584–3592

Godavarty A, Rodriguez S, Jung YJ, Gonzalez S (2015) Optical imaging for breast cancer prescreening. Breast Cancer 7:193–209

Goel H, Sinha VR (2016) P-204 Toxicological and biochemical assessment of specific COX-2 inhibitor as adjuvant in fluoropyrimidine therapy in chemically induced colon cancer. Ann Oncol 27:ii59

Goel H, Tikoo K, Karpe PA, Singla R, Sinha VR (2016) Ameliorated chemotherapeutic potential of diverse dose flouropyrimidine therapy by Etodolac via NF-KB pathway, PPAR-γ expression and COX-II inhibition in DMH induced colon. Clin Cancer Drugs 3(2):109–120

Greenwalt I, Zaza N, Das S, Li BD (2020) Precision medicine and targeted therapies in breast cancer. Surg Oncol Clin N Am 29:51–62

He Y, Zhang L, Song C, Zhu D (2014) Design of multifunctional magnetic iron oxide nanoparticles/mitoxantrone loaded liposomes for both magnetic resonance imaging and targeted cancer therapy. Int J Nanomedicine 9:4055–4066

Heimes AS, Schmidt M (2018) Atezolizumab for the treatment of triple-negative breast cancer. Exp Opin Investig Drugs 28:1–5

Hyseni FK, Beijnen JH, Schellens JH (2010) Bevacizumab. Oncologia 15:819–825

Infante M, Arranz-Ledo M, Lastra E, Olaverri A, Ferreira R, Orozco M, Hernández L, Martínez N, Durán M (2024) Profiling of the genetic features of patients with breast, ovarian, colorectal and extracolonic cancers: association to CHEK2 and PALB2 germline mutations. Clin Chim Acta 552:117695

Ishii K, Morii N, Yamashiro H (2019) Pertuzumab in the treatment of HER_2 positive breast cancer: an evidence-based review of its safety, efficacy, and place in therapy. Core Evid 14:51–70

Jablonowski LJ, Conover D, Teraphongphom NT, Wheatley MA (2018) Manipulating multifaceted microbubble Shell composition to target both TRAIL-sensitive and resistant cells. J Biomed Mater Res 106:1903–1915

Ji W, Wang B, Fan Q, Xu C, He Y, Chen Y (2017) Chemosensitizing indomethacin-conjugated dextran-based micelles for effective delivery of paclitaxel in resistant breast cancer therapy. PLoS One 12:0180037

Jones RH, Casbard A, Carucci M, Cox C, Butler R, Alchami F, Madden TA, Bale C, Bezecny P, Joffe J, Moon S, Twelves C, Venkitaraman R, Waters S, Foxley A, Howell SJ (2020) Fulvestrant plus capivasertib versus placebo after relapse or progression on an aromatase inhibitor in metastatic, oestrogen receptor-positive breast cancer: a multicentre, randomised, controlled, phase 2 trial. Lancet Oncol 21(3):345–357

Kato K, Chin K, Yoshikawa T, Yamaguchi K, Tsuji Y, Esaki T, Sakai K, Kimura M, Hamaguchi T, Shimada Y, Matsumura Y, Ikeda R (2012) Phase II study of NK105, a paclitaxel-incorporating micellar nanoparticle, for previously treated advanced or recurrent gastric cancer. Investig New Drugs 30(4):1621–1627

Kavosi A, Hosseini Ghale Noei S, Madani S, Khalighfard S, Khodayari S, Khodayari H, Mirzaei M, Kalhori MR, Yavarian M, Alizadeh AM, Falahati M (2018) The toxicity and therapeutic effects of single and multiwall carbon nanotubes on mice breast cancer. Sci Rep 8(1):8375

Kelly LPS, Sun SH, Ren C, Rampersaud IV, Albertson D, Duggan MC, Noel TC, Courtney N, Buteyn NJ, Moritz C, Yu L, Yildiz VO, Butchar JP, Tridandapani S, Rampersaud AA, Carson WE (2021) Antibody conjugation of fluorescent Nanodiamonds for targeted innate immune cell activation. ACS Appl Nano Mater 4(3):3122–3139

Kesharwani SS, Dachineni R, Bhat GJ, Tummala H (2019) Hydrophobically modified inulin-based micelles: transport mechanisms and drug delivery applications for breast cancer. J Drug Deliv Sci Technol 54:101254

Ko NR, Nafiujjaman M, Lee JS, Lim HN, Lee YK, Kwon IK (2017) Graphene quantum dot based theranostic agents for active targeting of breast cancer. RSC Adv 7:11420–11427

Kwapisz D (2017) Cyclin-dependent kinase 4/6 inhibitors in breast cancer: Palbociclib, ribociclib, and abemaciclib. Breast Cancer Res Treat 166:41–54

Lahart IM, Metsios GS, Nevill AM, Carmichael AR (2015) Physical activity, risk of death and recurrence in breast cancer survivors: a systematic review and meta-analysis of epidemiological studies. Acta Oncol 54(5):635–654

Li M, Tang Z, Zhang Y, Lv S, Li Q, Chen X (2015) Targeted delivery of cisplatin by LHRH-peptide conjugated dextran nanoparticles suppresses breast cancer growth and metastasis. Acta Biomater 18:132–143

Li B, Qi F, Zhu F, Lu Z, Wang M, Chu T, Wu S, Wei J, Song Z, Sukumar S, Zhang C, Xu J, Li S, Nie G (2023) Nanoparticle-based combination therapy enhances Fulvestrant efficacy and overcomes tumor resistance in ER-positive breast cancer. Cancer Res 83(17):2924–2937

Liao WS, Ho Y, Lin YW, Raj EN, Liu KK, Chen C, Zhou XZ, Lu KP, Chao JI (2019) Targeting EGFR of triple-negative breast cancer enhances the therapeutic efficacy of paclitaxel and cetuximab conjugated nanodiamond nanocomposite. Acta Biomater 86:395–405

Liu Y, Nguyen N, Colditz GA (2015) Links between alcohol consumption and breast cancer: a look at the evidence. Womens Health 11(1):65–77

Liu D, Zhang Q, Wang J, Fan L, Zhu W, Cai D (2019a) Hyaluronic acid-coated single walled carbon nanotubes loaded with doxorubicin for the treatment of breast cancer. Int J Pharm Sci Rev Res 74(2):83–90

Liu R, Hu C, Yang Y, Zhang J, Gao H (2019b) Theranostic nanoparticles with tumor-specific enzyme-triggered size reduction and drug release to perform photothermal therapy for breast cancer treatment. Acta Pharm Sin B 9:410–420

Liu J, Chen Z, Li Y, Zhao W, Wu J, Zhang Z (2021) PD-1/PD-L1 checkpoint inhibitors in tumor immunotherapy. Front Pharmacol 12:731798

Loibl S, Poortmans P, Morrow M, Denkert C, Curigliano G (2021) Breast cancer. Lancet 397:1750–1769

Luo X, Wang H, Ji D (2021) Carbon nanotubes loaded ginsenosides Rb3 suppress the PD-1/PD-L1 pathway in triple-negative breast cancer. Aging 13(13):17177

Ma M, Zhang X, Hao Y, Liu N, Yin Z, Wang L (2012) A novel lipid based nanomicelle of docetaxel: evaluation of antitumor activity and biodistribution. Int J Nanomedicine 7:3389–3398

Ma M, Lei M, Tan X, Tan F, Li N (2016) Theranostic liposomes containing conjugated polymer dots and doxorubicin for bio imaging and targeted therapeutic delivery. RSC Adv 6:1945–1957

Maji R, Dey NS, Satapathy BS, Mukherjee B, Mondal S (2014) Preparation and characterization of tamoxifen citrate loaded nanoparticles for breast cancer therapy. Int J Nanomedicine 25(9):3107–3118

Manaspon C, Pasuwat KV, Pimpha N (2012) Preparation of folate-conjugated Pluronic F127/chitosan Core-Shell nanoparticles encapsulating doxorubicin for breast cancer treatment. J Nanomat 2012:593878

Martorana F, Motta G, Pavone G, Motta L, Stella S, Vitale SR, Manzella L, Vigneri P (2021) AKT inhibitors: new weapons in the fight against breast cancer. Front Pharmacol 12:662232

Maximiano S, Magalhaes P, Guerreiro MP, Morgado M (2016) Trastuzumab in the treatment of breast cancer. BioDrugs 30:75–86

McCann KE, Goldfarb SB, Traina TA, Regan MM, Vidula N, Kaklamani V (2024) Selection of appropriate biomarkers to monitor effectiveness of ovarian function suppression in premenopausal patients with ER+ breast cancer. NPJ. Breast Cancer 10(1):8

Mehnati P, Tirtash MJ (2015) Comparative efficacy of four imaging instruments for breast cancer screening. Asian Pac J Cancer Prev 16(15):6177–6186

Miele E, Spinelli GP, Miele E, Tomao F, Tomao S (2009) Albumin-bound formulation of paclitaxel (Abraxane ABI-007) in the treatment of breast cancer. Int J Nanomedicine 4:99–105

Mukama T, Fallah M, Brenner H, Xing X, Sundquist K, Sundquist J, Kharazmi E (2020) Risk of invasive breast cancer in relatives of patients with breast carcinoma in situ: a prospective cohort study. BMC Med 18(1):295

Murthy RK, Loi S, Okines A, Paplomata E, Hamilton E, Hurvitz SA, Lin NU, Borges V, Abramson V, Anders C, Bedard PL, Oliveira M, Jakobsen E, Bachelot T, Shachar SS, Müller V, Braga S, Duhoux FP, Greil R, Cameron D, Carey LA, Curigliano G, Gelmon K, Hortobagyi G, Krop I, Loibl S, Pegram M, Slamon D, Wessels MCP, Walker L, Feng W, Winer EP (2020) Tucatinib, Trastuzumab, and Capecitabine for HER_2 -positive metastatic breast cancer. N Engl J Med 382(7):597–609

Nguyen X, Hooper M, Borlagdan JP, Palumbo A (2021) A review of fam-Trastuzumab Deruxtecan-nxki in HER_2 positive breast cancer. Ann Pharmacother 55(11):1410–1418

Nicolini A, Carpi A, Ferrari P, Biava PM, Rossi G (2016) Immunotherapy and hormone-therapy in metastatic breast cancer: a review and an update. Curr Drug Targets 17:1127–1139

Nie Z, Luo N, Liu J, Zeng X, Zhang Y, Su D (2020) Multi-mode biodegradable tumour microenvironment sensitive nanoparticles for targeted breast cancer imaging. Nanoscale Res Lett 15(1):81

Norat T, Scoccianti C, Boutron-Ruault MC, Anderson A, Berrino F, Cecchini M, Espina C, Key T, Leitzmann M, Powers H, Wiseman M, Romieu I (2015) European code against cancer 4th edition: diet and cancer. Cancer Epidemiol 1:S56–S66

Odeh F, Ismail SI, Dahab RA, Mahmoud IS, Bawab A (2012) Thymoquinone in liposomes: a study of loading efficiency and biological activity towards breast cancer. Drug Deliv 19(8):371–377

Osborne CK (1998) Tamoxifen in the treatment of breast cancer. N Engl J Med 339:1609–1618
Ozgen PSO, Atasoy S, Kurt BZ, Durmus Z, Yigit G, Dag A (2020) Glycopolymer decorated multiwalled carbon nanotubes for dual targeted breast cancer therapy. J Mater Chem B 8(15):3123–3137
Patel HK, Bihani T (2018) Selective estrogen receptor modulators and selective estrogen receptor degraders in cancer treatment. Pharmacol Ther 186:1–24
Pawar A, Singh S, Rajalakshmi S, Shaikh K, Bothiraja C (2018) Development of fisetin loaded folate functionalized pluronic micelles for breast cancer targeting. Artif Cells Nanomed Biotechnol 46:347–361
Peng J, Sengupta S, Jordan VC (2009) Potential of selective estrogen receptor modulators as treatments and preventives of breast cancer. Anti Cancer Agents Med Chem 9:481–499
Porta C, Paglino C, Mosca A (2014) Targeting PI3K/Akt/mTOR signaling in cancer. Front Oncol 4:64
Qiu Y, Ren K, Zhao W, Yu Q, Guo R, He J, Mei L, Liu Y, Tang J, Xu S (2021) A dual-guide bioinspired drug delivery strategy of a macrophage based carrier against postoperative triple negative breast cancer recurrence. J Control Release 329:191–204
Rau KM, Lin YC, Chen YY, Chen JS, Lee KD, Wang CH, Chang HK (2015) Pegylated liposomal doxorubicin combined with cyclophosphamide and 5-fluorouracil is effective and safe as salvage chemotherapy in taxane-treated metastatic breast cancer: an open-label, multi-center, non-comparative phase II study. BMC Cancer 15:423
Reeder J, Vogel V (2008) Breast cancer prevention. Cancer Treat Res 41:149–164
Rizzitelli S, Giustetto P, Cutrin J, Castelli DD, Boffa C, Ruzza M, Menchise V, Molinari F, Aime S, Terreno E (2015) Sonosensitive theranostic liposomes for preclinical in vivo MRI-guided visualization of doxorubicin release stimulated by pulsed low intensity non-focused ultrasound. J Control Release 202:21–30
Rizzitelli S, Giustetto P, Faletto D, Castelli DD, Aime S, Terreno E (2016) The release of doxorubicin from liposomes monitored by MRI and triggered by a combination of US stimuli led to a complete tumor regression in a breast cancer mouse model. J Control Release 230:57–63
Rockville MD (2002) U.S. preventive services task force. Chemoprevention of breast cancer: recommendations and rationale. Agency for Healthcare Research and Quality. http://www.ahrq.gov/clinic/3rduspstf/breastchemo/breastchemorr.Htm. Accessed 8 Feb 2010
Rodriguez-Miguel C, Moral R, Escrich R, Vela E, Solanas M, Escrich E (2015) The role of dietary Extra virgin olive oil and corn oil on the alteration of epigenetic patterns in the rat DMBA-induced breast cancer model. PLoS One 10(9):1–16
Sabatier R, Jauffret EC, Pierga JY, Cure H, Lambaudie E, Genre D, Houvenaeghel G, Viens P, Ginestier C, Bertucci F, Sfumato P, Extra JM, Goncalves A (2019) Stem cells inhibition by bevacizumab in combination with neoadjuvant chemotherapy for breast cancer. J Clin Med 8(5):612
Schneble E, Jinga DC, Peoples G (2015) Breast cancer immunotherapy. Maedica 10:185–191
Schwarz EB, Hoyt-Austin A, Fix M, Kair LR, Iwuagwu C, Chen MJ (2024) Prenatal counseling on the maternal health benefits of lactation: a randomized trial. Breastfeed Med 19(1):52–58
Shah AN, Gradishar WJ (2018) Adjuvant anthracyclines in breast cancer: what is their role. Oncologist 23:1153–1161
Sheng D, Deng L, Li P, Wang Z, Zhang Q (2021) Perfluorocarbon Nanodroplets with deep tumor penetration and controlled drug delivery for ultrasound/fluorescence imaging guided breast cancer therapy. ACS Biomater Sci Eng 7(2):605–616
Shirley AJ, Schweeberg S, Waag T, Peindl M, Dandekar G, Walles H, Jakob F, Kruegger A, Ebert R (2021) The influence of differently functionalized nanodiamonds on proliferation, apoptosis and EMT/MET phenomena in 2D and 3D tumor cell cultures. J Mater Chem B 9(45):9395–9405
Siegel R, Naishadham D, Jemal A (2013) Cancer statistics. CA Cancer J Clin 63:11–30
Siegel RL, Miller KD, Fuchs HE, Jemal A (2021) Cancer statistics. CA Cancer J Clin 71(1):7–33
Singhai NJ, Maheshwari R, Ramteke S (2020) CD44 receptor targeted smart multi walled carbon nanotubes for synergistic therapy of triple negative breast cancer. Colloids Interface Sci Commun 35:100235

Sinha VR, Honey (2012a) Critical aspects in rationale design of fluorouracil-based adjuvant therapies for the management of colon cancer. Crit Rev Ther Drug Carrier Syst 29(2):89–148

Sinha VR, Honey (2012b) Multidimensional role of inulin as synbiotic, drug carrier and synergistic adjuvant for the colonic delivery of 5-FU. Drug Deliv Lett 2(1):60–69

Slingerland M, Guchelaar HJ, Rosing H, Scheulen ME, van Warmerdam LJC, Beijnen JH, Gelderblom H (2013) Bioequivalence of liposome-entrapped paclitaxel easy-to-use (LEP-ETU) formulation and paclitaxel in Polyethoxylated Castor oil: a randomized, two-period crossover study in patients with advanced cancer. Clin Ther 35:1946–1954

Sobczak KD, Wroblewska HP, Klimonda Z, Karwat P, Purska KR, Clauser P, Baltzer PAT, Litniewski J (2021) Multiparametric ultrasound examination for response assessment in breast cancer patients undergoing neoadjuvant therapy. Sci Rep 11:2501

Sood A, Lang DK, Kaur R, Saini B, Arora S (2021) Relevance of aromatase inhibitors in breast cancer treatment. Curr Top Med Chem 21:1319–1336

Soyemi TO, Ayandipo OO, Ademola AF, Obajimi GO, Obajimi MO, Ogundiran TO (2018) Role of breast ultrasound in evaluating the response of locally advanced breast cancer to neoadjuvant anthracycline chemotherapy in Ibadan. Glob Surg 4:5–56

Starek-Świechowicz B, Budziszewska B, Starek A (2023) Alcohol and breast cancer. Pharmacol Rep 75(1):69–84

Stathopoulos GP, Boulikas T (2011) Lipoplatin formulation review article. J Drug Deliv 2012:581363

Sun JY, Zhang D, Wu S, Xu M, Zhou X, Lu XJ, Ji J (2020) Resistance to PD-1/PD-L1 blockade cancer immunotherapy: mechanisms, predictive factors, and future perspectives. Biomark Res 8:35

Taran Z, Yektaniroumand Digehsaraei S, Salouti M, Amini B, Mahmazi S, Kalantari M (2023) Methotrexate loaded in alginate beads for controlled drug release against breast cancer. Gene 30(851):146941

Tarantino P, Morganti S, Curigliano G (2020) Biologic therapy for advanced breast cancer: recent advances and future directions. Exp Opin Biol Ther 20:1009–1024

Thavendiranathan P, Shalmon T, Fan CS, Houbois C, Amir E, Thevakumaran Y, Somerset E, Malowany JM, Urzua-Fresno C, Yip P, McIntosh C, Sussman MS, Brezden-Masley C, Yan AT, Koch CA, Spiller N, Abdel-Qadir H, Power C, Hanneman K, Wintersperger BJ (2023) Comprehensive cardiovascular magnetic resonance tissue characterization and cardiotoxicity in women with breast cancer. JAMA Cardiol 8(6):524–534

Tian Y, Carrillo-Malani N, Feng K, Miller J, Busch TM, Sundaram KM, Cheng Z, Amirshaghaghi A, Tsourkas A (2024) Theranostic Phthalocyanine and Naphthalocyanine nanoparticles for photoacoustic imaging and Photothermal therapy of tumors. Nano 8(1):100–111

Ueno Y, Sonoda S, Suzuki R, Yokouchi M, Kawasoe Y, Tachibana K, Maruyama K, Sakamoto T, Komiya S (2011) Combination of ultrasound and bubble liposome enhance the effect of doxorubicin and inhibit murine osteosarcoma growth. Cancer Biol Ther 12(4):270–277

Van der Kolk DM, de Bock GH, Leegte BK, Schaapveld M, Mourits MJ, de Vries J, van der Hout AH, Oosterwijk JC (2010) Penetrance of breast cancer, ovarian cancer and contralateral breast cancer in BRCA1 and BRCA2 families: high cancer incidence at older age. Breast Cancer Res Treat 124(3):643–651

Wang Y, Wang Y, Chen G, Li Y, Xu W, Gong S (2017) Quantum-dot-based Theranostic micelles conjugated with an anti-EGFR Nanobody for triple-negative breast cancer therapy. ACS Appl Mater Interf 9:30297–30305

Wang J, De G, Yue Q, Ma H, Cheng J, Zhu G, Du M, Yi H, Zhao Q, Chen Y (2018) PH responsive polymer micelles enhances inhibitory efficacy on metastasis of murine breast cancer cells. Front Pharmacol 9:543

Wang B, Jiang T, Huang M, Wang J, Chu Y, Zhong L, Zheng S (2019) Evaluation of the response of breast cancer patients to neoadjuvant chemotherapy by combined contrast-enhanced ultrasonography and ultrasound Elastography. Exp Ther Med 17:3655–3663

Wang T, Shang M, Chow KM (2023) Effects of breastfeeding training programmes for midwives on breastfeeding outcomes: a systematic review and meta-analysis. BMC Pregnancy Childbirth 23(1):262

Weigel S, Heindel W, Heidrich J, Heidinger O, Hense HW (2015) Reduction of advanced breast cancer stages at subsequent participation in mammography. Screening 188(1):33–37

Willson ML, Burke L, Ferguson T, Ghersi D, Nowak AK, Wilcken N (2019) Taxanes for adjuvant treatment of early breast cancer. Cochrane Database Syst Rev 9:CD004421

World Health Organization (n.d.). https://www.who.int/news-room/fact-sheets/detail/breast-cancer. Accessed 23 Jan 2024

Xia Y, Deng X, Cao M, Liu S, Zhang X, Xiao X, Shen S, Hu Q, Sheng W (2018) Nanodiamond-based layer-by-layer nanohybrids mediate targeted delivery of miR-34a for triple negative breast cancer therapy. RSC Adv 8(25):13789–13797

Yallapu MM, Othman SF, Curtis ET, Bauer NA, Chauhan N, Kumar D, Jaggi M, Chauhan SC (2012) Curcumin loaded magnetic nanoparticles for breast cancer therapeutics and imaging applications. Int J Nanomedicine 7:1761–1779

Yang TJ, Ho AY (2013) Radiation therapy in the Management of Breast Cancer. Surg Clin N Am 93:455–471

Yang PJ, Hou MF, Yang F et al (2022) Association of early onset breast cancer with body mass index, menarche, and menopause in Taiwan. BMC Cancer 22(1):259

Yu H, Cui Z, Yu P, Guo C, Feng B, Jiang T, Wang S, Yin Q, Zhong D, Yang X (2015) PH- and NIR light-responsive micelles with hyperthermia-triggered tumor penetration and cytoplasm drug release to reverse doxorubicin resistance in breast cancer. Adv Funct Mater 25:2489–2500

Zhang C, Xu C, Gao X, Yao Q (2022) Platinum-based drugs for cancer therapy and anti-tumor strategies. Theranostic 12:2115–2132

Zhao YN, Cao YN, Sun J, Wu Q, Cui SH, Zhi DF, Zhang SB, Liang Z, Zhen YH, Guo ST (2019) Anti-breast cancer activity of resveratrol encapsulated in liposomes. J Mater Chem B 8:27–37

Zheng YR, Suntharalingam K, Johnstone TC, Yoo H, Lin W, Brooks JG, Lippard SJ (2014) Prodrugs designed to bind non-covalently to human serum albumin for drug delivery. J Am Chem Soc 136(24):8790–8798

Zhong Y, Zhang J, Cheng R, Deng C, Meng F, Xie F, Zhong Z (2015) Reversibly crosslinked hyaluronic acid nanoparticles for active targeting and intelligent delivery of doxorubicin to drug resistant CD44+ human breast tumor xenografts. J Cont Rel 205:144–154

Nanoscale Theranostics for Hepatocellular Carcinoma

7

Mayuresh Ramesh Redkar
and Manojkumar Mukundrao Nitalikar

Abstract

Hepatocellular carcinoma (HCC) has a major influence on individuals throughout the world. The use of nanomaterials in the diagnosis and treatment of HCC demonstrated a promising future. Because of their enormous surface area and compact size, nanomaterials with unique features provide a limitless number of design options in the rapidly increasing field of nanotechnology. Scientists have used nanotechnology to create drug delivery systems that may be used for diagnosis, therapy, and as a tool for imaging procedures. This paper offers an overview of the broad idea and prospects of cancer nanotechnology; it focuses specifically on HCC targeting through therapeutic nanomaterials. Challenges in the development of nanotheranostics for HCC treatment will be highlighted as well.

Keywords

Nanocarriers · Nanoparticles · Hepatocellular cancer · Nanotheranostics · Nanomedicines · Nanodrug delivery

M. R. Redkar (✉)
Department of Pharmaceutics, Yashwantrao Bhonsale College of Pharmacy Sawantwadi, Sindhudurg, Maharashtra, India

Department of Pharmaceutics, Rajarambapu College of Pharmacy Kasegaon, Sangli, Maharashtra, India

M. M. Nitalikar
Department of Pharmaceutics, Rajarambapu College of Pharmacy Kasegaon, Sangli, Maharashtra, India

M. A. Barkat et al. (eds.), *Nanotheranostics for Diagnosis and Therapy*,
https://doi.org/10.1007/978-981-97-3115-2_7

7.1 Introduction

Hepatocellular carcinoma is the most common primary liver tumor and the second highly observed cause of cancer death worldwide. Roughly 80% of cirrhosis cases linked to chronic viral hepatitis B or C are the primary cause of HCC. The most frequent type of primary liver cancer worldwide, HCC is the fifth most widespread cancer disease having a death rate of about 54% (Sangiovanni et al. 2004) to 70% (Benvegnu 2004) in those suffering from compensatory cirrhosis (Wong et al. 2018). HCC accounts for about 90% of all cases of primary liver cancer. Although liver transplantation and surgical resection are thought to be the curative treatments for long-term HCC control, most HCC patients are discovered at advanced stages that are not amenable to normal surgical care. Its high fatality rate results in more than 600,000 deaths globally each year. Owing to HCC's stealthy growth pattern, most patients receive a diagnosis of the disease at an advanced stage, when few and inefficient treatment choices remain (Dhanasekaran et al. 2012; Llovet et al. 2003). Due to its poor prognosis and the fact that liver resection is now the most effective treatment, HCC is considered a major global health issue, particularly in recently industrialized nations (Harshita et al. 2019).

Nanomedicine is a new discipline that combines nanoscale experimentation in science, engineering, and technology with life sciences to produce novel outcomes for the healthcare community and society (Harshita et al. 2019; Siafaka et al. 2016). A variety of nanomedicine platforms, but nanotechnology-based medication delivery systems and imaging nano-agents are of particular interest (Siafaka et al. 2015). Nanoscale theranostics entail the use of nanosized theranostics with wide purposes such as site-specific delivery, modified release, increased drug delivery efficiency via endocytosis, stimuli-responsive systems, and the combination of therapeutic approaches such as multimodality diagnosis and therapy (Chen et al. 2020). The term "theranostics" refers to systems that can serve as both medicines and diagnostic agents (Qi et al. 2020). Nanoscale theranostics are techniques that use carriers having nanosize. Polymeric nanoparticles (NPs), liposomes, dendrimers, metal nanocarriers, carbon nanotubes, and nanospheres and systems that include all these categories, i.e., polymeric layered nanocarriers, can all be used to create nanotheranostics. Carbon-based nanomaterials, such as carbon nanotubes alone or coated with other materials, fullerenes, graphene oxide, and carbon quantum dots, have been used for drug detection or imaging in biological samples. Octahedral Prussian blue cubes with conductive and magnetic characteristics that have been created as detecting instruments (Siafaka et al. 2021).

An effective nanotheranostic should circulate in the body for a longer period and possess sufficient release behavior, excellent tissue target penetrability and selectivity, imaging possibility, and an excellent target-to-background ratio. Nanotheranostics can target diagnostic and therapeutic substances to specific disease locations, reducing unwanted side effects. Nanoparticles can readily extravagate from circulation into tumor tissues and become trapped due to inadequate lymphatic drainage (Dasgupta et al. 2020; Filippousi et al. 2013a, b). Nuclear medicine imaging has been studied extensively as a sophisticated tool for the diagnosis of disease, and it

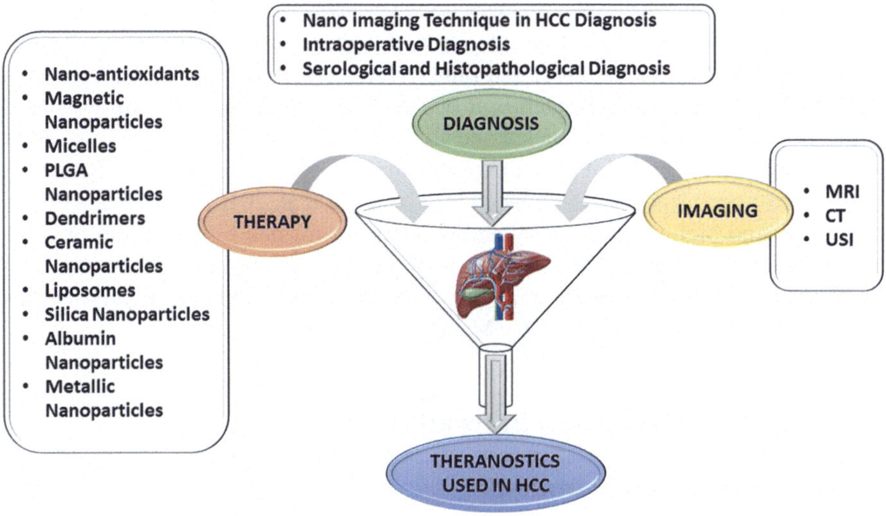

Fig. 7.1 Diagrammatic representation of theranostics used in the treatment of HCC

consists of introducing radioactive elements into the body, detecting gamma rays emitted by the radionuclides, and generating images that provide detailed radionuclide distribution as well as physiological characterization of organs and tissues. Many unique tactics have been used to enhance the pharmacokinetics and biodistribution of these drugs. As diagnostic and clinical diagnostic instruments, nanomaterials exhibit dual behavior. They offer comprehensive radionuclide distribution, gamma-ray detection, and physiological characterization of organs and tissues (Han et al. 2013; Ge et al. 2020). The applications of nanotheranostics or NP-based theranostics have improved rapidly in recent years. Reviewing the existing applications of imaging modalities may potentially result in development in the field of nanomedicine. Nanotheranostics used in the HCC treatment are represented diagrammatically in Fig. 7.1. This article might help medical experts provide greater progress in this subject by summarizing current advances in Nanotechnology.

7.2 Chemicals Inducing HCC

During the experimental investigation, various chemically active carcinogenic agents are used to induce carcinogenesis. HCC-inducing chemical agents including diethyl nitrosamine (DNS), thioacetamide, B1 aflatoxin, carbon tetrachloride, anabolic steroids, and alcohol also contribute to HCC. Hepatitis viruses like hepatitis B and hepatitis C are also involved in the pathogenesis of HCC. Fried meals, hard water, cigar smoke, cheddar, alcoholic brews, cheese, pharmaceutical, and cosmetics formulations contain DNS which is well known carcinogenic agent having a wide potential to induce HCC (Barbara et al. 1992).

In the animal administered with DNS, the liver is the most affected organ, and rat animal is used for experimental investigation of pathogenesis and treatment of HCC because rat metabolizes the DNS and other related chemical moieties as the livers of the human being. Most frequently it has been observed that DNS administered animal shows hepatocarcinogenesis and in very few cases it has been found that other organs were affected. For the experimental investigation rat model is used to study the pathogenesis of DNS-induced HCC (Gupta et al. 2010). DNS metabolized in the liver and highly reactive oxygen moieties are formed by the action of cytochrome isoform 2E1 (CYP 2E1) which may lead to impaired cellular function by exerting the oxidative force (Ha et al. 2001). Alpha-fetoprotein is generally used as a diagnostic tumor marker for HCC (Mandal et al. 2008).

7.3 Current Challenges in Treating HCC

7.3.1 Tumor Microenvironment

The latest research studies have emphasized the relationship between tumor cells and the milieu around them as well as the critical function of the tumor microenvironment in the pathogenicity of HCC (Yang et al. 2011). Hepatic stellate cells, fibroblasts, helper or destructive cells and tumor-associated macrophages (TAMs), endothelial cells, proteolytic enzymes like matrix metalloproteinases (MMPs), and tissue inhibitors of metalloproteinases (TIMPs), growth factors, such as platelet-derived growth factor (PDGF) and transforming growth factor b1 (TGF-1), inflammatory cytokines, and extracellular matrix (ECM) are the constituents of the tumor microenvironment (Friedman et al. 1985; Pietras and Ostman 2010; Benetti et al. 2008; Kessenbrock et al. 2010). Tumor cells consist of poor alignment of endothelial cells, deficiency of smooth muscles, and lymphatic drainage all these factors contribute to impaired fluid and molecular transport this specialized phenomenon is referred to as enhanced permeability and retention (EPR), is in charge of creating holes and leaky channels with a diameter of 100 nm to 2 μm, liposomes, nanoparticles, and therapeutic macromolecules frequently build up in tumor tissue much more than in healthy tissues This benefits the development of new anticancer nanoformulations for tumor targeting in all tumor types (Kessenbrock et al. 2010; Whiteside 2008; Balkwill et al. 2012; Kalyane et al. 2019). Because of membrane hyperpermeability and lack of extracellular matrix in the tumor tissues in comparison to normal tissue blood vessels, the EPR effect increases nanoparticle accumulation at the tumor site, allowing for more precise therapeutic targeting as well as helping to lower the toxicity and improve the biocompatibility of the therapeutic agents (Byrne et al. 2008; Siemann 2011; Ruoslahti 2002). Although the liver vasculature naturally has leaky arteries, anomalies in the capillary structure are common in the presence of chronic liver disorders like cirrhosis (Maeda et al. 2000). To provide an efficient drug delivery system in the context of liver illnesses, several nanosystems with particle size within the vascular structure would need to be designed and developed. Low extracellular pH, which ranges between 6.0 and 7.0

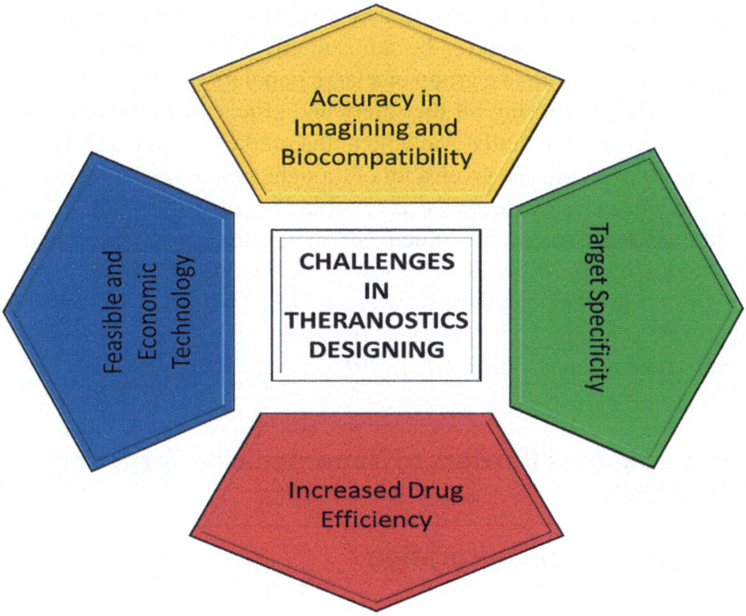

Fig. 7.2 Challenges in theranostics designing

as opposed to normal tissues and blood pH of 7.4, is another feature of the HCC tumor environment. This is because the rate of glycolysis has risen, causing lactic acid to build up in hypoxic cancerous cells (Urano et al. 2009; Gallagher et al. 2008; Nichols and Bae 2012; Trédan et al. 2007). The distribution of drug moiety to hepatocarcinogenic cells is affected due to changes in pH; an acidic pH promotes the cellular absorption of weakly acidic drugs and inhibits the uptake of weakly basic treatment drugs. This factor may also be taken into account while creating nanoparticles for the best possible HCC tumor targeting (Bakrania et al. 2021). Challenges in designing the theranostics are diagrammatically represented in Fig. 7.2.

7.4 The Gap Between Preclinical and Clinical Trials

Nanoformulations used to treat HCC have the major hurdle of regulatory (FDA) approval and its commercialization. So that it becomes the gap between preclinical and clinical studies. There is no known therapeutic nanomaterial for HCC that has been authorized by the US FDA. Before actually moving on to in vitro and in vivo evaluations, the technique is normally required to be both physically and chemically evaluated for drug-loaded NPs to be available commercially. After a protracted process of clinical testing, patients are ultimately given those NPs (Malik 2017).

During the clinical trial studies safety and efficacy of therapeutic agents are evaluated. This study comprises four phases, phase one involves 25–100 subjects who participated, and safety and efficacy are tested whereas phase two consists of

100–500 participants, and phase three involves 500–3000 participants. Once the formulation and or drug is found to be safe and effective in all three phases then it has been experimented with in a group of a large population during phase four of the clinical study. After qualifying all the safety and efficacy parameters of the phase four study product commercially available in the market (Resnik and Tinkle 2007). Pre-clinical trials must thus be given a lot of thought and effort for the lengthy clinical trial procedure to go as smoothly as feasible. Given all of that, the move from pre-clinical to clinical trials is not anticipated to be simple, especially when factors like immunotoxicity are involved. According to reports, from 1969 to 2005, 10% to 20% of FDA-approved medications were taken off the market due to immunotoxicity events, such as anaphylaxis, allergies, hypersensitivity, idiosyncratic reactions, and immunosuppression (Dobrovolskaia 2015; Tingting et al. 2023).

7.5 Physiological Barriers to Nanomedicine Targeting HCC Cells

7.5.1 Coronal Protein Adsorption

NPs surrounded by coupled or adsorbed proteins it is called protein corona. Numerous biological fluids and proteins like transferrin, fibrinogen, albumin, complement C3, haptoglobin,-2-macroglobulin, and immunoglobulins A, M, and G are exposed to them in the bloodstream (Nguyen and Lee 2017). Their entire pharmacological and toxicological profile will unavoidably change and result in an unexpected therapeutic effect. The hydrophilic or hydrophobic nature, zeta potential, size, and shape of nanoparticle materials are some of the physicochemical elements in charge of coronal protein adsorption (Nguyen and Lee 2017; Rampado et al. 2020; Ke et al. 2017; Cedervall et al. 2007; Cai and Chen 2019). During the experimental investigation, it was revealed that reducing the size of NPs diminishes the surface contact that affects protein conformation. In a similar vein, increasing the mass-to-surface ratio of nanoparticles significantly decreases interaction with surrounding proteins (García-Álvarez et al. 2018; Magro et al. 2019; Xu et al. 2016). The formation of the protein corona is greatly influenced by the NP's composition. Electrostatic forces are affected by lipophobic NPs to attract charged proteins whereas van der waal forces or π-π interactions are used by lipophilic NPs to form a connection with lipophilic proteins (Rampado et al. 2020; Saha et al. 2016).

In recent studies it has been observed that NPs surrounded by biological fluids promote the loos aggregation of proteins to their surrounding referred to as smooth protein corona which facilitates the replacement of bounteous proteins throughout the circulation these soft aggregates are further layered by high-affinity proteins to form a hard aggregate also referred as hard corona (Vroman et al. 1980; Vroman and Lukosevicius 1964; Monopoli et al. 2012). This interaction lowers the hydration layer and the enthalpy of the NPs which leads to enhance the aqueous solubility of NPs and improves protein aggregation and misfolding, proteins undergo conformational changes that trigger phagocytosis and conceal the elements of the nanoparticles that define their pharmacological action (Ke et al. 2017).

Coronal protein adsorption is also significantly influenced by the HCC milieu, which includes its composition, period of exposure, pH and temperature condition, and shear stress. These elements have the potential to affect the protein corona's composition as well. Recently, several methods have been used to combat this phenomenon. For instance, carbohydrate moieties, dysopsonic proteins, zwitterions, and hydrophilic "stealth" polymers like PEG might enable nanoparticles to avoid opsonization, resulting in effective nanoparticle transport (Yan et al. 2013; Dai et al. 2014; Kang et al. 2015; Mizuhara et al. 2015; Oh et al. 2018; Maiolo et al. 2015; Wilhelm et al. 2016; Hadjidemetriou and Kostarelos 2017). Given the aforementioned difficulties in HCC targeting, it is crucial to improve NPs targeting in HCC even without the interference of the protein corona to achieve the most effective and ideal targeted treatment. This can be done by tailoring the protein corona utilizing particular non-covalent antibodies for HCC or novel targeted ligands considering disease heterogeneity (Barui et al. 2020; Ren et al. 2019; Corbo et al. 2017).

7.5.2 Reticulo Endothelial System

The reticuloendothelial system (RES) of the human liver, is a specialized component of the cellular network meant to separate the vital strains from blood circulation. This system is recently referred to as a mononuclear phagocyte system made up of monocytes, macrophages, and dendritic cells and elucidates the immune response. Once the NPs entered the liver RES seized the NPs (Geissmann et al. 2010; Jenne and Kubes 2013; Strauss et al. 2015; Crispe 2011; Allen 1994; Parmar and Patel 2019). Research studies presented that the previously existing hepatic macrophages engulf the NPs and clear from the blood circulation and this interaction is based on the molecular size and charge of NPs (Lu et al. 2018; Wake and Karl Wilhelm Kupffer and his contributions to modern hepatology 2004). It becomes the reason for complexity in HCC targeting so it is considered a basic need of NPs to protect them from the Kupffer cells to prevent its non-selective uptake before they reach the HCC cells. Research studies prove that bypassing the effect of Kuffer cells in nanodrug delivery results in improved tumor targeting, and effective nanoparticle delivery (Sadauskas et al. 2007; Tavares et al. 2017; Zhang et al. 2016). By modifying the NPs surface with the coating of zwitterion or neutral charge and constructing them with very much hydrophilic, "stealth" polymers like PEG and poly(vinyl alcohol) (PVA) we can prevent the effect of phagocytosis (García et al. 2014; Karakoti et al. 2011; Owensiii and Peppas 2006).

7.5.3 Biocompatibility and Toxicity of Nanocarriers

It is essential to study the toxicity and biocompatibility profile, especially in the case of nanoformulations like nanoparticles to ascertain the drug delivery to the targeted site is safely and smoothly devoid of any toxic effect. NPs have the potential to elucidate the immune response by stimulating or suppressing the immune system. Morphological phenomena and the surrounding environment of NPs significantly matter for their safety and biocompatibility (Naahidi et al. 2013). Opsonins are the specialized proteins involved in the phagocytosis process which help to remove the foreign matter from the body with the help of macrophages it is difficult to remain in the body by the NPs due to their inability to develop surface modification so NPs are easily and rapidly cleared from body system by the macrophages (Gref et al. 1994). Silver nanoparticles are considered highly toxic because they cause the outflow of lactate dehydrogenase and they can produce highly reactive oxygen moieties. Deformations in cell organelles, metallic toxicity, oxidative stress, and platelet aggregation are the problems associated with carbon nanotubes (Naahidi et al. 2013). Gold NPs show the ability to cross the blood-brain barrier and also have toxicity issues like hemolysis and platelet aggregation (de Jong. 2008). Amongst all the NPs formulations PLGA [poly (lactic-co-glycolic acid)] enabled NPS are found to be safer and have maximum biocompatibility (Bahadar et al. 2016).

7.5.4 Limitations of Nanoparticles

Due to their very tiny structure, NPs limited their use. Because of the reduction in particle size surface area of NPs increases which escalates the physicochemical interactions. NPs become more toxic due to their potential to generate highly reactive oxygen molecules and also increase oxidative stress which may lead to destructing the proteins, DNA, and membrane structure. Various factors contribute to the toxicity profile of NPs which includes morphological characters, the surrounding environment, the composition of chemically active components and functional groups, etc. NPs facilitate the numerous interactions inside the body and also cross the blood-brain barriers. Biodegradable NPs excreted through various excretory channels of the body but other NPs which are not biodegradable start accumulating in the body and initiate the inflammatory reaction. Cationic NPs are cardiotoxic and escalate platelet aggregation and hemolysis (Shubhika 2013).

7.5.5 Active Targeting and Nanocarriers

The receptor-specific interaction between ligands and nanocarriers (NCs) on the cell surface is referred to as active targeting or NCs functionalization. The functionality of the NCs depends upon the type of material used for the development of NCs. Active targeting is based on the combination of nanocarriers and ligands that interact with specific receptors on the cell surface. This process is called nanocarrier functionalization. Functionalization modalities differ according to the nature of the materials involved. In reality, a variety of carriers may be classified into two groups depending on where they came from inorganic NPs and organic NPs (Martinelli et al. 2019). Recent advancements in inorganic NPs include gold particles, iron oxide, nanoshells, silica NPs, and quantum dots (Hauck and Chan 2007; Brown et al. 2010). Gold NPs are found to be safer and biocompatible (Vines et al. 2019). Liposomes, polymersomes, micelles, polymer nanoparticles, dendrimers, and solid lipid particles are examples of organic particles. Here, we list a few properties of organic NPs in short. Although polymersomes and liposomes have similar morphologies, polymersomes are more stable than typical liposomes. Natural like albumin, gelatin, or biodegradable polymers, such as polylactide-co-glycolide, can be used to create polymer NPs (Zhang and Zhang 2017; Rabanel et al. 2015). The identification of dendrimers is highly accurate. Salinomycin-based nanoparticles also benefit from great stability and sustained release (Akhoond Zardini et al. 2018). This variety of nanocarriers provides a wide range of possibilities for focusing on over-expressed receptors, resulting in more effective cancer medicines and diagnostics.

7.5.6 Passive Targeting Strategies

Cancerous cell organelles that have porous vascular structures are intended to target the NCs by the EPR mechanism this mode of drug delivery is referred to as passive targeting of NCs. Through these channels, these tiny carriers travel by passive transport mechanism. The EPR effect helps to accumulate drugs in the liver tissues for a longer duration of time so that the requisite concentration of therapeutic agent builds up in hepatic sites and becomes the benefit of passive targeting (Rabanel et al. 2015). Drug accumulation at the hepatic site by the EPR effect is due to nonselective interactions of NCs when they enter into the systemic circulation (Bae and Park 2011). Passive targeting is subject to molecular characteristics of NCs like molecular weight, surface charge, and hydrophilicity or hydrophobicity of NCs as well as cancer cell characteristics including pH and surface charge of the cancerous cell, temperature, vascularity, and membrane permeability of cancerous cells (Ferreira et al. 2013; Joshi et al. 2015; Xiao et al. 2011). The sinusoid passages in the liver endothelium wall include 100–200 nm fenestrations that allow passive NCs accumulation to reach malignant areas. NCs smaller than 200 nm make passive targeting into the liver easier because greater sinusoidal fenestrations make it simple to discharge these particles (Xia et al. 2017). As these NCs reach the sinusoidal membrane, the liver's kuffer cells absorb them, causing an accumulation in the liver. The medication is then enough delivered to the liver cancer cells. Because of this, it may be necessary to add proteins, stabilizers, or polymers to the surface of NCs to increase their affinity for cells and decrease protein corona (Wilhelm et al. 2016; Nkansah et al. 2013; Walkey and Chan 2012).

7.5.7 Nanomedicine Potential in HCC

The creation of medicines and diagnostics in nanomedicine is confined to the nanoscale ranging from 1 nm to 100 nm (Guo et al. 2019). Proliferated hepatic cells are treated with nanoformulation is a trustworthy line of attack to treat HCC. Chemotherapeutics that have been encapsulated into NPs enable the medicine to concentrate within the targeted cells with prolonged bodily circulation (Pranatharthiharan et al. 2017). Tumor cell-associated overexpressed receptors are precisely identified by the NPs and help to target the drug delivery at a specific site. In research experiments, many nanocarrier systems, including graphene (Rahman et al. 2015a; Turato et al. 2017), Polylactic-co-glycolic acid PLGA-based materials (Pandey et al. 2018a; Rahman et al. 2015b), nanolipid (Kumar et al. 2017), systems (Akhter et al. 2011; Beg et al. 2017; Rahman et al. 2019), and nanometal have shown potential in the treatment of HCC. Liposomes, quantum dots, dendrimers, micelles, nanoshells, metallic NPs, and carbon nanotubes are only a few examples of the various types of nanostructures that have been employed for medical treatment. When creating nano-targeting drug delivery, effective medicines, NPs, and surface ligands that facilitate effective NPs targeting should all be taken into account. NPs should be between 30 to 200 nm in size (Usmani et al. 2018; Ungureanu et al. 2016; Ahmad et al. 2010). The renal system will eliminate less than 10 nm and the reticuloendothelial system of the spleen or liver will eliminate 100 nm. To safeguard gastrointestinal medication, NPs must be biocompatible and very stable. For NPs to be protected from the immune system and Kupffer cells, their surface must be hydrophilic (Ahmad et al. 2010). Passive targeting, active targeting, and stimuli-assisted NCs targeting are the three main forms of NPs targeting that have been employed for diagnostic and therapeutic applications. The vascular structure of tumor cells in HCC serves as the foundation for passive targeting, commonly referred to as the "Enhanced Permeability and Retention (EPR) effect." Cancerous cells will release the components of a nano-targeted drug delivery system. The most popular kind of HCC treatment is active targeting, wherein markers or ligands on the drug delivery system connect to particular corresponding receptors on the target cell, causing some cellular deformation (Varshosaz 2015; Zhong et al. 2014; Li et al. 2016a). NPs directed by stimulations such as ultrasound, radio waves, microwaves, magnetic fields like an external field, or the body's stimulations like pH, body temperature, and cellular microenvironment help to release the drug at targeted cells (Li et al. 2016a).

7.5.8 Nano-Antioxidants

Reactive oxygen radicals (ROR) consist of a single unpaired electron in their outermost orbital and are considered the most reactive species. Tumor cells show high potential for metabolic, peroxisome, and oncogene activity which is due to the high proportion of ROR present in tumor cells (Liou and Storz 2010). The phenomenon of neurocarcinogenesis and alterations in signaling pathways take place due to the disproportion between antioxidants and ROR. ROR concentration in a body is regulated with the help of antioxidants and used to produce an immune response against viruses and bacteria (Yaswen et al. 2015; Amin et al. 2015; Thyagarajan and Sahu 2018). Antioxidants can be employed as a cancer treatment by removing ROR from the tumor microenvironment. However, barriers to employing antioxidants as an anticancer therapeutic include low membrane permeability, poor solubility in water, degradation rate, and storage issues. Nanotechnology has been used to find solutions to these problems. The so-called "nano-antioxidants" are NPs that can be employed as nano-carriers for antioxidants. Nano-antioxidants include carbon nanotubes, and metal and metal oxide nanoparticles (Usmani et al. 2018). DEN-induced HCC in rats treated with PLGA NCs of Rutin shows improvement in the physiology of liver tissues which is analyzed by color staining technique. Mild photodegradation and a total lack of nodules prove the effectiveness of Rutin NCs (Pandey et al. 2018b). Curcumin NPs were used to treat DEN-induced HCC. Formulation shows positive outcomes against HCC which are confirmed by microscopic and biochemical examination of hepatic cells (Eftekhari et al. 2018).

7.5.9 Magnetic Nanoparticles

In liver tumors, magnetic NPs like magnetite and ferrimagnetic were discovered to work well in combination with ultrasound treatment. The contrast of magnetic scanning for liver cancer has been improved by the introduction of magnetically sensitive NPs. Controlled medication distribution can also be accomplished using NPs (Ebrahimi Fard et al. 2015). The magnetic properties of NPs can trigger physical reactions. Magnetic tweezers can be used for direct mechanical stimulation of cells in vitro. Magnetic field manipulation is used to achieve magnetothermal energy conversion and anticancer mechanical effects (Kilinc et al. 2016). Drug delivery using magnetic NPs to handle anticancer medicines to treat liver cancer has sparked a lot of attention. Researchers have designed the magnetically driven NCs consisting of Fe3O4 enabled head portion and a body composed of SiO_2 to deliver Doxorubicin. Selectively inhibitory action produced by these nanocomposites utilizing endocytosis action elucidated by cancerous cells (Shao et al. 2016).

7.5.10 Micelles

Colloidal composition of amphiphilic polymer with particle size ranging from 5 nm to 100 nm referred to as micelles. Micelles consist of both water-loving and lipid-loving components which act as a building block that helps to exhibit affinity for hydrophilic and lipophilic drugs (Husseini and Pitt 2008; Torchilin 2007). Micelles are the potential carriers for improving the solubility of lipid-loving bioactive with prolonged residence in the bloodstream of the body. Drug degradation is guarded against by the hydrophobic core of micelles, while opsonization is avoided by the hydrophilic surface. Nanomicelles composition of Curcumin was developed using Gum Arabica. This conjugate of Curcumin and Gum Arabica (CurGA) has shown a significant increase in the solubility of curcumin. Efficiently drug delivered to HepG2 cells by CurGA complex (Sarika et al. 2015).

7.5.11 Polylactic-Co-Glycolic Acid (PLGA) Nanoparticles

Polylactic-Co-Glycolic Acid (PLGA) is the synthetic polymeric composition of glycolic and lactic acid monomers. US and European drug regulatory authorities have approved the use of PLGA in therapeutic formulations (Danhier et al. 2012). PLGA NPs are found to be more effective in HCC and have good targeting, biocompatibility, and drug-loading capacity. Sorafenib NPs developed using PLGA polymer coated with lipids and CXCR4 antagonist to treat HCC. Sorafenib PLGA nano composite shows that lowering the proliferation of tumor cells and blood vessel progression leads to producing an angiogenesis inhibition effect. Also observed that increase in the survival rate of the orthotopic HCC mouse model (Gao et al. 2015). Dangi et al. (2014) have developed lactobionic acid-coated PLGA-based nanodrug delivery of 5 Fluorouracil. During the study, it has been observed that the improvement in bioavailability and drug loading of 5 Fluorouracil NPs (Dangi et al. 2014).

7.5.12 Dendrimers

Dendrimers are synthetic branched macromolecules having efficient renal filtration capacity (Peer et al. 2007; Nazemi and Gillies 2013). Solubility enhancement, smaller particle size and shape, uniformity in particle size distribution, and high drug loading capacity are the benefits offered by the dendrimer-based drug composite (Noriega-Luna et al. 2014). Drugs used to treat HCC are lipophilic in nature. Dendrimer provides the opportunity to target the lipophilic drug in the liver to kill the tumor affected cells because the inner central portion and branches of dendrimers have an affinity for lipophilic drugs. Researchers have developed doxorubicin dendrimer conjugates using lacto bionic acid and polyethylene glycol to treat the HCC and results proved that dendrimer-based doxorubicin complex magnificently targeted the drug to prevent the progression of tumor affected liver cells (Fu et al. 2014). The surface of dendrimers consists of hydrophilic functional groups which make the dendrimers an efficient nanocarrier to deliver water soluble, ionizing therapeutic agents to treat HCC (Chopdey et al. 2015; Sharma et al. 2017).

7.5.13 Ceramic Nanoparticles

Metal oxide-based biomaterials are considered ceramics which include titanium dioxide, aluminum oxide, magnetite, silica, and zinc oxide. Ceramic NPs are porous in nature replacing the organic NPs and becoming a prominent alternative that offers extensive biomedical applications. Amongst all the ceramic materials silica and magnetite have widespread applications in drug delivery to treat cancer patients. Silica materials have very good functional stability with most anticancer drugs (Immordino et al. 2006; Baig et al. 2019).

7.5.14 Liposomes

This system of drug delivery is referred to as the lipid vesicular system of drug transport. Consists of spherically arranged individual or group of lipid layers usually made up of phospholipids with amphiphilic in nature. Liposomes are biodegradable, stable, and hollow vesicular structures. The liposomal structure is divided into two parts hydrophilic head and lipophilic tail. The drug is enclosed in a hollow portion of the vesicles according to its affinity (Quagliariello et al. 2019; Hann and Prentice 2001). Liposomes have widespread utility as a carrier in drug delivery and contribute remarkably to enhancing the effectiveness of anticancer agents.

Studied the effectiveness of hepatoblastoma as an anticancer agent using HepG2 cell lines in the formulation of liposomes coated with chitosan and butyric acid enclosed uncoated liposomes. The study revealed that chitosan coated liposomes show maximum cytotoxic effect (Lombardi et al. 2011). Researchers have developed Doxorubicin enclosed liposomes coated with Polyethylene Glycol (DLP) to treat HCC. Study proves the effectiveness of DLP over the use of a pure drug to target the tumor affected cells of the liver. The effectiveness of PLD was confirmed by enhanced liver uptake and its drug targeting potential to the spleen as well as hepatic cells (He et al. 2018). Sorafenib enclosed liposomes (SEL) was prepared by He et al. (2018) and characterized its effectiveness in Hep3B tumor affected xenograft mice (Nguyen et al. 2019). The study proved the effectiveness of SEL for its biocompatibility and activity against HCC. From the results, it has been concluded that liposomes are the most suitable drug carrier to treat HCC.

7.5.15 Silica Nanoparticles

Silica NPs (SNPs) have excellent biocompatibility and capacity for drug loading, SNPs have been created in a variety of ways to investigate the entire tumor diagnostics and therapy in recent years (Chen et al. 2018; Jiang et al. 2021; Liu et al. 2020). SNPs have a lot of promise for use in HCC treatment offering the benefits of their distinctive form, modifiable particle size, and explicit surface area (Chen et al. 2015; Li et al. 2018). The interfaces of SNPs are typically altered and substituted with other NPs or chemically active compounds for the targeted release at the HCC region to take greater advantage of the diverse design of SNPs (Ashley et al. 2011; Wang et al. 2017). Numerous tumor-targeted delivery technologies have emerged over the past decade. These include lipid bilayers anchored by mesoporous silica nanoparticles that have been modified with the SP94 peptide HCC and VLPs from the bacteriophage MS2 virus (Li et al. 2016b; Ashley et al. 2012). Due to SP94 peptide multivalent interaction 10,000 times increase in aversion to human HCC compared to hepatocytes, endothelial cells, T-lymphocytes, T-lymphocytes, or B-lymphocytes from peripheral blood (Jeong et al. 2016). SP94 peptide's exquisite targeting specificity in combination with improved multi-component cargo delivery to tumors allows for sensitivity differentiation between the targeted tissue and healthy tissue Therefore, SP94 peptide is a good model to study how active tumor targeting works (Ashley et al. 2012).

7.5.16 Albumin Nanoparticles

It is the most prevalent protein composite with a molecular weight of 66.5 kDa present in the human body referred to as Human Serum Albumin (HSA) produced by the liver having very good biocompatible and sufficient target specificity (Dockal et al. 1999; Kratz 2008; Abbasi et al. 2012; Sebak et al. 2010). X-ray imaging confirmed that HSA consists of three domains each domain having two subunits (Elzoghby et al. 2012). HSA molecules assemble in solution by the effect of intermolecular disulfide bonds, which results in the formation of HSA NPs (Su et al. 2009; Lomis et al. 2016). Albumin is a dynamic choice as a carrier in chemotherapeutics due to its capacity to stay in inflammatory tumor cells and a variety of groups for drug binding. These NPs might also be combined with different dyes to track their mobility and localization within cancer cells. Albumin complexed Paclitaxel NPs were formulated by Zhou et al. and clear destruction of the hepatic malignant tumor was seen. It was seen from the HCC xenograft models' higher effectiveness and reduced cytotoxicity results. By using a passive targeting strategy, the controlled release of significant amounts of the medicines cinobufagin and sodium ferulate from albumin NPs was assessed (Amina and Guo 2020; Li et al. 2008).

7.5.17 Metallic Nanoparticles

Metallic nanoparticles can be utilized for both medicinal and diagnosis applications including gold (GNPs), silver (SNPs), and platinum (PNsP). Typically, the production of GNPs involves chemical reduction as well as stabilization with the use of a capping agent, electrochemical layering on the surface of electrodes especially in electro-sensor-based NPs, or recently developed biological fabrication methods (Amina and Guo 2020; Bhattacharya et al. 2022). Particle size and surface characteristics of the GNPs can be easily modified. GNPs possess the ability to absorb visible light rays which makes them the most suitable diagnostic agent (Gavas et al. 2021). At ordinary concentrations, level GNPs are safe and non-toxic but at higher concentrations lead to gene toxicity (Desai et al. 2021). Concerning HCC diagnostics and therapeutics for the detection of biomarkers like AFP or glypican-3 as well as to image cancers and photo thermally destroy them (Gobin et al. 2007) GNPs (Li et al. 2021), gold-platinum (Li et al. 2020a), or platinum-palladium NCs (Shi et al. 2021) have been utilized in combination with carbon-based NCs.

7.6 Nanoparticles as an HCC Diagnostics

7.6.1 Nanoimaging Technique in HCC Diagnosis

Usually, MRI and CT scan technique is employed for the diagnosis of HCC. Patients having high-risk HCC development are supervised by a modern diagnostic method such as ultrasound imaging (USI) technique. Although diagnosis is reasonably precise due to the presence of the radio marker of HCC (Llovet et al. 2021). USI is an affordable, non-invasive, and effective method to supervise HCC development at an early stage. However, the difficulty with this technique is that it does not always reveal the necrotic region, tumor surface areas, and outer lines clearly (Xu et al. 2022). It was discovered that the distinction between the neighboring tissues and the HCC cells is noticeable, owing to the specific metabolism and breakdown products of HCC cells. Furthermore, when compared to other abdominal malignant tumors, early detection of HCC is less susceptible to gas interference. So it makes more sense to formulate an ultrasound contrast agent to identify and differentiate the tumor affected hepatic cells and normal growth of liver tissues. The use of nanomaterials in USI makes the HCC diagnosis improvised and simplified (Li et al. 2020b).

Rats with HCC were imaged using both MRI and visual near-infrared fluorescence simultaneously using a fifth-generation dendrimer coupled with an aptamer. In order to increase the nanoprobe's selectivity to tumor tissue, the aptamer was selected for endoglin, a biomarker present on the outermost layers of endothelial cells that line the vessels in tumors. To detect near-infrared fluorescence, the customized dendrimers have been imprinted with fluorophores. Additionally, paramagnetic Gd3 + -DTPA chelators were used to generate T1 graded relaxation and an MRI response. When compared to aptamer-free dendrimers, the generated nanoprobes fluoresced strongly in tumor cells. Tumor boundaries were clearly revealed by in vivo MRI scanning, implying that this method might be utilized to guide surgical operations for the removal of tiny hepatic tumors (Yan et al. 2018).

Lee et al. revealed yet another dual-imaging technique. Nd3+ doped up-conversion nanoparticles (UCNP) were used in their investigation for MRI and real-time up-conversion luminescence imaging of HCC in a simple rat model. Nd3+ was utilized because it has a lesser propensity for overheating during irradiation than the more often employed Yb3+. The thermal breakdown was used to create the core-shell particles, which were then shell-coated through a seed-mediated method. To boost specificity for tumor cells, the surface of the nanoprobe was activated with an anti-CD44 antibody. MRI imaging and the strong UCL signal recorded in tumor tissue relative to surrounding healthy tissue both revealed preferential aggregation of the nanoprobes in tumor tissue (Lee et al. 2016).

7.6.2 Intraoperative Diagnosis

An intraoperative diagnostic is a novel approach that has lately evolved. This aspect pertains to the outlining of the residual tumor tissue, which is generally on the hepatocellular resection area and might be challenging to see without additional procedures.

There was also talk of intraoperative visualization using fluorescence dye. Tsunda et al. created indocyanine green loaded lactosomal vesicles, a photosensitive chemical used in near-infrared fluorescence imaging. Lactosomes are a kind of core surrounded shell polymeric micelle created when sarcosine and lactic acid polymerize together. These NPs were delivered through an intravenous route to naked mice and displayed extended retention in liver tumors as compared to indocyanine green alone. This is advantageous in the event of lengthy surgical operations when a single dosage would suffice for the whole treatment. Another benefit is the capacity of ICG-loaded lactosomes to perform photodynamic therapy on tumors when exposed to laser light (Andreou et al. 2016).

Surface-enhanced Raman scattering (SERS) silica NPs were created by Andreou et al. for the intraoperative screening of HCC in mice as a contrast medium. With the help of a modified Stöber approach, gold NPs were produced by chemical reduction and subsequently layered with silica in the presence of a Raman reporter. The identification of HCC utilizing these NPs was based on their increased concentration in healthy liver tissue vs malignancies, indicating the tumor margins. When compared to MRI intraoperative imaging, the SERS-based technology was able to detect smaller tumors. Furthermore, as compared to fluorescence approaches, the SERS-based methodology is more sensitive and photostable (Tsuda et al. 2017).

7.6.3 Serological and Histopathological Diagnosis

Amongst all the available biomarkers very few are nanoscale biomarkers used in the HCC diagnosis. This might result in the development of methods that will significantly enhance HCC diagnosis (Wu et al. 2021; Parikh et al. 2020; Pan et al. 2020; Zacharakis et al. 2018; Abreu et al. 2020). Although AFP is the most often used HCC biomarker and the sole one with phase 5 validation evidence, multiple additional biomarkers osteopontin, Golgiprotein 73, glypican-3 and PIVKA have recently been discovered (Parikh et al. 2020).

Serological tumor markers employed in the identification and tracking of HCC are often less, and detecting these biomarkers in biological materials necessitates the development of sensitive technologies. Metallic nanoparticles or nanocarbon materials can be employed to enhance the surface area of sensors, improve their detecting capacities, and so ultimately boost sensitivity. Furthermore, they can be designed and synthesized with biological elements such as enzymes and antibodies or bioinspired material such as aptamers and molecularly imprinted polymer recognition systems, increasing the detection method's specificity towards a specific target. Many nanomaterial-based detection techniques for cancer biomarkers, such as proteins, in the blood have been described (Sun et al. 2021; Kaya et al. 2022; Dai et al. 2020).

For the diagnosis of HCC, an AFP sensing methodology employs an electrochemical method and an AFP aptamer. To boost the electrical conductivity of an electrode, gold-platinum NPs were coated on its surface, along with a reduced form of graphene oxide, chitosan, and a redox mediator. Finally, the aptamer was introduced to assure AFP sensitivity. The approach had some interferences with other proteins, but it was sensitive enough to detect minimal amounts of AFP (Li et al. 2020a).

Metallic NPs such as platinum-palladium nanoconjugates, gold NPs, etc. used in the diagnosis of glypican 3 by electrochemical mode. Metallic NPs along with a reduced form of graphene oxide, ferrocene, hemin, and selective aptamer of glypican 3 deposited on the surface of the electrode. When glypican-3 was introduced to the aptasensor, the reduction in the differential-pulse voltammetry oxidation potential of the redox mediator was measured. The platinum-palladium NPs offered greater stability and a higher linear correlation coefficient than the AuNP approach, however, it has very little sensitivity (Li et al. 2021; Shi et al. 2021).

7.7 Conclusion

HCC is a complicated illness with a rising incidence and fatality rate. Management of HCC using nanomedicines is highly encouraged since it makes use of the appealing properties given by the EPR effect, as well as the biodistribution of a drug in the specified body locations and the release of the loaded medications into the target areas. This review summarized the challenges in the treatment of HCC, biocompatibility, and toxicity as well as the limitations of NPs. A nanotheranostics-based drug delivery system with active targeting and effective endocytosis might be employed as a potential treatment for hepatocellular carcinoma.

In the future nanomedicine, particularly nanocarriers have tremendous potential in the holistic care of liver carcinomas. However, numerous hurdles in drug delivery systems must be addressed, including extensive carrier buildup in the liver, limited antitumor activity, and various barriers around tumor sites or in vascular areas that provide impediments to penetration into liver neoplasms. To accomplish active targeting of nanodrugs, the ligands or targeting moieties must translocate the nanocarriers to a specific location of tumors.

Only a few clinically approved nanodrugs for liver cancer have been created by existing nanotherapeutic techniques. To attain acceptable therapeutic payload and imaging sensitivity, several new therapeutic medicines with various physicochemical characteristics of NCs must be created.

References

Abbasi S, Paul A, Shao W, Prakash S (2012) Cationic albumin nanoparticles for enhanced drug delivery to treat breast cancer: preparation and in vitro assessment. J Drug Deliv 2012:686108

Abreu P, Ferreira R, Mineli V, Ribeiro MA, Ferreira FG, De Mello Vianna RM et al (2020) Alternative biomarkers to predict tumor biology in hepatocellular carcinoma. Anticancer Res 40(12):6573–6784

Ahmad MZ, Akhter S, Jain GK, Rahman M, Pathan SA, Ahmad FJ et al (2010) Metallic nanoparticles: technology overview and drug delivery applications in oncology. Expert Opin Drug Deliv 7(8):927–942

Akhoond Zardini A, Mohebbi M, Farhoosh R, Bolurian S (2018) Production and characterization of nanostructured lipid carriers and solid lipid nanoparticles containing lycopene for food fortification. J Food Sci Technol 55(1):287–298

Akhter S, Zaki Ahmad M, Singh A, Ahmad I, Rahman M, Anwar M et al (2011) Cancer targeted metallic nanoparticle: targeting overview, recent advancement and toxicity concern. Curr Pharm Des 17(18):1834–1850

Allen TM (1994) The use of glycolipids and hydrophilic polymers in avoiding rapid uptake of liposomes by the mononuclear phagocyte system. Adv Drug Deliv Rev 13(3):285–309

Amin ARMR, Karpowicz PA, Carey TE, Arbiser J, Nahta R, Chen ZG et al (2015) Evasion of anti-growth signaling: a key step in tumorigenesis and potential target for treatment and prophylaxis by natural compounds. Semin Cancer Biol 35:S55–S77

Amina SJ, Guo B (2020) A review on the synthesis and functionalization of gold nanoparticles as a drug delivery vehicle. Int J Nanomedicine 15:9823–9857

Andreou C, Neuschmelting V, Tschaharganeh DF, Huang CH, Oseledchyk A, Iacono P et al (2016) Imaging of liver tumors using surface-enhanced Raman scattering nanoparticles. ACS Nano 10(5):5015–5026

Ashley CE, Carnes EC, Phillips GK, Padilla D, Durfee PN, Brown PA et al (2011) The targeted delivery of multicomponent cargos to cancer cells by nanoporous particle-supported lipid bilayers. Nat Mater 10(5):389–397

Ashley CE, Carnes EC, Epler KE, Padilla DP, Phillips GK, Castillo RE et al (2012) Delivery of small interfering RNA by peptide-targeted mesoporous silica nanoparticle-supported lipid bilayers. ACS Nano 6(3):2174–2188

Bae YH, Park K (2011) Targeted drug delivery to tumors: myths, reality and possibility. J Control Release 153(3):198–205

Bahadar H, Maqbool F, Niaz K, Abdollahi M (2016) Toxicity of nanoparticles and an overview of current experimental models. Iran Biomed J 20(1):1–11

Baig B, Halim SA, Farrukh A, Greish Y, Amin A (2019) Current status of nanomaterial-based treatment for hepatocellular carcinoma. Biomed Pharmacother 116:108852

Bakrania A, Zheng G, Bhat M (2021) Nanomedicine in hepatocellular carcinoma: a new frontier in targeted cancer treatment. Pharmaceutics 14(1):41

Balkwill FR, Capasso M, Hagemann T (2012) The tumor microenvironment at a glance. J Cell Sci 125(Pt 23):5591–5596

Barbara L, Benzi G, Gaiani S, Fusconi F, Zironi G, Siringo S et al (1992) Natural history of small untreated hepatocellular carcinoma in cirrhosis: a multivariate analysis of prognostic factors of tumor growth rate and patient survival. Hepatology 16(1):132–137

Barui AK, Oh JY, Jana B, Kim C, Ryu J (2020) Cancer-targeted nanomedicine: overcoming the barrier of the protein corona. Adv Ther (Weinh) 3(1):1900124

Beg S, Rahman M, Jain A, Saini S, Midoux P, Pichon C et al (2017) Nanoporous metal organic frameworks as hybrid polymer–metal composites for drug delivery and biomedical applications. Drug Discov Today 22(4):625–637

Benetti A, Berenzi A, Gambarotti M, Garrafa E, Gelati M, Dessy E et al (2008) Transforming growth factor-β1 and CD105 promote the migration of hepatocellular carcinoma–derived endothelium. Cancer Res 68(20):8626–8634

Benvegnu L (2004) Natural history of compensated viral cirrhosis: a prospective study on the incidence and hierarchy of major complications. Gut 53(5):744–749

Bhattacharya T, Das D, Borges e Soares GA, Chakrabarti P, Ai Z, Chopra H et al (2022) Novel green approaches for the preparation of gold nanoparticles and their promising potential in oncology. PRO 10(2):426

Brown SD, Nativo P, Smith JA, Stirling D, Edwards PR, Venugopal B et al (2010) Gold nanoparticles for the improved anticancer drug delivery of the active component of oxaliplatin. J Am Chem Soc 132(13):4678–4684

Byrne JD, Betancourt T, Brannon-Peppas L (2008) Active targeting schemes for nanoparticle systems in cancer therapeutics. Adv Drug Deliv Rev 60(15):1615–1626
Cai R, Chen C (2019) The crown and the scepter: roles of the protein corona in nanomedicine. Adv Mater 31(45):e1805740
Cedervall T, Lynch I, Lindman S, Berggård T, Thulin E, Nilsson H et al (2007) Understanding the nanoparticle–protein corona using methods to quantify exchange rates and affinities of proteins for nanoparticles. Proc Natl Acad Sci 104(7):2050–2055
Chen F, Goel S, Valdovinos HF, Luo H, Hernandez R, Barnhart TE et al (2015) In vivo integrity and biological fate of chelator-free zirconium-89-labeled mesoporous silica nanoparticles. ACS Nano 9(8):7950–7959
Chen X, Zhang Q, Li J, Yang M, Zhao N, Xu FJ (2018) Rattle-structured rough nanocapsules with in-situ -formed gold nanorod cores for complementary gene/chemo/Photothermal therapy. ACS Nano 12(6):5646–5656
Chen X, Wang T, Le W, Huang X, Gao M, Chen Q et al (2020) Smart sorting of tumor phenotype with versatile fluorescent ag nanoclusters by sensing specific reactive oxygen species. Theranostics 10(8):3430–3450
Chopdey PK, Tekade RK, Mehra NK, Mody N, Jain NK (2015) Glycyrrhizin conjugated dendrimer and multi-walled carbon nanotubes for liver specific delivery of doxorubicin. J Nanosci Nanotechnol 15(2):1088–1100
Corbo C, Molinaro R, Tabatabaei M, Farokhzad OC, Mahmoudi M (2017) Personalized protein corona on nanoparticles and its clinical implications. Biomater Sci 5(3):378–387
Crispe IN (2011) Liver antigen-presenting cells. J Hepatol 54(2):357–365
Dai Q, Walkey C, Chan WCW (2014) Polyethylene glycol backfilling mitigates the negative impact of the protein corona on nanoparticle cell targeting. Angew Chem Int Ed Engl 53(20):5093–5096
Dai Y, Han B, Dong L, Zhao J, Cao Y (2020) Recent advances in nanomaterial-enhanced biosensing methods for hepatocellular carcinoma diagnosis. TrAC Trends Anal Chem 130:115965
Dangi R, Hurkat P, Jain A, Shilpi S, Jain A, Gulbake A et al (2014) Targeting liver cancer via ASGP receptor using 5-FU-loaded surface-modified PLGA nanoparticles. J Microencapsul 31(5):479–487
Danhier F, Ansorena E, Silva JM, Coco R, Le Breton A, Préat V (2012) PLGA-based nanoparticles: an overview of biomedical applications. J Control Release 161(2):505–522
Dasgupta A, Biancacci I, Kiessling F, Lammers T (2020) Imaging-assisted anticancer nanotherapy. Theranostics 10(3):956–967
de Jong. (2008) Drug delivery and nanoparticles: applications and hazards. Int J Nanomedicine 3(2):133–149
Desai N, Momin M, Khan T, Gharat S, Ningthoujam RS, Omri A (2021) Metallic nanoparticles as drug delivery system for the treatment of cancer. Expert Opin Drug Deliv 18(9):1261–1290
Dhanasekaran R, Limaye A, Cabrera R (2012) Hepatocellular carcinoma: current trends in worldwide epidemiology, risk factors, diagnosis, and therapeutics. Hepat Med 4:19–37
Dobrovolskaia MA (2015) Pre-clinical immunotoxicity studies of nanotechnology-formulated drugs: challenges, considerations and strategy. J Control Release 220:571–583
Dockal M, Carter DC, Rüker F (1999) The three recombinant domains of human serum albumin. Structural characterization and ligand binding properties. J Biol Chem 274(41):29303–29310
Ebrahimi Fard A, Zarepour A, Zarrabi A, Shanei A, Salehi H (2015) Synergistic effect of the combination of triethylene-glycol modified Fe3O4 nanoparticles and ultrasound wave on MCF-7 cells. J Magn Magn Mater 394:44–49
Eftekhari A, Dizaj SM, Chodari L, Sunar S, Hasanzadeh A, Ahmadian E et al (2018) The promising future of nano-antioxidant therapy against environmental pollutants induced-toxicities. Biomed Pharmacother 103:1018–1027
Elzoghby AO, Samy WM, Elgindy NA (2012) Albumin-based nanoparticles as potential controlled release drug delivery systems. J Control Release 157(2):168–182
Ferreira DD, Lopes SC, Franco MS, Oliveira MC (2013) pH-sensitive liposomes for drug delivery in cancer treatment. Ther Deliv 4(9):1099–1123

Filippousi M, Altantzis T, Stefanou G, Betsiou M, Bikiaris DN, Angelakeris M et al (2013a) Polyhedral iron oxide core–shell nanoparticles in a biodegradable polymeric matrix: preparation, characterization and application in magnetic particle hyperthermia and drug delivery. RSC Adv 3(46):24367

Filippousi M, Papadimitriou SA, Bikiaris DN, Pavlidou E, Angelakeris M, Zamboulis D et al (2013b) Novel core-shell magnetic nanoparticles for Taxol encapsulation in biodegradable and biocompatible block copolymers: preparation, characterization and release properties. Int J Pharm 448(1):221–230

Friedman SL, Roll FJ, Boyles J, Bissell DM (1985) Hepatic lipocytes: the principal collagen-producing cells of normal rat liver. Proc Natl Acad Sci USA 82(24):8681–8685

Fu F, Wu Y, Zhu J, Wen S, Shen M, Shi X (2014) Multifunctional lactobionic acid-modified dendrimers for targeted drug delivery to liver cancer cells: investigating the role played by PEG spacer. ACS Appl Mater Interfaces 6(18):16416–16425

Gallagher FA, Kettunen MI, Day SE, Hu DE, Ardenkjaer-Larsen JH, Zandt R et al (2008) Magnetic resonance imaging of pH in vivo using hyperpolarized 13C-labelled bicarbonate. Nature 453(7197):940–943

Gao DY, Lin TT, Sung YC, Liu YC, Chiang WH, Chang CC et al (2015) CXCR4-targeted lipid-coated PLGA nanoparticles deliver sorafenib and overcome acquired drug resistance in liver cancer. Biomaterials 67:194–203

García KP, Zarschler K, Barbaro L, Barreto JA, O'Malley W, Spiccia L et al (2014) Zwitterionic-coated "stealth" nanoparticles for biomedical applications: recent advances in countering biomolecular corona formation and uptake by the mononuclear phagocyte system. Small 10(13):2516–2529

García-Álvarez R, Hadjidemetriou M, Sánchez-Iglesias A, Liz-Marzán LM, Kostarelos K (2018) In vivo formation of protein corona on gold nanoparticles. The effect of their size and shape. Nanoscale 10(3):1256–1264

Gavas S, Quazi S, Karpiński TM (2021) Nanoparticles for cancer therapy: current progress and challenges. Nanoscale Res Lett 16(1):173

Ge J, Zhang Q, Zeng J, Gu Z, Gao M (2020) Radiolabeling nanomaterials for multimodality imaging: new insights into nuclear medicine and cancer diagnosis. Biomaterials 228:119553

Geissmann F, Gordon S, Hume DA, Mowat AM, Randolph GJ (2010) Unravelling mononuclear phagocyte heterogeneity. Nat Rev Immunol 10(6):453–460

Gobin AM, Lee MH, Halas NJ, James WD, Drezek RA, West JL (2007) Near-infrared resonant nanoshells for combined optical imaging and photothermal cancer therapy. Nano Lett 7(7):1929–1934

Gref R, Minamitake Y, Peracchia MT, Trubetskoy V, Torchilin V, Langer R (1994) Biodegradable long-circulating polymeric nanospheres. Science (1979) 263(5153):1600–1603

Guo R, Wu Z, Wang J, Li Q, Shen S, Wang W et al (2019) Development of a non-coding-RNA-based EMT/CSC inhibitory nanomedicine for in vivo treatment and monitoring of HCC. Adv Sci 6(9):1801885

Gupta C, Vikram A, Tripathi DN, Ramarao P, Jena GB (2010) Antioxidant and antimutagenic effect of quercetin against DEN induced hepatotoxicity in rat. Phytother Res 24(1):119–128

Ha WS, Kim CK, Song SH, Kang CB (2001) Study on mechanism of multistep hepatotumorigenesis in rat: development of hepatotumorigenesis. J Vet Sci 2(1):53–58

Hadjidemetriou M, Kostarelos K (2017) Nanomedicine: evolution of the nanoparticle corona. Nat Nanotechnol 12(4):288–290

Han N, Yang YY, Wang S, Zheng S, Fan W (2013) Polymer-based cancer nanotheranostics: retrospectives of multi-functionalities and pharmacokinetics. Curr Drug Metab 14(6):661–674

Hann IM, Prentice HG (2001) Lipid-based amphotericin B: a review of the last 10 years of use. Int J Antimicrob Agents 17(3):161–169

Harshita, Barkat MA, Rizwanullah M, Beg S, Pottoo FH, Siddiqui S, Ahmad FJ (2019) Paclitaxel-loaded nanolipidic carriers with improved oral bioavailability and anticancer activity against human liver carcinoma. AAPS PharmSciTech 20(2):87. https://doi.org/10.1208/s12249-019-1304-4

Hauck TS, Chan WC (2007) Gold nanoshells in cancer imaging and therapy: towards clinical application. Nanomedicine 2(5):735–738

He Q, He X, Deng B, Shi C, Lin L, Liu P et al (2018) Sorafenib and indocyanine green co-loaded in photothermally sensitive liposomes for diagnosis and treatment of advanced hepatocellular carcinoma. J Mater Chem B 6(36):5823–5834

Husseini GA, Pitt WG (2008) Micelles and nanoparticles for ultrasonic drug and gene delivery. Adv Drug Deliv Rev 60(10):1137–1152

Immordino ML, Dosio F, Cattel L (2006) Stealth liposomes: review of the basic science, rationale, and clinical applications, existing and potential. Int J Nanomedicine 1(3):297–315

Jenne CN, Kubes P (2013) Immune surveillance by the liver. Nat Immunol 14(10):996–1006

Jeong K, Kang CS, Kim Y, Lee YD, Kwon IC, Kim S (2016) Development of highly efficient nanocarrier-mediated delivery approaches for cancer therapy. Cancer Lett 374(1):31–43

Jiang H, Guo Y, Wei C, Hu P, Shi J (2021) Nanocatalytic innate immunity activation by mitochondrial DNA oxidative damage for tumor-specific therapy. Adv Mater 33(20):2008065

Joshi MD, Patravale V, Prabhu R. Polymeric nanoparticles for targeted treatment in oncology: current insights. Int J Nanomedicine 2015;101001–18

Kalyane D, Raval N, Maheshwari R, Tambe V, Kalia K, Tekade RK (2019) Employment of enhanced permeability and retention effect (EPR): nanoparticle-based precision tools for targeting of therapeutic and diagnostic agent in cancer. Mater Sci Eng C Mater Biol Appl 98:1252–1276

Kang B, Okwieka P, Schöttler S, Winzen S, Langhanki J, Mohr K et al (2015) Carbohydrate-based nanocarriers exhibiting specific cell targeting with minimum influence from the protein corona. Angew Chem Int Ed Engl 54(25):7436–7440

Karakoti AS, Das S, Thevuthasan S, Seal S (2011) PEGylated inorganic nanoparticles. Angew Chem Int Ed 50(9):1980–1994

Kaya SI, Ozcelikay G, Mollarasouli F, Bakirhan NK, Ozkan SA (2022) Recent achievements and challenges on nanomaterial based electrochemical biosensors for the detection of colon and lung cancer biomarkers. Sens Actuat B Chem 351:130856

Ke PC, Lin S, Parak WJ, Davis TP, Caruso F (2017) A decade of the protein corona. ACS Nano 11(12):11773–11776

Kessenbrock K, Plaks V, Werb Z (2010) Matrix metalloproteinases: regulators of the tumor microenvironment. Cell 141(1):52–67

Kilinc D, Dennis CL, Lee GU (2016) Bio-nano-magnetic materials for localized mechanochemical stimulation of cell growth and death. Adv Mater 28(27):5672–5680

Kratz F (2008) Albumin as a drug carrier: design of prodrugs, drug conjugates and nanoparticles. J Control Release 132(3):171–183

Kumar V, Bhatt P, Rahman M, Kaithwas G, Choudhry H, Al-Abbasi F et al (2017) Fabrication, optimization, and characterization of umbelliferone β-D-galactopyranoside-loaded PLGA nanoparticles in treatment of hepatocellular carcinoma: in vitro and in vivo studies. Int J Nanomedicine 12:6747–6758

Lee J, Gordon AC, Kim H, Park W, Cho S, Lee B et al (2016) Targeted multimodal nano-reporters for pre-procedural MRI and intra-operative image-guidance. Biomaterials 109:69–77

Li FQ, Su H, Wang J, Liu JY, Zhu QG, Fei YB et al (2008) Preparation and characterization of sodium ferulate entrapped bovine serum albumin nanoparticles for liver targeting. Int J Pharm 349(1–2):274–282

Li M, Zhang W, Wang B, Gao Y, Song Z, Zheng QC (2016a) Ligand-based targeted therapy: a novel strategy for hepatocellular carcinoma. Int J Nanomedicine 11:5645–5669

Li Y, Hu Y, Xiao J, Liu G, Li X, Zhao Y et al (2016b) Investigation of SP94 peptide as a specific probe for hepatocellular carcinoma imaging and therapy. Sci Rep 6(1):33511

Li Z, Zhang H, Han J, Chen Y, Lin H, Yang T (2018) Surface nanopore engineering of 2D MXenes for targeted and synergistic multitherapies of hepatocellular carcinoma. Adv Mater 30(25):1706981

Li W, Chen M, Liang J, Lu C, Zhang M, Hu F et al (2020a) Electrochemical aptasensor for analyzing alpha-fetoprotein using RGO–CS–Fc nanocomposites integrated with gold–platinum nanoparticles. Anal Methods 12(41):4956–4966

Li J, Wang J, Yu S, Yuan G, He S (2020b) Letter to the editor: hepatocellular carcinoma surveillance in high-risk patients with cirrhosis: contrast-enhanced ultrasound may be a choice. Hepatology 71(1):392

Li G, Feng H, Shi X, Chen M, Liang J, Zhou Z (2021) Highly sensitive electrochemical aptasensor for Glypican-3 based on reduced graphene oxide-hemin nanocomposites modified on screen-printed electrode surface. Bioelectrochemistry 138:107696

Liou GY, Storz P (2010) Reactive oxygen species in cancer. Free Radic Res 44(5):479–496

Liu Q, Wang C, Zheng Y, Zhao Y, Wang Y, Hao J et al (2020) Virus-like nanoparticle as a co-delivery system to enhance efficacy of CRISPR/Cas9-based cancer immunotherapy. Biomaterials 258:120275

Llovet JM, Burroughs A, Bruix J (2003) Hepatocellular carcinoma. Lancet 362(9399):1907–1917

Llovet JM, Kelley RK, Villanueva A, Singal AG, Pikarsky E, Roayaie S et al (2021) Hepatocellular carcinoma. Nat Rev Dis Primers 7(1):6

Lombardi G, Zustovich F, Farinati F, Cillo U, Vitale A, Zanus G et al (2011) Pegylated liposomal doxorubicin and gemcitabine in patients with advanced hepatocellular carcinoma: results of a phase 2 study. Cancer 117(1):125–133

Lomis N, Westfall S, Farahdel L, Malhotra M, Shum-Tim D, Prakash S (2016) Human serum albumin nanoparticles for use in cancer drug delivery: process optimization and in vitro characterization. Nano 6(6):116

Lu J, Wang J, Ling D (2018) Surface engineering of nanoparticles for targeted delivery to hepatocellular carcinoma. Small 14(5):1702037

Maeda H, Wu J, Sawa T, Matsumura Y, Hori K (2000) Tumor vascular permeability and the EPR effect in macromolecular therapeutics: a review. J Control Release 65(1–2):271–284

Magro M, Baratella D, Bonaiuto E, de Almeida RJ, Chemello G, Pasquaroli S et al (2019) Stealth iron oxide nanoparticles for organotropic drug targeting. Biomacromolecules 20(3):1375–1384

Maiolo D, Del Pino P, Metrangolo P, Parak WJ, Baldelli Bombelli F (2015) Nanomedicine delivery: does protein corona route to the target or off road? Nanomedicine (Lond) 10(21):3231–3247

Malik S (ed) (2017) Biotechnology and production of anti-cancer compounds. Springer International Publishing, Cham

Mandal AK, Das S, Mitra M, Chakrabarti RN, Chatterjee M, Das N (2008) Vesicular flavonoid in combating diethylnitrosamine induced hepatocarcinoma in rat model. J Exp Ther Oncol 7(2):123–133

Martinelli C, Pucci C, Ciofani G (2019) Nanostructured carriers as innovative tools for cancer diagnosis and therapy. APL Bioeng 3(1):011502

Mizuhara T, Saha K, Moyano DF, Kim CS, Yan B, Kim YK et al (2015) Acylsulfonamide-functionalized zwitterionic gold nanoparticles for enhanced cellular uptake at tumor pH. Angew Chem Int Ed Engl 54(22):6567–6570

Monopoli MP, Åberg C, Salvati A, Dawson KA (2012) Biomolecular coronas provide the biological identity of nanosized materials. Nat Nanotechnol 7(12):779–786

Naahidi S, Jafari M, Edalat F, Raymond K, Khademhosseini A, Chen P (2013) Biocompatibility of engineered nanoparticles for drug delivery. J Control Release 166(2):182–194

Nazemi A, Gillies ER (2013) Dendritic surface functionalization of nanomaterials: controlling properties and functions for biomedical applications. Brazilian. J Pharm Sci 49(spe):15–32

Nguyen VH, Lee BJ (2017) Protein corona: a new approach for nanomedicine design. Int J Nanomedicine 12:3137–3151

Nguyen TL, Choi Y, Kim J (2019) Mesoporous silica as a versatile platform for cancer immunotherapy. Adv Mater 31(34):1803953

Nichols JW, Bae YH (2012) Odyssey of a cancer nanoparticle: from injection site to site of action. Nano Today 7(6):606–618

Nkansah P, Antipas A, Lu Y, Varma M, Rotter C, Rago B et al (2013) Development and evaluation of novel solid nanodispersion system for oral delivery of poorly water-soluble drugs. J Control Release 169(1–2):150–161

Noriega-Luna B, Godínez LA, Rodríguez FJ, Rodríguez A, Zaldívar-Lelo de Larrea G, Sosa-Ferreyra CF et al (2014) Applications of dendrimers in drug delivery agents, diagnosis, therapy, and detection. J Nanomater 2014:1–19

Oh JY, Kim HS, Palanikumar L, Go EM, Jana B, Park SA et al (2018) Cloaking nanoparticles with protein corona shield for targeted drug delivery. Nat Commun 9(1):4548

Owensiii D, Peppas N (2006) Opsonization, biodistribution, and pharmacokinetics of polymeric nanoparticles. Int J Pharm 307(1):93–102

Pan Y, Chen H, Yu J (2020) Biomarkers in hepatocellular carcinoma: current status and future perspectives. Biomedicines 8(12):576

Pandey P, Rahman M, Bhatt PC, Beg S, Paul B, Hafeez A et al (2018a) Implication of nano-antioxidant therapy for treatment of hepatocellular carcinoma using PLGA nanoparticles of rutin. Nanomedicine 13(8):849–870

Pandey P, Rahman M, Bhatt PC, Beg S, Paul B, Hafeez A et al (2018b) Implication of nano-antioxidant therapy for treatment of hepatocellular carcinoma using PLGA nanoparticles of rutin. Nanomedicine (Lond) 13(8):849–870

Parikh ND, Mehta AS, Singal AG, Block T, Marrero JA, Lok AS (2020) Biomarkers for the early detection of hepatocellular carcinoma. Cancer Epidemiol Biomarkers Prev 29(12):2495–2503

Parmar K, Patel JK (2019) Surface modification of nanoparticles to oppose uptake by the mononuclear phagocyte system. In: Surface modification of nanoparticles for targeted drug delivery. Springer International Publishing, Cham, pp 221–236

Peer D, Karp JM, Hong S, Farokhzad OC, Margalit R, Langer R (2007) Nanocarriers as an emerging platform for cancer therapy. Nat Nanotechnol 2(12):751–760

Pietras K, Ostman A (2010) Hallmarks of cancer: interactions with the tumor stroma. Exp Cell Res 316(8):1324–1331

Pranatharthiharan S, Patel MD, Malshe VC, Pujari V, Gorakshakar A, Madkaikar M et al (2017) Asialoglycoprotein receptor targeted delivery of doxorubicin nanoparticles for hepatocellular carcinoma. Drug Deliv 24(1):20–29

Qi B, Crawford AJ, Wojtynek NE, Talmon GA, Hollingsworth MA, Ly QP et al (2020) Tuned near infrared fluorescent hyaluronic acid conjugates for delivery to pancreatic cancer for intraoperative imaging. Theranostics 10(8):3413–3429

Quagliariello V, Masarone M, Armenia E, Giudice A, Barbarisi M, Caraglia M et al (2019) Chitosan-coated liposomes loaded with butyric acid demonstrate anticancer and anti-inflammatory activity in human hepatoma HepG2 cells. Oncol Rep 41(3):1476–1486

Rabanel JM, Faivre J, Tehrani SF, Lalloz A, Hildgen P, Banquy X (2015) Effect of the polymer architecture on the structural and biophysical properties of PEG–PLA nanoparticles. ACS Appl Mater Interfaces 7(19):10374–10385

Rahman M, Akhter S, Ahmad MZ, Ahmad J, Addo RT, Ahmad FJ et al (2015a) Emerging advances in cancer nanotheranostics with graphene nanocomposites: opportunities and challenges. Nanomedicine 10(15):2405–2422

Rahman M, Ahmad M, Ahmad J, Firdous J, Ahmad F, Mushtaq G et al (2015b) Role of graphene nano-composites in cancer therapy: theranostic applications, metabolic fate and toxicity issues. Curr Drug Metab 16(5):397–409

Rahman M, Al-Ghamdi SA, Alharbi KS, Beg S, Sharma K, Anwar F et al (2019) Ganoderic acid loaded nano-lipidic carriers improvise treatment of hepatocellular carcinoma. Drug Deliv 26(1):782–793

Rampado R, Crotti S, Caliceti P, Pucciarelli S, Agostini M (2020) Recent advances in understanding the protein corona of nanoparticles and in the formulation of "stealthy" nanomaterials. Front Bioeng Biotechnol 8:166

Ren J, Cai R, Wang J, Daniyal M, Baimanov D, Liu Y et al (2019) Precision nanomedicine development based on specific opsonization of human cancer patient-personalized protein coronas. Nano Lett 19(7):4692–4701

Resnik DB, Tinkle SS (2007) Ethical issues in clinical trials involving nanomedicine. Contemp Clin Trials 28(4):433–441

Ruoslahti E (2002) Antiangiogenics meet nanotechnology. Cancer Cell 2(2):97–98

Sadauskas E, Wallin H, Stoltenberg M, Vogel U, Doering P, Larsen A et al (2007) Kupffer cells are central in the removal of nanoparticles from the organism. Part Fibre Toxicol 4(1):10

Saha K, Rahimi M, Yazdani M, Kim ST, Moyano DF, Hou S et al (2016) Regulation of macrophage recognition through the interplay of nanoparticle surface functionality and protein corona. ACS Nano 10(4):4421–4430

Sangiovanni A, Del Ninno E, Fasani P, De Fazio C, Ronchi G, Romeo R et al (2004) Increased survival of cirrhotic patients with hepatocellular carcinoma detected during surveillance☆. Gastroenterology 126(4):1005–1014

Sarika PR, James NR, Kumar PRA, Raj DK, Kumary TV (2015) Gum arabic-curcumin conjugate micelles with enhanced loading for curcumin delivery to hepatocarcinoma cells. Carbohydr Polym 134:167–174

Sebak S, Mirzaei M, Malhotra M, Kulamarva A, Prakash S (2010) Human serum albumin nanoparticles as an efficient noscapine drug delivery system for potential use in breast cancer: preparation and in vitro analysis. Int J Nanomedicine 5:525–532

Shao D, Li J, Zheng X, Pan Y, Wang Z, Zhang M et al (2016) Janus "nano-bullets" for magnetic targeting liver cancer chemotherapy. Biomaterials 100:118–133

Sharma AK, Gothwal A, Kesharwani P, Alsaab H, Iyer AK, Gupta U (2017) Dendrimer nanoarchitectures for cancer diagnosis and anticancer drug delivery. Drug Discov Today 22(2):314–326

Shi X, Chen M, Feng H, Zhou Z, Wu R, Li W et al (2021) Glypican-3 electrochemical aptasensor based on reduced graphene oxide-chitosan-ferrocene deposition of platinum–palladium bimetallic nanoparticles. J Appl Electrochem 51(5):781–794

Shubhika K (2013) Nanotechnology and medicine—the upside and the downside. Int J Drug Dev Res 5(1):1–10

Siafaka P, Betsiou M, Tsolou A, Angelou E, Agianian B, Koffa M et al (2015) Synthesis of folate-pegylated polyester nanoparticles encapsulating ixabepilone for targeting folate receptor overexpressing breast cancer cells. J Mater Sci Mater Med 26(12):275

Siafaka PI, Üstündağ Okur N, Karavas E, Bikiaris DN (2016) Surface modified multifunctional and stimuli responsive nanoparticles for drug targeting: current status and uses. Int J Mol Sci 17(9):1440

Siafaka PI, Okur NÜ, Karantas ID, Okur ME, Gündoğdu EA (2021) Current update on nanoplatforms as therapeutic and diagnostic tools: a review for the materials used as nanotheranostics and imaging modalities. Asian J Pharm Sci 16(1):24–46

Siemann DW (2011) The unique characteristics of tumor vasculature and preclinical evidence for its selective disruption by tumor-vascular disrupting agents. Cancer Treat Rev 37(1):63–74

Strauss O, Dunbar PR, Bartlett A, Phillips A (2015) The immunophenotype of antigen presenting cells of the mononuclear phagocyte system in normal human liver—A systematic review. J Hepatol 62(2):458–468

Su YH, Zhang JG, Shen J, Li FQ, Su H, Ling CQ (2009) Preparation of cinobufagin-loaded bovine serum albumin nanoparticles for hepatocarcinoma therapy. Nano 4(1):47–54

Sun ZF, Chang Y, Xia N (2021) Recent development of nanomaterials-based cytosensors for the detection of circulating tumor cells. Biosensors (Basel) 11(8):281

Tavares AJ, Poon W, Zhang YN, Dai Q, Besla R, Ding D et al (2017) Effect of removing Kupffer cells on nanoparticle tumor delivery. Proc Natl Acad Sci 114(51):E10871–E10880

Thyagarajan A, Sahu RP (2018) Potential contributions of antioxidants to cancer therapy: immunomodulation and radiosensitization. Integr Cancer Ther 17(2):210–216

Tingting L, Li L, Li L, Jing C (2023) Exploiting targeted nanomedicine for surveillance, diagnosis, and treatment of hepatocellular carcinoma. Mater Today Bio 22:10766

Torchilin VP (2007) Micellar nanocarriers: pharmaceutical perspectives. Pharm Res 24(1):1–16

Trédan O, Galmarini CM, Patel K, Tannock IF (2007) Drug resistance and the solid tumor microenvironment. J Natl Cancer Inst 99(19):1441–1454

Tsuda T, Kaibori M, Hishikawa H, Nakatake R, Okumura T, Ozeki E et al (2017) Near-infrared fluorescence imaging and photodynamic therapy with indocyanine green lactosome has antineoplastic effects for hepatocellular carcinoma. PLoS One 12(8):e0183527

Turato C, Balasso A, Carloni V, Tiribelli C, Mastrotto F, Mazzocca A et al (2017) New molecular targets for functionalized nanosized drug delivery systems in personalized therapy for hepatocellular carcinoma. J Control Release 268:184–197

Ungureanu BS, Teodorescu CM, Săftoiu A (2016) Magnetic nanoparticles for hepatocellular carcinoma diagnosis and therapy. J Gastrointestin Liver Dis 25(3):375–383

Urano Y, Asanuma D, Hama Y, Koyama Y, Barrett T, Kamiya M et al (2009) Selective molecular imaging of viable cancer cells with pH-activatable fluorescence probes. Nat Med 15(1):104–109

Usmani A, Mishra A, Ahmad M (2018) Nanomedicines: a theranostic approach for hepatocellular carcinoma. Artif Cells Nanomed Biotechnol. 46(4):680–690

Varshosaz J (2015) Nanoparticles for targeted delivery of therapeutics and small interfering RNAs in hepatocellular carcinoma. World J Gastroenterol 21(42):12022

Vines JB, Yoon JH, Ryu NE, Lim DJ, Park H (2019) Gold nanoparticles for photothermal cancer therapy. Front Chem 7:167

Vroman L, Lukosevicius A (1964) Ellipsometer recordings of changes in optical thickness of adsorbed films associated with surface activation of blood clotting. Nature 204(4959):701–703

Vroman L, Adams AL, Fischer GC, Munoz PC (1980) Interaction of high molecular weight kininogen, factor XII, and fibrinogen in plasma at interfaces. Blood 55(1):156–159

Wake K, Karl Wilhelm Kupffer and his contributions to modern hepatology (2004) Comp Hepatol 3(Suppl 1):S2

Walkey CD, Chan WCW (2012) Understanding and controlling the interaction of nanomaterials with proteins in a physiological environment. Chem Soc Rev 41(7):2780–2799

Wang Z, Shao D, Chang Z, Lu M, Wang Y, Yue J et al (2017) Janus gold nanoplatform for synergetic chemoradiotherapy and computed tomography imaging of hepatocellular carcinoma. ACS Nano 11(12):12732–12741

Whiteside TL (2008) The tumor microenvironment and its role in promoting tumor growth. Oncogene 27(45):5904–5912

Wilhelm S, Tavares AJ, Dai Q, Ohta S, Audet J, Dvorak HF et al (2016) Analysis of nanoparticle delivery to tumours. Nat Rev Mater 1(5):16014

Wong CM, Tsang FHC, Ng IOL (2018) Non-coding RNAs in hepatocellular carcinoma: molecular functions and pathological implications. Nat Rev Gastroenterol Hepatol 15(3):137–151

Wu H, Wang MD, Liang L, Xing H, Zhang CW, Shen F et al (2021) Nanotechnology for hepatocellular carcinoma: from surveillance, diagnosis to management. Small 17(6):e2005236

Xia Q, Li L, Zhao L (2017) Silica nanoparticle-based dual-responsive nanoprodrug system for liver cancer therapy. Exp Ther Med 14(3):2071–2077

Xiao K, Li Y, Luo J, Lee JS, Xiao W, Gonik AM et al (2011) The effect of surface charge on in vivo biodistribution of PEG-oligocholic acid based micellar nanoparticles. Biomaterials 32(13):3435–3446

Xu F, Reiser M, Yu X, Gummuluru S, Wetzler L, Reinhard BM (2016) Lipid-mediated targeting with membrane-wrapped nanoparticles in the presence of corona formation. ACS Nano 10(1):1189–1200

Xu M, Yang L, Lin Y, Lu Y, Bi X, Jiang T et al (2022) Emerging nanobiotechnology for precise theranostics of hepatocellular carcinoma. J Nanobiotechnol 20(1):427

Yan Y, Gause KT, Kamphuis MMJ, Ang CS, O'Brien-Simpson NM, Lenzo JC et al (2013) Differential roles of the protein corona in the cellular uptake of nanoporous polymer particles by monocyte and macrophage cell lines. ACS Nano 7(12):10960–10970

Yan H, Gao X, Zhang Y, Chang W, Li J, Li X et al (2018) Imaging tiny hepatic tumor xenografts via endoglin-targeted paramagnetic/optical nanoprobe. ACS Appl Mater Interfaces 10(20):17047–17057

Yang JD, Nakamura I, Roberts LR (2011) The tumor microenvironment in hepatocellular carcinoma: current status and therapeutic targets. Semin Cancer Biol 21(1):35–43

Yaswen P, MacKenzie KL, Keith WN, Hentosh P, Rodier F, Zhu J et al (2015) Therapeutic targeting of replicative immortality. Semin Cancer Biol 35 Suppl(Suppl):S104–S128

Zacharakis G, Aleid A, Aldossari KK (2018) New and old biomarkers of hepatocellular carcinoma. Hepatoma Res 4(10):65

Zhang X, Zhang P (2017) Polymersomes in nanomedicine—a review. Curr Nanosci 13(2):124–129

Zhang YN, Poon W, Tavares AJ, McGilvray ID, Chan WCW (2016) Nanoparticle–liver interactions: cellular uptake and hepatobiliary elimination. J Control Release 240:332–348

Zhong Y, Meng F, Deng C, Zhong Z (2014) Ligand-directed active tumor-targeting polymeric nanoparticles for cancer chemotherapy. Biomacromolecules 15(6):1955–1969

Progress and Prospect of Nanocarriers: Design, Concept, and Recent Advances

8

Mohd Ahmar Rauf, Saadullah Khattak, Mohammad Oves, and Mohammad Azam Ansari

Abstract

In the realm of contemporary scientific inquiry, nanomedicine and nanodelivery systems represent a burgeoning and dynamic sector. This discipline harnesses nanoscale materials for dual purposes: as diagnostic instruments and as vehicles for the targeted and controlled conveyance of therapeutic agents. The utilization of nanotechnology in this context is pivotal for facilitating the precision-targeted administration of pharmaceuticals for the management of chronic human ailments. Recent years have witnessed the proliferation of significant deployments of nanomedicine, encompassing a spectrum of chemotherapeutic, biological, and immunotherapeutic agents, applied to a variety of disease conditions. This chapter delves into recent advancements in nanomedicine and nano-based drug delivery systems. It scrutinizes both the discovery and application of nanomaterials, focusing on their role in augmenting the effectiveness of both emergent and established pharmaceuticals, as well as in selective diagnosis through disease marker molecules. Furthermore, it conducts a comprehensive examination of the merits and demerits of nanomedicines, from their synthesis or natural derivation to their clinical applications. This chapter also integrates insights into the prevail-

M. A. Rauf (✉)
Department of Surgery, Miller School of Medicine, University of Miami, Miami, FL, USA
e-mail: mxr2481@miami.edu

S. Khattak
The 5th Affiliated Hospital of Wenzhou Medical University, Lishui, Zhejiang, China

M. Oves
Center of Excellence in Environmental Studies, King Abdulaziz University, Jeddah, Saudi Arabia

M. A. Ansari
Department of Epidemic Disease Research, Institute for Research and Medical Consultations, Imam Abdulrahman Bin Faisal University, Dammam, Saudi Arabia

M. A. Barkat et al. (eds.), *Nanotheranostics for Diagnosis and Therapy*,
https://doi.org/10.1007/978-981-97-3115-2_8

ing trends and prospective trajectories in the domain of nanomedicine, providing a holistic overview of its current state and future potential.

Keywords

Drug delivery · Nanomedicine · Polymeric · Inorganic · Pharmaceuticals

8.1 Introduction

Throughout ancient history, humans have extensively depended on botanical-based traditional treatments as a primary means of addressing diverse health conditions. A significant proportion of contemporary pharmaceuticals are sourced from botanical sources through the utilization of traditional knowledge and methodologies. Natural resources account for more than 25% of the essential pharmaceutical compounds and their derivatives that are now accessible (Dahl et al. 2007). Natural compounds with diverse molecular architectures serve as a foundation for the exploration and identification of novel pharmaceutical agents. In contemporary natural product-based medication development, there has been a notable shift towards prioritizing the construction of lead compounds that possess synthetic feasibility and closely resemble the chemical properties of their natural product equivalents (Slocik et al. 2006). Natural products are widely recognized for their extensive range of chemical compounds as well as their ability to exhibit unique interactions with macromolecules and possess low levels of toxicity. Consequently, they exhibit potential as viable candidates for the exploration and development of novel pharmaceuticals (Kharissova et al. 2016; Shukla and Iravani 2016). Moreover, computational investigations have played a crucial role in the anticipation of drug-drug interactions and the advancement of innovative drug discovery, including target-oriented drug exploration and delivery methods.

Despite the various benefits, pharmaceutical companies are reluctant to allocate more resources towards the development and implementation of natural-product-based drug discovery and delivery systems (Brinker 2006). Instead, they tend to prioritize the exploration of existing chemical compound libraries to discover new pharmaceuticals. Currently, there is ongoing research on the potential use of natural chemicals in the treatment of significant diseases such as cancer, diabetes, cardiovascular disorders, inflammatory conditions, and microbial infections. The primary reason for this phenomenon can be attributed to the several advantages associated with natural drugs, such as minimal toxicity and side effects, affordability, and significant therapeutic potential. Nevertheless, concerns regarding the biocompatibility and toxicity of natural compounds pose challenges for their application in the field of medicine. Numerous natural compounds have encountered challenges that hinder their advancement in various phases of clinical trials (Kingston 2011; Ray et al. 2009). The utilization of macroscopic materials in the context of drug administration poses notable challenges such as in vivo instability, limited bioavailability and solubility, inadequate absorption within the organism, difficulties in achieving target-specific delivery and desired therapeutic efficacy, and the potential occurrence of undesired pharmacological effects. Therefore, an effective method for addressing these fundamental challenges was proposed by Rehman et al. (2020) and Tatiparti et al. (2020a). Nanotechnology plays a crucial role in the advancement of sophisticated medicine and drug formulations as well as in the targeted distribution and controlled release of drugs. This has been demonstrated with significant success in the field of nanomedicine and nano-based drug delivery systems, which have garnered particular attention (Kumar Teli et al. 2010; Lakhal and Wood 2011; Peng et al. 2011). Nanomaterials encompass substances with diameters smaller than 100 nm, and they play a significant role in advancing various aspects of nanomedicine. These include biosensors, microfluidics, drug delivery systems, microarray tests, and tissue engineering, as demonstrated by studies conducted by Seo et al. (2020); Shakir et al. (2016); Handa et al. 2021; and Wang (2005). Nanotechnology is employed in the development of nanomedicines, which involves the utilization of therapeutic agents at a minuscule level. Nanoparticles have significantly contributed to the advancement of the field of biomedicine, including several areas, such as nanobiotechnology, drug delivery, biosensors, and tissue engineering. Nanoparticles are frequently characterized by their composition, which consists of materials engineered at the atomic or molecular scale, resulting in the formation of diminutive nanospheres (Rehman et al. 2016, 2018). Consequently, smaller materials exhibit enhanced mobility throughout the human body compared with larger substances. Nanoparticles possess a distinctive amalgamation of structural, chemical, mechanical, magnetic, electrical, and biological functionalities. The utilization of nanomedicines has experienced a surge in popularity in recent times, owing to their capacity to serve as delivery vehicles for therapeutic drugs. This is achieved through the encapsulation or attachment of medications, enabling more precise and controlled release to specific tissues (Alsaab et al. 2018; Tatiparti et al. 2020b; Tatiparti et al. 2020a). Nanomedicine is an emerging field that utilizes the principles and methodologies of nanoscience to advance medical biology as well as enhance disease prevention and treatment strategies. The utilization of nanoscale materials, including nanorobots and nano-sensors, is involved in several applications, such as diagnostics, drug transport, sensory functions, and the manipulation of materials within living cells (see Fig. 8.1). An innovative strategy utilizing nanoparticles has been developed to integrate cancer diagnosis with both treatment and imaging techniques (Duce and Bush 2010; Laroui et al. 2013; Matthews et al. 2010). The first iteration of nanoparticle-based therapy involved the utilization of lipid systems, specifically liposomes and micelles, which have subsequently received approval from the FDA (Bozzuto and Molinari 2015; Felnerova et al. 2004). Inorganic nanoparticles, such as gold or magnetic nanoparticles, have been effectively encapsulated within liposomes and micelles (Chen et al. 2008; Felnerova et al. 2004; Wang et al. 2009). These characteristics have led to heightened utilization of inorganic nanoparticles for drug delivery, imaging, and therapeutic purposes.

Furthermore, nanostructures are believed to play a role in preventing the contamination of pharmaceuticals within the gastrointestinal tract and facilitating the targeted distribution of drugs with low water solubility to their intended sites.

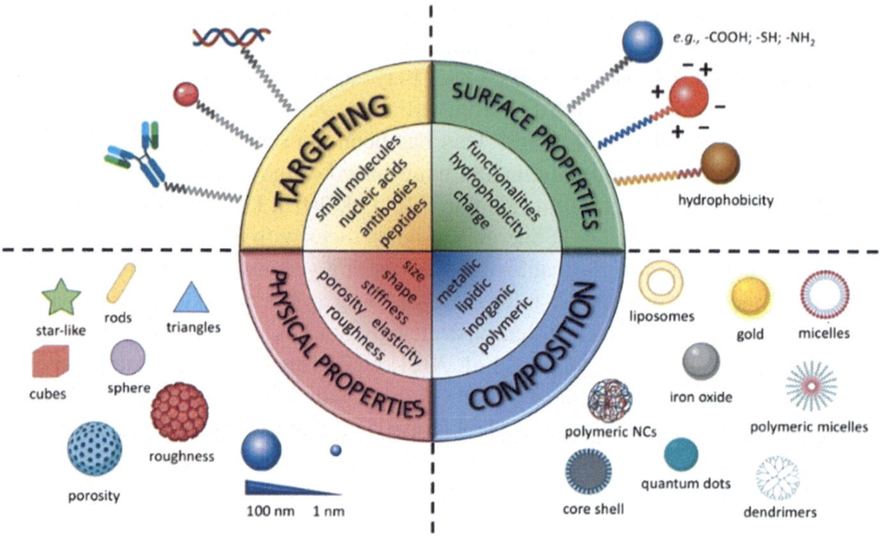

Fig. 8.1 Different types of nanoparticles for targeting drugs. Image with reprint permission (Salvioni et al. 2019)

Nanodrugs exhibit enhanced oral bioavailability compared with conventional drugs owing to their characteristic absorptive endocytosis mechanisms of absorption.

Nanostructures exhibit prolonged retention inside the bloodstream, thus facilitating precise administration of coalesced therapeutic agents at the prescribed dosage. Consequently, these phenomena lead to a reduction in plasma fluctuations and adverse outcomes. The nanoscale dimensions of these structures allow their penetration into the tissue system, facilitating the cellular uptake of medication, promoting effective drug delivery, and ensuring targeted action at the desired location. According to recent studies conducted by Rauf et al. (2019b) and Oves et al. (2019), it has been observed that nanostructures exhibit a higher degree of cellular absorption compared to larger particles within the size range of 1 10 m. Consequently, these entities collaborate harmoniously to enhance the efficacy of cell repair while minimizing or eliminating detrimental consequences.

The efficacy of nanoparticles in acquiring information across all stages of clinical therapy has been demonstrated owing to their utilization in a diverse range of pioneering tests for disease treatment and diagnosis. The principal benefits of these nanoparticles are associated with their surface characteristics, which facilitate the binding of various proteins. AuNPs are commonly employed as biomarkers and tumor labels in various assays for biomolecule detection. The selection of a nanoparticle for medication administration in the context of nanomaterials is determined by the physicochemical characteristics of the drug, as discussed in previous studies (Elbakry et al. 2009). Peng et al. 2014; Sivanesan and Rajeshkumar 2019). The integration of nanoscience with bioactive natural substances has garnered significant interest and witnessed rapid growth in recent years. There are several benefits associated with the use of this approach for administering natural compounds for the treatment of cancer and various other ailments. Natural chemicals have been extensively investigated for their therapeutic properties because of their various qualities, including their ability to induce tumor-suppressive autophagy and function as antimicrobial agents. Previous studies have shown that curcumin and caffeine induce autophagy (Akram et al. 2010; Tiwari et al. 2014). Additionally, the antimicrobial activities of cinnamaldehyde, carvacrol, curcumin, and eugenol have been previously reported (Alam et al. 2009; Rauf et al. 2018).

The incorporation of nanoparticles resulted in the augmentation of various characteristics, including improved bioavailability, greater targeting capabilities, and enhanced control over release mechanisms. One area of current research involves the examination of thymoquinone, a biologically active compound derived from Nigella sativa, which is being studied for its incorporation into lipid nanocarriers. The studies conducted by Al-Haj et al. (2010) and Yaghmur et al. (2019) revealed a significant enhancement in bioavailability, namely a sixfold increase, when thymoquinone was encapsulated as opposed to being in its free form. The encapsulation also exhibited a protective effect on the gastrointestinal system. Furthermore, it enhances the pharmacokinetic features of natural substances, leading to enhanced therapeutic advantages.

8.2 Nano-Based Drug Delivery Systems

In recent years, significant advancements have been achieved in the domain of delivery systems used for the administration of therapeutic agents or naturally derived active compounds to specific target sites with the aim of treating various medical conditions (Dahl et al. 2007; Gunalan et al. 2012; Kharissova et al. 2016). Although some drug delivery technologies have demonstrated efficacy in recent years, certain challenges persist, necessitating the development of advanced technologies to ensure the precise and efficient delivery of pharmaceuticals to their intended destinations. Nano-based drug delivery systems are currently being investigated to facilitate the advancement of improved medication delivery systems.

8.3 Fundamentals of Nanotechnology-Based Drug Design Methodologies

Nanomedicine is a field within the medical domain that employs nanotechnology principles to address the prevention and treatment of diverse diseases. This is achieved through the utilization of nanoscale materials, such as biocompatible nanoparticles and nanorobots, for a range of applications, including disease diagnosis, drug delivery, sensory applications, and actuation. These applications have been extensively studied and documented by various researchers (Khan et al. 2015; Rauf et al. 2019b; Sivanesan and Rajeshkumar 2019; Croy and Kwon 2006; Vyas et al. 2005; Alhodieb et al. 2023; Hervé et al. 2008). Pharmaceutical compounds characterized by significantly low solubility present various challenges in terms of biopharmaceutical delivery. These challenges include restricted bioavailability after oral administration, reduced ability to permeate the outer membrane, heightened dosage requirements for intravenous administration, and adverse effects preceding conventional formulation-based vaccination procedures. The aforementioned limitations can be effectively addressed through the integration of nanotechnology into the drug delivery mechanism.

Extensive research has been conducted on drug design at the nanoscale, which represents a highly sophisticated technology in the realm of nanoparticle applications. This area of study holds significant promise in terms of its ability to adjust several aspects, such as solubility, drug release patterns, diffusivity, bioavailability, and immunogenicity, hence offering major potential benefits. This phenomenon can lead to the emergence of easily accessible administration techniques, a reduction in toxicity levels, a decrease in side effects, an improvement in biodistribution, and an extension of the drug's life cycle. Customized drug delivery systems can be targeted towards a specific site or engineered to achieve a controlled distribution of medicinal chemicals to that particular spot. The construction of these structures or patterns is achieved through a process known as self-assembly, wherein building blocks undergo spontaneous organization (Gref et al. 2000). In addition, it is necessary for them to surmount obstacles such as opsonization/sequestration by the mononuclear phagocyte system (Klajnert et al. 2007; Maeda et al. 2000).

There are two primary methods via which nanostructures can deliver medications: passive and autonomous. In the previous scenario, medications were predominantly incorporated into the internal cavity of the structure through a hydrophobic effect. Controlled release of medication can be achieved by directing nanostructured materials to certain regions, where the hydrophobic nature of the drug leads to its release in the desired quantity, owing to its low concentration. In contrast, the latter approach facilitates convenient delivery of drugs through direct conjugation to the carrier nanostructure material. The timing of drug release is of the utmost importance in this technology because of the inability of the drug to effectively reach the intended target site and its tendency to quickly disassociate from the carrier. Conversely, if a drug is released from its nanocarrier system at an appropriate moment, its bioactivity and efficacy will be diminished. Drug targeting is a crucial aspect that involves the utilization of nanomaterials or nanoformulations as vehicles for drug delivery. This approach can be classified into two categories: active and passive targeting. Active targeting entails the attachment of molecules such as antibodies and peptides to a drug delivery system. This conjugation enables the drug delivery system to specifically bind to receptor structures expressed at the desired target site. Passive targeting involves circulation of the drug carrier complex inside the bloodstream, whereby it is transported to the desired site through affinity or binding mechanisms. This process is regulated by various factors such as pH, temperature, molecular site, and shape, as shown in Fig. 8.2. Within the organism, the main focal points reside in the cellular framework, encompassing receptors situated on cell membranes, the lipid constituents of cell membranes, and the antigens or proteins present on cell surfaces (Ansari et al. 2020; Cetin et al. 2005; Davis et al.

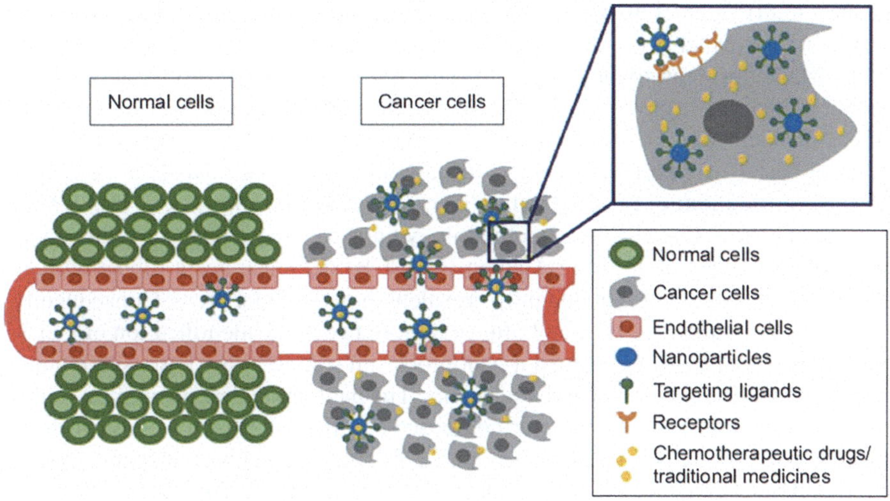

Fig. 8.2 Active and passive targeting of nanoparticles. Image presented with permission (Muhamad et al. 2018)

2008; Krishnamurthy et al. 2016; Moreira et al. 2008). Currently, the predominant focus of nanotechnology-based drug delivery systems is on cancer and its therapeutic interventions.

The term "theranostics" is frequently used within the domain of cancer therapy to denote the amalgamation of therapeutic measures with diagnostic methodologies (Krishnamurthy et al. 2016; Tortorella and Karagiannis 2014). Theranostic nanoparticles can enhance the process of diagnosing, localizing, identifying the stage of disease, and monitoring the response to therapy. Furthermore, prior research conducted by Chen et al. (2014) and Li et al. (2019) has provided evidence to support the notion that nanoparticles exhibit the capacity to selectively deliver therapeutic medicines for tumor treatment. This facilitates precise administration of medicinal substances at the required levels via molecular and/or environmental cues. Chitosan, a biopolymer, demonstrates a notable level of biocompatibility and is characterized by the presence of functional groups (Aktaş et al. 2005; Anitha et al. 2009; Lu et al. 2017; Shakir et al. 2018). Encapsulation or coating techniques on various nanoparticles have been employed to produce multifunctional particles that have the potential to be utilized in the detection and diagnosis of a diverse array of diseases.

The study conducted by Yang et al. (2011) developed nanoparticles that exhibited a significant ability to selectively target cells affected by colorectal cancer (CC) using a light-mediated mechanism. The aforementioned outcome was attained through the physical conjugation of alginate with folic acid-modified chitosan. This process led to the creation of nanoparticles that demonstrated an improved release of 5-aminolevulinic acid (5-ALA) within the cell lysosome. The results of the investigation indicate that the nanoparticles, following modification, were efficiently internalized by CC cells via an endocytic mechanism facilitated by the folate receptor (Korthagen et al. 2019; Xia et al. 2009). Following this, 5-aminolevulinic acid (5-ALA), which has a positive charge, was transported into the lysosome with the assistance of deprotonated alginate. This transport process results in the accumulation of protoporphyrin IX (PpIX) within the cellular milieu, serving as a photodynamic detector. The study's findings demonstrate that the incorporation of chitosan-based nanoparticles, along with alginate and folic acid, is a successful strategy for facilitating the targeted delivery of 5-aminolevulinic acid (5-ALA) to cervical cancer (CC) cells. This methodology enhances the efficiency of the endoscopic fluorescence monitoring. Cathepsin B (CB) demonstrates a robust correlation with metastasis and is abundantly prevalent in pericellular areas. Consequently, circulating biomarkers (CB) are of utmost importance in the identification of metastasis. Ryu et al. (2014) developed a nanoprobe referred to as a CB-sensitive nanoprobe (CB-CNP) in their research endeavors. The nanoprobe described in this study consists of self-assembled structures composed of CB-CNP, which are glycol chitosan nanoparticles (CNPs) that incorporate a fluorogenic peptide attached to their exterior. These CNPs have been engineered to target tumors. The nanoprobe had a spherical structure with a diameter of 280 nm. The object exhibited spherical morphology and did not demonstrate fluorescence under biological conditions. The utilization of a nanoprobe that is sensitive to carbon black (CB) in three rat models that display metastasis has shown the effectiveness of these nanoprobes in distinguishing metastatic cells from healthy cells using non-invasive imaging techniques. Hyaluronic acid (HA) is a distinct biopolymer. Glycosaminoglycan is biocompatible and possesses a negative charge. It functions as a substantial component of the extracellular matrix, as evidenced by investigations undertaken by Lee et al. (2014) and Rauf et al. (2019b). The relationship between the CD44 receptor and HA binding is facilitated through the receptor linker interaction, predominantly observed in malignant cells where the CD44 receptor is highly expressed. Consequently, there has been a notable surge in interest in the application of hyaluronic acid (HA)-modified nanoparticles in the realm of cancer diagnosis and treatment. This is evidenced by the attention received in the studies conducted by Kamat et al. (2010) and Lapcık Jr et al. (1998). In a study conducted by Wang et al. (2018), dopamine-modified hyaluronic acid (HA) was utilized as a coating material for iron oxide nanoparticles (IONP). The nanoparticles comprise a surface with hydrophilic characteristics, whereas their core is hydrophobic and contains the anticancer drug homocamptothecin. The biopotential of this process was examined under controlled laboratory conditions as well as in live cellular contexts. Na et al. (2009) reported an increase in the uptake of nanoparticles by malignant cells when an external magnetic field was utilized in conjunction with MRI. After the intravenous injection of the nano-vehicle at a dosage of 3 mg/kg, significant tumor ablation was noted, leading to nearly total elimination of the tumors following the therapy, in comparison to the delivery of the unbound drug.

Dextran, a neutral polymer, is widely regarded as a pioneering example of a microbial exopolysaccharide utilized in medical contexts. One noteworthy characteristic of dextran is its remarkable tolerance, lack of toxicity, and biodegradability in the human body, which results in the absence of any negative reactions (de Groot et al. 2001; Naeye et al. 2010). Photodynamic therapy (PDT) is a targeted therapeutic approach for cancer that causes minimal harm to healthy cells. Ding et al. (2017) developed a composite system consisting of nanoparticles that exhibited multifunctionality in near-infrared (NIR) and magnetic resonance (MR) imaging. This was achieved by enclosing Fe_3O_4 nanoparticles within dextran nanoparticles coupled with redox-responsive chlorine 6 (C6).

The nanoparticles exhibited a redox biological response that enabled precise tumor imaging by an "on/off" modulation of the fluorescent signal. Furthermore, a notable magnetic targeting capability was discovered both in vitro and in vivo, resulting in the enhanced efficacy of photodynamic therapy. Hong et al. (2016) synthesized theranostic nanoparticles for the purpose to study their effects on C6 murine glioma cells. The nanoparticles were coated with folic acid-conjugated dextran (FA) or paclitaxel (PTX). The bioprotective effects of dextran coating and the chemotherapeutic effect of PTX on C6 glioma cells were assessed using an MTT assay. The nanoparticles generated demonstrated the ability to enter C6 tumor cells via receptor-mediated endocytosis. Furthermore, these nanoparticles display enhanced contrast activity in magnetic resonance imaging (MRI), which is dependent on their concentration. This increase in contrast activity can be attributed to the paramagnetic properties of the gadolinium nanoparticles. The efficacy of multifunctional nanoparticles in lowering cell viability was superior to that of uncoated gadolinium nanoparticles. Therefore, the use of nanoparticles coupled with folic acid (FA) and paclitaxel (PTX) is promising for their application as theragnostic agents, possessing both paramagnetic and chemotherapeutic properties.

8.4 Process and Mechanism of Drug Designing and Delivery

The emergence of nanomedicine, advancements in drug discovery/design, and drug delivery technologies have prompted extensive research into enhancing drug specificity and diagnostic accuracy through various therapeutic procedures and traditional clinical diagnostic methodologies. Novel approaches for drug delivery are currently under investigation, with a particular focus on targeted delivery to specific locations. This emphasis aims to reduce the toxicity of drugs and improve their availability within an organism (Juillerat-Jeanneret 2008; Lalatsa et al. 2014; Leonor Pinzon-Daza et al. 2013).

Drug design has emerged as a promising approach in this domain, involving the discovery of new therapeutic agents by understanding a specific biological target. Advances in computer science and refinement of experimental protocols for the categorization and refinement of proteins, peptides, and biological targets play a pivotal role in fostering the expansion and progress of the field (Araujo-Lima et al. 2017; Gacche et al. 2011). Furthermore, a plethora of research studies and literature reviews have been published in this field. These studies mostly focused on the logical development of diverse compounds and underscored the importance of comprehending the distinct mechanisms associated with drug release (Loo et al. 2016; Mun et al. 2014; Packiavathy et al. 2013). Furthermore, natural products have the potential to offer pragmatic and captivating solutions to challenges in drug design while also serving as a valuable source of inspiration for the discovery of medications possessing desirable physicochemical characteristics.

Furthermore, there has been a notable increase in the importance of pharmaceutical delivery technologies in recent years. The development of these systems is quite complicated, and they possess the ability to provide a controlled release of bioactive compounds within the human body. Chen et al. (2016) conducted a comprehensive review that explored the utilization of nanocarriers in imaging and sensory applications, along with an examination of the therapeutic benefits associated with these systems. Furthermore, Pelaz et al. (2017) provided a comprehensive overview of the various uses of nanocarriers in the field of nanomedicine, presenting current advancements and addressing emerging opportunities and challenges within this sector.

It is noteworthy that each of these drug delivery systems possesses unique chemical, physical, and morphological characteristics, and may demonstrate an inclination towards various drug polarities through chemical interactions such as covalent bonds and hydrogen bonds, or physical interactions such as electrostatic and van der Waals interactions. For example, Mattos et al. (2017) discovered that neem bark extract-grafted biogenic silica nanoparticles exhibited a reduced release profile in terms of chemical interactions compared to neem bark extract-loaded biogenic silica nanoparticles. Therefore, the aforementioned elements influence the interplay between nanocarriers and biological systems, as well as the kinetics of release of the active ingredient within the organism. Furthermore, Sethi et al. (2014) conducted a study in which they developed a lipid shell known as cross linkable lipid shell (CLS) that contained docetaxel and wortmannin. The lipid shell was designed as a prototype medicine to manipulate the kinetics of drug discharge. The researchers then examined the discharge profile of this lipid shell and observed that it was influenced by both in vivo and in vitro conditions.

Furthermore, it is important to consider additional factors that significantly influence the drug-delivery properties of nanocarriers. These factors include the composition of the carriers, which may consist of organic, inorganic, or hybrid materials, as well as the specific approach employed to attach pharmaceutical substances to the carriers. Examples of such attachment methods include the utilization of a core-shell system or a matrix system (Liong et al. 2008; Na et al. 2009; Slocik et al. 2006; Sokolova and Epple 2008). Extensive studies have been conducted to investigate the mechanisms underlying the release of medicines from nanocarriers. Various techniques can be employed to characterize the release of pharmaceuticals from nanocarriers, including diffusion, solvent-mediated release, chemical-reaction-based release, and stimuli-responsive release. Kamaly et al. (2016) undertook a comprehensive assessment of controlled-release systems, with a specific emphasis on investigations related to the release of pharmaceuticals from polymeric nanocarriers.

Numerous methods are currently under development with the aim of augmenting the specificity of nanocarriers towards particular locations inside the organism, while concurrently mitigating their immunogenicity. The implementation of these tactics encompasses the use of many substances, including polymers, natural polysaccharides, antibodies, cell membranes, tunable surfactants, peptides, and other compounds, to coat or chemically modify nanostructures. In situations where drugs demonstrate inadequate binding and affinity for a specific target or encounter difficulties in crossing physiological barriers, such as the blood-brain or blood-cerebrospinal fluid barriers, scientists have utilized ligand-modified nanocarriers to enhance cellular membrane penetration and accomplish targeted drug delivery within a particular milieu (Bradbury 1985; Duan et al. 2016; Pan et al. 1998). Researchers have employed hyaluronic acid, a polysaccharide found in the extracellular matrix, as a ligand-appended constituent in diverse nanocarriers. The aforementioned investigations have demonstrated encouraging results in enhancing antitumor efficacy against melanoma stem-like cells (Maeda et al. 2013), breast cancer cells (Bhise et al. 2020), and lung adenocarcinoma cells (Hsieh et al. 2011; Maeda et al. 2013). Moreover, previous studies have demonstrated that hyaluronic acid plays a crucial role in enhancing the efficacy of intravitreal drug delivery in retinal gene therapy (Martens et al. 2015). By reducing the immunogenicity however, the development of drug delivery systems incorporating ligand attachments is a laborious process that necessitates the implementation of multiple targeted designs. The consideration of multiple physiological parameters, such as blood circulation, sickness status, and tissue structure, is crucial for the development of these designs.

Furthermore, few studies have been conducted to elucidate the interplay between ligands conjugated to nanocarriers and cellular membranes, and the mechanism by which they are absorbed remains elusive. Furthermore, previous studies have confirmed that cells internalize nanoparticles through either phagocytic or non-phagocytic pathways, such as clathrin-mediated endocytosis and caveolae-mediated endocytosis (Gref et al. 2000). However, establishing a standardized understanding of how these delivery systems function and interact within cells has been challenging because of the distinctive physicochemical properties of nanoparticles. Salatin and Khosroushahi provided a comprehensive overview of the primary endocytosis mechanisms implicated in the internalization of polysaccharide nanoparticles carrying bioactive compounds into cellular structures.

In contrast, stimuli-responsive nanocarriers have demonstrated the ability to control the rate of release of pharmaceutical agents, referred to as triggered release, by utilizing external stimuli, such as ultrasound, heat, magnetism, light, pH, and ionic strength (Chen et al. 2016; Hong et al. 2016; Leonor Pinzon-Daza et al. 2013; Yang 2010). The aforementioned capability exhibits potential in augmenting drug targeting and facilitating accurate regulation of dosage. An instance can be observed whereby superparamagnetic iron oxide nanoparticles are covalently linked to polymeric nanocarriers or lipids, thereby facilitating the establishment of a controlled-release mechanism via the utilization of an external magnetic field. Ulbrich et al. (2016) conducted a scholarly investigation that critically analyzed recent advancements in drug delivery methods, specifically focusing on polymeric and magnetic nanoparticles. This study further examined the influence of drugs that are either covalently or noncovalently bound on the effectiveness of cancer treatment. In addition, Chen et al. (2017) documented the production of Au/Fe_3O_4@polymer nanoparticles, which exhibit promise for application in chemo-photothermal therapy triggered by near-infrared radiation. Hybrid nanocarriers have become a very promising technology in the field of nanomedicine owing to their capacity to amalgamate many system features inside a single organism. The integration process described above results in the creation of materials that demonstrate improved functionality for both therapeutic and diagnostic purposes. These materials are known as theranostic systems. Nevertheless, there is a lack of understanding of the specific processes that drive the actions and toxicological characteristics of drug delivery systems. Therefore, additional research is required to further explore this field. Moreover, there has been a surge in scholarly endeavors centered on the development of nanocarriers that utilize plant extracts and microorganisms.

8.5 Nanoparticles Used in Drug Delivery System

8.5.1 Polymeric Nanoparticles

Various biopolymeric materials are used in pharmaceutical delivery systems. The subsequent sections examine these materials and their respective characteristics.

8.5.2 Chitosan

Chitosan exhibits mucoadhesive properties, rendering it suitable for targeting epithelial junctions that are characterized by high tightness. Chitosan-based nanoparticles are commonly utilized in the development of sustained drug delivery systems for various epithelial tissues such as buccal, intestinal, nasal, ophthalmic, and pulmonary tissues (Alhodieb et al. 2022). In a study conducted by Silva et al. (2017), the researchers developed and evaluated the effectiveness of a 0.75% w/w isotonic solution of hydroxypropyl methylcellulose (HPMC) that incorporated chitosan/sodium tripolyphosphate/hyaluronic acid nanoparticles. The purpose of this formulation is to facilitate the transport of ceftazidime to the eye. The rheological synergism parameter was assessed by measuring the viscosity of nanoparticles in the presence of mucin at different mass ratios. The combination of chitosan nanoparticles with mucin results in a reduction in viscosity. Nevertheless, the nanoparticles demonstrated mucoadhesive properties, leading to beneficial contact with the ocular mucosa and extended antibiotic release. This suggests that nanoparticles have the potential to enhance ocular drug retention time. The cytotoxicity of the nanoparticles was assessed in two cell lines, ARPE-19 and HEK 239 T, revealing their non-cytotoxic nature. Furthermore, the nanoparticles exhibited sustained antibacterial efficacy, rendering them a promising formulation for the delivery of ocular drugs with improved mucoadhesive properties.

Pistone et al. (2017) synthesized chitosan, alginate, and pectin nanoparticles with the aim of explore their potential as drug delivery systems for oral administration. The biocompatibility of the formulations was assessed by evaluating the solubility of the nanoparticles in a simulated salivary environment, whereas the cytotoxicity potential was determined by examining their effects on an oral cell line. According to previous studies by Anitha et al. (2009) and Yang et al. (2011) observed that alginate nanoparticles exhibited stability in artificial saliva for a minimum duration of 2 h. In contrast, pectin and chitosan nanoparticles, especially the latter, were unstable under the same conditions.

In this study, chitosan nanoparticles were observed to exhibit the highest cytocompetitive properties. In contrast, both alginate and pectin nanoparticles demonstrated cytotoxic effects at various doses and time intervals. The observed cytotoxicity may be attributed to the presence of Zn2+ as a cross-linking agent, and every formulation exhibited both advantages and disadvantages with regard to oral cavity release, thus requiring further adjustment.

In a separate investigation, Jain and Jain (2016) conducted a study investigated the release pattern of 5-fluorouracil (5-FU) from chitosan nanoparticles coated with hyaluronic acid, with a focus on its release into the gastrointestinal tract after oral administration. The trials conducted under simulated conditions replicating the transition from the stomach to the colon demonstrated a release profile of 5-FU that exhibited protection against discharge in the stomach and small intestine. Moreover, the localized abundance of drugs can effectively extend the period of exposure, thereby enhancing the effectiveness of anticancer treatment and reducing overall toxicity to the body in the context of colon cancer therapy.

8.5.3 Liposomes

Liposomes are widely utilized in the pharmaceutical and cosmetic industries as carriers for a wide variety of substances. They are considered one of the most extensively studied vehicles for drug delivery. Liposomes have gained significant recognition as a well-established method for enhancing the efficacy of drug delivery. Previous studies have documented the existence of spherical vesicles consisting of phospholipids and steroids, exhibiting a typical size range of 50–450 nm (Onyeji et al. 1994; Tandrup Schmidt et al. 2016; Wei et al. 2015). Drug delivery vehicles are highly regarded for their exceptional properties, mostly attributed to their membrane structure, which closely matches cellular membranes, and their remarkable capacity to efficiently incorporate medicinal compounds. Moreover, prior research has provided empirical support for the capacity of these compounds to augment the stability of therapeutic molecules, boost their systemic dispersion, demonstrate compatibility with both hydrophilic and hydrophobic medications, and undergo biodegradation. Liposomes can be classified into four categories:

1. The term "traditional liposomes" refers to lipid bilayers composed of cholesterol and phospholipids with anionic, cationic, or neutral properties, encompassing an aqueous core material. Materials that demonstrate hydrophobicity or hydrophi-

licity can be used to occupy the lipid bilayer or aqueous gap, respectively (see Fig. 8.3).

2. PEGylated liposomes are subject to modification through the introduction of polyethylene glycol (PEG) onto their surface, with the purpose of achieving steric equilibrium.
3. Ligand-targeted liposomes encompass the process of affixing ligands, such as antibodies, polysaccharides, and peptides, to the outside of liposomes or at the end of pre-attached PEG chains.
4. The theranostic liposome type is a liposome composite that incorporates components from the three previously stated liposome types. The composition generally comprises nanoparticles, accompanied by other components designed for targeting, imaging, and therapeutic objectives.

Liposomes are typically generated via a sequential process that includes several steps: thin-layer hydration, mechanical agitation, solvent evaporation, solvent injection, and surfactant solubilization. It is imperative to emphasize that liposomes exhibit the property of encapsulate pharmaceutical compounds, leaving them inaccessible until release. Therefore, it is imperative to carefully consider the distribution of these substances in specific regions to optimize the accessibility of medications within the intended therapeutic range while also ensuring optimal rates and timings. Pharmaceutical loading into liposomes encompasses both active and passive approaches. The active approach entails drug encapsulation subsequent to liposome synthesis, whereas the passive approach involves drug encapsulation during liposome production. Medications that possess hydrophilic characteristics, such

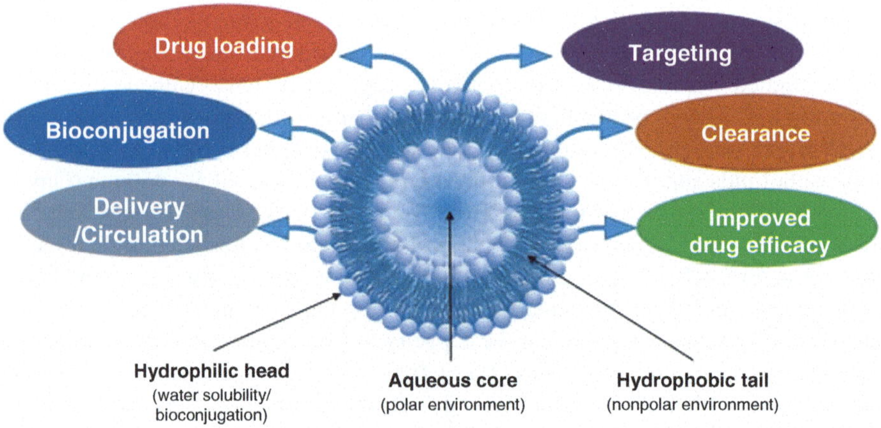

Fig. 8.3 Structural characteristics of liposomes and their advantages for enhanced drug delivery and therapeutic efficacy. Liposomes are composed of an aqueous core surrounded by a lipid bilayer, which provides an environment conducive to the loading of hydrophilic or hydrophobic medicines. The functional groups on the lipid head enable conjugation to polymers, peptides, and proteins, among others, to improve medication targeting, circulation, and efficacy. Image represented with permission from (Bunggulawa et al. 2018)

as ampicillin and 5-fluoro-deoxyuridine, are commonly localized within the water core of liposomes. Therefore, encapsulation of these drugs does not rely on altering their drug-to-lipid ratio. However, previous studies have reported the presence of hydrophobic antibiotics, such as Amphotericin B and Indomethacin, within the acyl hydrocarbon chain of liposomes. Therefore, the encapsulation of these antibiotics into liposomes is contingent upon the distinct properties of the acyl chain. In the context of passive loading procedures, it is noteworthy to mention the mechanical and solvent dispersion methods, as well as the detergent removal strategy.

Despite various challenges, including the reticuloendothelial system (RES), opsonization, and immunogenicity, liposomes have demonstrated the ability to improve the efficiency of drug delivery. This is achieved through mechanisms such as enhanced permeability and the enhanced permeability and retention (EPR) effect (Klajnert et al. 2007; Maeda et al. 2000, 2013; Ansari et al. 2021; Ansari et al. 2022). When liposomes are introduced into the human body, they come into contact with opsonins, high-density lipoproteins (HDLs), and low-density lipoproteins (LDLs) as they circulate in the bloodstream. Opsonins, including immunoglobulins and fibronectin, are essential for the recognition and elimination of liposomes by the reticuloendothelial system (RES). The interaction between high-density lipoproteins (HDLs) and low-density lipoproteins (LDLs) has been reported to induce the formation of complexes with liposomes, potentially compromising their stability.

Liposomes demonstrate a tendency to selectively accumulate in organs such as the liver and spleen, which offers distinct advantages in terms of enabling the targeted administration of therapeutic medicines to these particular locations. Nevertheless, the accumulation of lipophilic anticancer medicines could potentially cause difficulties in the realm of cancer treatment by hindering their prompt elimination. This is the rationale behind the formulation of many types of liposomes, particularly PEGylated liposomes. Dimov et al. (2017) introduced a methodological approach that employs a continuous flow system for the synthesis, functionalization, and purification of liposomes. The current investigation utilized vesicles with a diameter of less than 300 nm in a lab-on-a-chip system, with the objective of enhancing the development of affordable pharmaceuticals and methodologies for encapsulating proteins. The importance of this issue is that the expenses associated with production have a decisive impact on the viability of bringing a particular pharmaceutical product into the commercial market. Liposome-based systems have received authorization from the Food and Drug Administration (FDA).

8.5.4 Micelles

Polymeric micelles are nanostructures that arise from the spontaneous organization of amphiphilic block copolymers in aqueous settings, leading to the development of core-shell shapes. The hydrophobic core can tolerate hydrophobic medicinal molecules, such as camptothecin, docetaxel, or paclitaxel. Simultaneously, the hydrophilic outer layer confers water solubility to the entire system and stabilizes the core. Polymeric micelles typically display an average diameter of less than 100 nm and exhibit a restricted distribution that hinders their rapid removal via the renal system. These characteristics facilitate the accumulation of these substances in tumor tissues by utilizing the increased permeability and retention (EPR) phenomenon. Additionally, the inclusion of a polymeric shell impedes any non-specific interactions that may occur between these entities and biological constituents. The interior core structure of these nanostructures enables the efficient assimilation of hydrophobic pharmaceuticals, thereby improving their stability and bioavailability. This characteristic makes them highly promising for hydrophobic drug administration, as highlighted by the studies conducted by Croy and Kwon (2006) and Wen and Li (2018).

Polymeric micelles can be formed using two distinct methods: (1) through the straightforward process of dissolving the polymer in a suitable solvent and subsequently subjecting it to dialysis or (2) precipitation of one block of the polymer through the addition of a solvent. The process of micelle production is subject to various factors, such as the hydrophobic chain length of the amphiphilic compound, concentration of amphiphiles, solvent system employed, and temperature conditions. Micelle formation is initiated upon the attainment of a critical micelle concentration (CMC) by amphiphilic molecules. Amphiphilic molecules exhibit diminutive sizes and manifest autonomously at reduced concentrations. Three methods are frequently employed for loading pharmaceuticals into polymeric micelles: direct dissolution, solvent evaporation, and dialysis. The copolymer and pharmaceutical compounds spontaneously mix in aqueous media, resulting in the formation of drug-loaded micelles. The solvent evaporation method involves dissolution of both the copolymer and the required medication in a volatile organic solvent. On the other hand, the dialysis technique entails the combination of the drug in solution and the copolymer in the organic solvent within a dialysis bag, followed by dialysis to produce the micelle.

The use of diverse polymeric micelles for drug targeting has been demonstrated through various mechanisms of action. These mechanisms include enhanced penetrability and retention effects, attachment of a specific targeting ligand molecule to the surface of the micelle, or incorporation of monoclonal antibodies into the outer layer of the micelle. Polymeric micelles have demonstrated efficacy in the delivery of cancer drugs as well as in the administration of drugs for eye conditions.

8.5.5 Dendrimers

Dendrimers are complex molecular structures characterized by a highly branched architecture that exhibits monodispersity, well-defined molecular weight, and three-dimensional geometry. The structures of these entities are spherical, and their exteriors may be readily modified in a controlled manner, rendering them suitable for use as vehicles for delivering pharmaceutical substances (Liu et al. 2014; Menjoge et al. 2010). The user does not provide any text. Dendrimers can be synthesized using these two methods. There are two distinct approaches to synthesizing dendrimers: the divergent route, characterized by the formation of the dendrimer from its core and subsequent outward growth, and the convergent method, characterized by the formation of the dendrimer from the periphery.

Dendrimers can be categorized into various types according to their functionalized moieties, including PAMAM, PPI, liquid crystalline, core-shell, chiral, peptide, glycodendrimers, and PAMAMOS. Among these, PAMAM dendrimers have received the most attention in the field of oral drug delivery because of their water solubility and capacity to traverse epithelial tissues, thereby facilitating their paracellular transfer (Luo et al. 2002). The therapeutic usefulness of dendrimers is limited by the presence of amine groups. Owing to their positive charge or cationic nature, these groups provide a potential hazard. Consequently, efforts have often been made to modify or eliminate the toxicity associated with dendrimers. The loading of pharmaceuticals into dendrimers involves many stages, namely simple encapsulation, electrostatic interactions, and covalent conjugation, as depicted in Fig. 8.4.

Dendrimers serve as primary vehicles for drug delivery via two distinct mechanisms. First, in vivo degradation of covalent bonds within the drug dendrimer occurs in the presence of appropriate enzymes or a conducive environment capable of cleaving these bonds. Second, drug release is triggered by alterations in the physical environment such as changes in pH and temperature. Dendrimers have been developed for application in drug delivery systems, specifically for transdermal, oral, ocular, and pulmonary routes of administration.

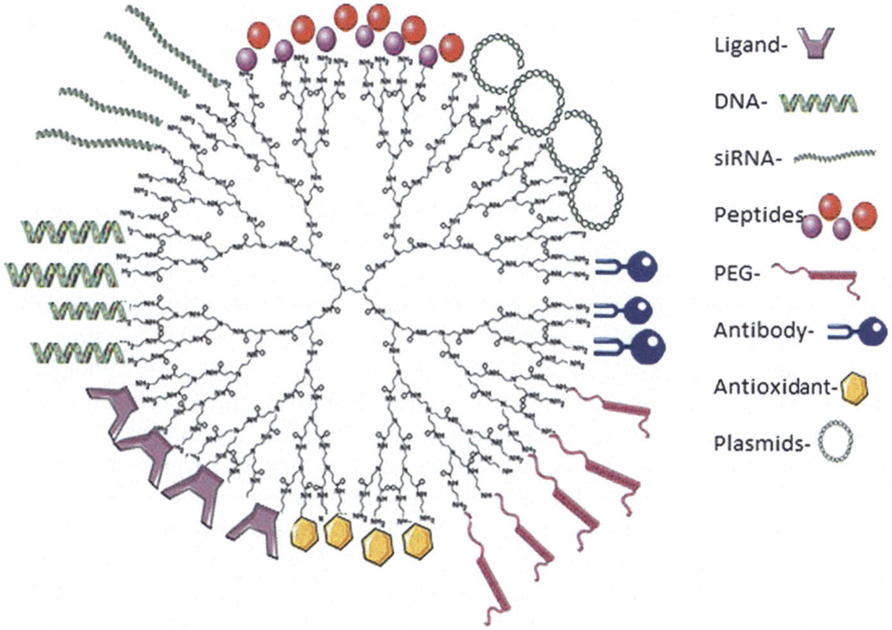

Fig. 8.4 A representation of G4-PAMAM dendrimer depicting some possible surface modifications on their surface. Image adapted from (Fana et al. 2020) with permission

In their study, Jain and Kumar (2010) described folate-conjugated poly-l-lysine dendrimers (doxorubicin hydrochloride) as a potential model for delivering cancer prevention drugs. The authors highlighted the ability of the model to release drugs in a pH-dependent manner, its specificity for targeting cancer cells, and its antiangiogenic and anticancer capabilities. The study provided evidence that the utilization of doxorubicin-folate-conjugated poly-l-lysine dendrimers resulted in elevated accumulation of doxorubicin within the tumor. In a similar vein, Kaur et al. (2017) conducted a study whereby they devised folate-conjugated polypropylene imine dendrimers (FA-PPI) as a pH-responsive nanocarrier for methotrexate (MTX), with the aim of selectively targeting cancer cells for the purpose of anticancer therapy. Researchers demonstrated the ability of the tested substances to achieve sustained release, improve cellular uptake, and exhibit little cytotoxicity in in vitro experiments conducted on MCF-7 cell lines (Kaur et al. 2017). Furthermore, it is important to acknowledge that the synthesized formulations, namely methotrexate (MTX)-loaded and folic acid-conjugated 5.0G PPI (MTX-FAPPI), exhibited enhanced tumor cell uptake compared to methotrexate alone (MTX).

8.5.6 Inorganic Nanoparticles

Nanoparticles composed of inorganic materials. Inorganic nanoparticles encompass a variety of materials, such as silver, gold, iron oxide, and silica nanoparticles. While the number of studies conducted on these nanoparticle types is comparatively limited in comparison to the other types discussed in this section, they do exhibit potential applications. Nevertheless, the application of nanoparticles in therapeutics is still limited, as only a small number of these particles have received approval for medical use. The majority of nanoparticles are still being evaluated in clinical studies. Silver and gold metal nanoparticles possess distinct characteristics, including surface plasmon resonance (SPR), which is not observed in liposomes, dendrimers, or micelles. The study conducted by Zhang et al. (2004) showed several notable benefits of the tested material, such as its remarkable biocompatibility and capacity to be easily modified on the surface. The investigation of their drug delivery efficacy has yielded inconclusive results regarding the potential toxicity associated with their particulate or ionized states. Additionally, despite the proposal of two transport mechanisms, paracellular transport and transcytosis, there is a lack of comprehensive knowledge regarding their in vivo transport and uptake mechanisms (Gericke and Pinches 2006; Kim et al. 2012; Lengke et al. 2006). The attachment of drugs to gold nanoparticles (AuNPs) can be achieved through ionic or covalent bonding as well as physical absorption. This allows for regulated delivery and release of drugs in response to biological stimuli or light activation. Despite the documented antibacterial activity of AgNPs, there is a paucity of research investigating their potential for use in medication delivery. In a study conducted by Prusty and Swain (2018), a hybrid system of interconnected and spongy polyacrylamide/dextran nanohydrogels was produced. This system incorporated covalently bonded AgNPs and was designed for the controlled release of ornidazole. The in vitro release of ornidazole from this hybrid system was found to be 98.5%. In a comparable investigation, iron oxide nanoparticles were produced using laser pyrolysis, followed by the application of coatings consisting of Violamycine B1 and anthracycline antibiotics. Subsequently, the specimens were subjected to cytotoxicity and anti-proliferation assessments using MCF-7 cells, and were further compared to commercially accessible iron oxide nanoparticles.

8.5.7 Metallic Nanoparticles

Recently, there has been increasing scientific interest in the application of metallic nanoparticles in many medical contexts. The applications discussed in this study include bioimaging, biosensors, targeted and sustained drug administration, hyperthermia, and photoablation treatment (Kaur et al. 2017; Prusty and Swain 2018; Ansari et al. 2024a) (see Fig. 8.5). Moreover, by modifying and functionalizing nanoparticles with certain functional groups, they can efficiently bind to antibodies, drugs, and other ligands. This improvement has made these systems highly sought-after for numerous biological applications (Laurent et al. 2010; Miller et al. 1994). Extensive research has been conducted on metallic nanoparticles, specifically gold, silver, iron, and copper nanoparticles. In recent years, there has been a surge in scholarly attention towards various metallic nanoparticles, including zinc oxide, titanium oxide, platinum, selenium, gadolinium, palladium, and cerium dioxide

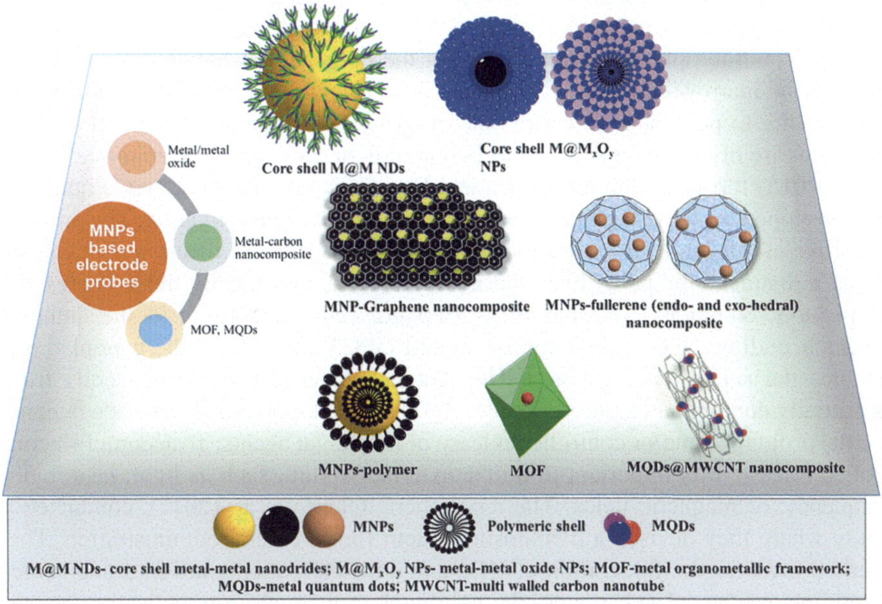

Fig. 8.5 Different types of metallic nanoparticles for development of biosensors. Figure reprinted with permission (Islam et al. 2020)

(Bao et al. 2015; Jahromi et al. 2016; Jia et al. 2017; Khan et al. 2015; Li et al. 2016; Patil et al. 2007; Shoeibi et al. 2017).

8.5.8 Quantum Dots

Quantum dots (QDs) are nanocrystals composed of semiconducting materials that typically fall within the range of 2–10 nm. The optical characteristics of these nanocrystals, including absorbance and photoluminescence, are dependent on their size (Mortensen et al. 2008; Jahangir et al. 2019; Ansari et al. 2024a). The utilization of QDs in the field of nanomedicine has garnered considerable attention because of their capacity to emit light within the near-infrared spectrum at a wavelength of 650 nm. This attribute confers significant benefits in the context of biomedical imaging applications, as it serves to mitigate tissue absorption and diminish light scattering. Consequently, QDs are preferred over conventional organic dyes in this domain. Furthermore, discrete emission colors covering a large range can be generated when QDs of varying sizes and/or compositions are excited using a common light source. In this particular situation, QDs have demonstrated a significant degree of appropriateness for multiplex imaging. Extensive research has been conducted on QDs in the field of medicine, with a primary focus on their application as carriers for drug administration, sensors, and agents for bioimaging. The current corpus of scholarly literature comprises a substantial amount of research on the application of QDs as contrast agents for in vivo imaging. Han et al. (2015) introduced a new fluorophore designed for intravital cytometric imaging. This fluorophore was created using QD antibody conjugates coated with polyimidazole ligands with norbornene activity. The fluorophore was employed for the purpose of achieving in vivo labeling of bone marrow cells. The researchers noted that the fluorophore demonstrated the capacity to permeate the entire bone marrow and efficiently label cell populations that are often untagged, such as hematopoietic stem and progenitor cells. Shi et al. (2015) formulated a multifunctional biocompatible graphene oxide quantum dot enveloped by a bright magnetic nanoplatform in their investigation. The primary objective of this innovative construct was to facilitate the identification and diagnosis of a distinct subtype of liver cancer tumor cells, specifically those expressing glypican-3, commonly referred to as Hep G2 cells. According to the authors' research findings, the conjugation of an anti-GPC3 antibody to the nanoplatform enables the selective isolation of Hep G2 hepatocellular carcinoma cells from potentially contaminated blood samples. QDs have the potential to provide benefits for the prolonged and/or controlled release of medicinal agents. To accomplish controlled release, it is important to utilize external stimuli, such as light, heat, radio frequency, or magnetic fields. The researchers, Olerile et al. (2017), conducted a study where they devised a theranostic system for parenteral administration. This method involves the integration of QDs and anticancer medicines into nanostructured lipid carriers. The nanoparticles exhibited a spherical form, indicating greater efficiency in the encapsulation of paclitaxel (80.7 ± 2.11%) and a substantial reduction in tumor growth (77.85%). Additionally, the researchers noted that the system demonstrated a noteworthy capacity to discern and pinpoint H22 tumor cells with exceptional accuracy. Cai et al. (2016) successfully produced quantum dots that exhibited pH-responsive behavior. This was achieved by coating polyethylene glycol (PEG) and hyaluronic acid (HA) onto ZnO quantum dots. The purpose of this coating was to improve the stability of the quantum dots under physiological conditions and facilitate their specific targeting towards cells that express the HA-receptor CD44. Prolonged release of doxorubicin (DOX) was also examined using this nanocarrier. The nanocarrier was stable when exposed to physiological pH conditions. DOX loading was accomplished by either forming complexes with Zn^{2+} ions or by attaching them to PEG. DOX is specifically released within the acidic intracellular environment of tumor cells owing to the disintegration of ZnO quantum dots (QDs).

8.6 Nanoparticles of Proteins and Carbohydrates

Natural biopolymers are composed of polysaccharides and proteins, and are derived from many biological sources, such as plants, animals, microbes, and marine species. Protein-based nanoparticles possess intrinsic properties that make them susceptible to biodegradation, metabolism, and convenient customization for targeted binding to certain medicinal drugs and other ligands. Protein-based products are frequently manufactured using two unique approaches: (a) employing water-soluble proteins, such as bovine and human serum albumin, and (b) utilizing insoluble proteins, such as zein and gliadin. The procedures frequently utilized for the synthesis of these substances include coacervation/desolvation, emulsion/solvent extraction, complicated coacervation, and electrospraying.

Chemical alterations are applied to protein-based nanoparticles to integrate targeting ligands that selectively recognize cells and tissues, thereby augmenting and facilitating their targeting mechanisms. In contrast, polysaccharides consist of monosaccharide molecules interconnected through O-glycosidic bonds. The distinctive physicochemical features of these polysaccharides result from the makeup of their monomers and their biological origin. One of the main drawbacks associated with the utilization of polysaccharides in the field of nanomedicine is their vulnerability to degradation, particularly oxidation, when exposed to high temperatures that exceed their melting point. This susceptibility to degradation is particularly relevant in industrial processes where elevated temperatures are often necessary. Moreover, a significant proportion of polysaccharides exhibit solubility in water, which imposes constraints on their utilization in specific areas of nanomedicine such as tissue engineering. Nevertheless, the process of crosslinking polymer chains has been employed to guarantee the durability of polysaccharide chains when exposed to watery environments. Figure 8.6 presents a compilation of the diverse polysaccharides employed in the field of nanomedicine, sourced from multiple references. The efficacy of biopolymers in the fields of nanomedicine and drug delivery can be attributed to their multifunctionality and distinctive characteristics. These include their capacity to exist in various forms, such as soft gels, flexible fibers, and rigid structures, as well as their ability to possess porous or nonporous

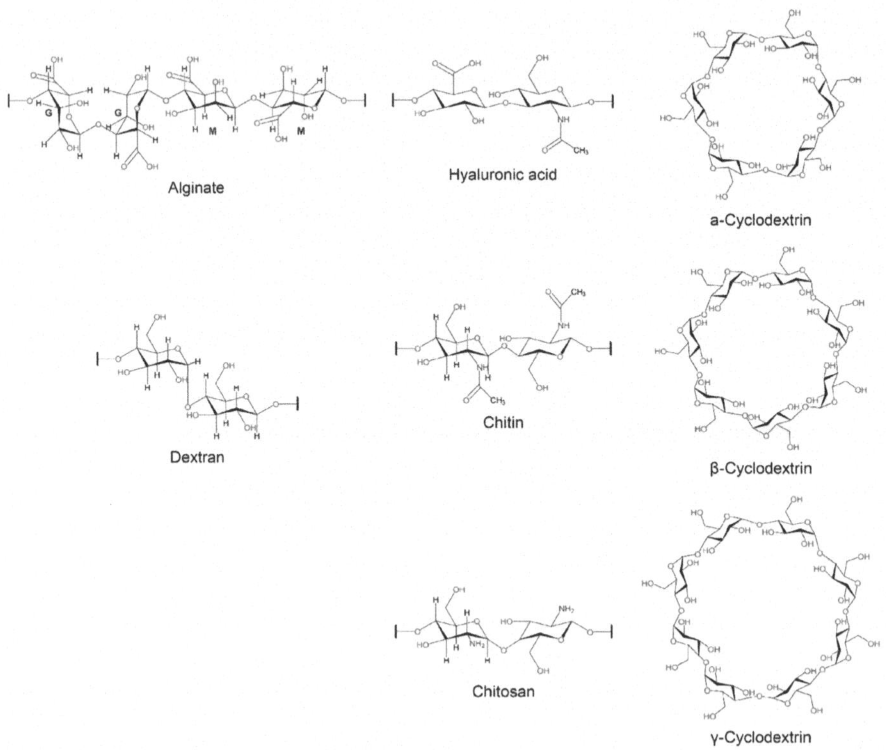

Fig. 8.6 Different types of polysaccharide materials used for the synthesis of nanoparticles. Image reprinted permission (Miao et al. 2018)

properties. Moreover, biopolymers exhibit a remarkable resemblance to the components found in the extracellular matrix, which potentially mitigates the risk of immunological responses.

8.7 Nanomedicine and Medication Delivery Systems in the Future

Nanomedicine is an emerging scientific discipline characterized by its dynamic nature and represents one of the most rapidly advancing fields in contemporary research. Over the course of the previous two decades, notable advancements have been observed in this field, as demonstrated by the acquisition of approximately 1500 patents and the implementation of numerous clinical trials. This chapter examines the significant impact of nanomedical breakthroughs in the field of cancer research, with a specific focus on their contributions to the detection and treatment of cancer. The precise utilization of diverse nanoparticles to deliver drugs specifically to cancer cells exemplifies the precision and potential of nanomedicine, thereby reducing its impact on healthy cells. Ongoing investigation and advancement in nano-drug delivery systems hold great potential as a prominent area of research in the foreseeable future.

The nanoparticles that have been shown exhibit a range of sizes, from the nanometer scale to the sub-micrometer scale. Subsequent investigations may derive advantages by directing their attention towards materials that exhibit heightened homogeneity and improved capacities for drug loading and release. This study also provides insights into the increased utilization of metal-based nanoparticles, specifically gold and silver, in the field of diagnostics. Gold nanoparticles have demonstrated potential in the context of soft tumor tissues, as they can improve the effectiveness of heat therapies that rely on radiation in the near-infrared band. This enhancement enables the targeted killing of tumors.

Although the potential of nanomedicine and nano-drug delivery in healthcare, particularly in the areas of cancer treatment and diagnosis, has been recognized, their current influence is constrained. The aforementioned constraint is indicative of the field's relatively nascent stage, characterized by a mere two decades of comprehensive investigation and numerous essential facets that remain unexplored. The identification of crucial biological markers for precise targeting, while minimizing disruption to normal cellular functioning, represents a critical area for future research. Future developments in nanomedicine are expected to be driven by advancements in the knowledge of illnesses at the molecular level and the identification of markers comparable in size to nanoparticles.

This chapter emphasizes the necessity for additional studies on many facets of nanomedicine. The refinement of controlled drug release principles, evaluation procedures, drug interactions at the tissue/cellular level, and prediction of theoretical models represent promising areas for further development. The current focus of research in the field of nanomedicine pertains to the study of biomaterials and formulations, which serve as the foundation for various biomedical applications. The integration of multiple disciplines in research and the examination of animals, albeit requiring substantial resources, plays a vital role in the progress of pharmacological treatment and diagnostic methods. The future of nanomedicine and nanodrug delivery technology appears to be promising owing to the global trend towards precision medicine and diagnostics. The potential of nanorobots and nanodevices in tissue diagnostics and repair, although still considered futuristic, is highly promising. The assessment of the risks associated with nanomedicines to both individuals and the environment has increased as the area continues to expand. Furthermore, the examination of pricing and regulations pertaining to nanomedicines is vital and will undergo development in tandem with technological progress within this domain.

8.8 Conclusion

This chapter provides an overview of current progress in the field of nanomedicine, encompassing breakthroughs in drug delivery techniques and innovative diagnostic strategies. A variety of nanoscale materials, including nanorobots and nanosensors, have been extensively studied owing to their potential use in diagnostics, targeted substance delivery, sensing capabilities, and the activation of elements within living systems. In its early stages, nanotechnology predominantly focused on enhancing the solubility, absorption, bioavailability, and controlled-release characteristics of medicinal substances. Despite the inherent uncertainties surrounding the discovery of nanodrugs and the diminished interest in identifying pharmacologically active compounds from natural sources in contemporary times relative to half a century ago, the utilization of nanotechnology to augment the effectiveness of established natural bioactive compounds has emerged as a prominent characteristic. The therapeutic uses of nanotechnology include the exploitation of several bioactive compounds such as berberine, curcumin, ellagic acid, resveratrol, curcumin, and quercetin. The effectiveness of naturally derived substances has been greatly improved by the utilization of nanocarriers composed of diverse materials, including gold, silver, cadmium sulfide, titanium dioxide polymeric nanoparticles, solid lipid nanoparticles, crystal nanoparticles, liposomes, micelles, superparamagnetic iron oxide nanoparticles, and dendrimers.

The persistent and enduring need for innovative natural biomaterials is driven by their intrinsic properties including biodegradability, biocompatibility, ease of procurement, renewability, and low toxicity. In addition to classifying polysaccharides and proteins as intrinsic biopolymers, ongoing research efforts are directed toward augmenting their stability in both industrial processing environments and biological matrices by means of procedures such as crosslinking. The discipline embodies a profoundly sophisticated realm of inquiry.

Polymeric nanoparticles, including nanocapsules and nanospheres, have been widely utilized by various methods such as solvent evaporation, emulsion polymerization, and surfactant-free emulsion polymerization. Recently, there has been notable focus on the integration of therapy and diagnosis in the field of nanomedicine, known as theranostics. This approach has garnered special attention in the context of cancer research, and serves as a representative model for disease investigation.

Numerous documented cases have highlighted the utilization of various nanoparticles for diagnostic purposes in the near-infrared range. These examples include the application of oleic acid-coated iron oxide nanoparticles, which have demonstrated efficacy in this regard. Additionally, the use of alginate and folic acid-based chitosan nanoparticles has been explored for photodynamic detection of colorectal cancer. Another notable example involves the conjugation of cathepsin B fluorogenic peptide probes to glycol chitosan nanoparticles, which enables the detection of metastatic processes. Furthermore, iron oxide-coated hyaluronic acid nanoparticles have been investigated for their potential diagnostic applications.

The prevalence of nanotechnology-based products approved by the Food and Drug Administration (FDA) and the number of clinical studies conducted in this field witnessed a substantial increase in the 1990s. This extension spans a diverse array of entities including synthetic polymer particles, liposome formulations, micellar nanoparticles, protein nanoparticles, and nanocrystals. Although the development of regulatory protocols and safety evaluations for nanomedicines is now underway, the discipline of nanomedicine has already had a profound impact on the discovery and delivery of pharmaceuticals in biological systems. Advancements in the field of nanomedicine have enabled the recognition of disorders and the combination of diagnostic and treatment methods.

References

Akram M, Shahab-Uddin AA, Usmanghani K, Hannan A, Mohiuddin E, Asif M (2010) Curcuma longa and curcumin: a review article. Rom J Biol Plant Biol 55(2):65–70

Aktaş Y, Yemisci M, Andrieux K, Gürsoy RN, Alonso MJ, Fernandez-Megia E, Novoa-Carballal R, Quiñoá E, Riguera R, Sargon MF (2005) Development and brain delivery of chitosan–PEG nanoparticles functionalized with the monoclonal antibody OX26. Bioconjug Chem 16(6):1503–1511

Alam M, Dwivedi V, Khan AA, Mohammad O (2009) Efficacy of niosomal formulation of diallyl sulfide against experimental candidiasis in Swiss albino mice. Nanomedicine 4(7):713–724

Al-Haj NA, Shamsudin MN, Alipiah NM, Zamri HF, Bustamam A, Ibrahim S, Abdullah R (2010) Characterization of nigella sativa L. essential oil-loaded solid lipid nanoparticles. Am J Pharmacol Toxicol 5(1):52. https://doi.org/10.3844/ajptsp.2010.52.57

Alhodieb FS, Barkat MA, Barkat HA, Hadi HA, Khan MI, Ashfaq F, Rahman MA, Hassan MZ, Alanezi AA (2022) Chitosan-modified nanocarriers as carriers for anticancer drug delivery: promises and hurdles. Int J Biol Macromol 217:457–469. https://doi.org/10.1016/j.ijbiomac.2022.06.201

Alhodieb FS, Rahman MA, Barkat MA, Alanezi AA, Barkat HA, Hadi HA, Harwansh RK, Mittal V (2023) Nanomedicine-driven therapeutic interventions of autophagy and stem cells in the management of Alzheimer's disease. Nanomedicine (Lond) 18(2):145–168. https://doi.org/10.2217/nnm-2022-0108

Alsaab HO, Sau S, Alzhrani RM, Cheriyan VT, Polin LA, Vaishampayan U, Rishi AK, Iyer AK (2018) Tumor hypoxia directed multimodal nanotherapy for overcoming drug resistance in renal cell carcinoma and reprogramming macrophages. Biomaterials 183:280–294

Anitha A, Rani VVD, Krishna R, Sreeja V, Selvamurugan N, Nair SV, Tamura H, Jayakumar R (2009) Synthesis, characterization, cytotoxicity and antibacterial studies of chitosan, O-carboxymethyl and N, O-carboxymethyl chitosan nanoparticles. Carbohydr Polym 78(4):672–677

Ansari MA, Almatroudi A, Alzohairy MA, AlYahya S, Alomary MN, Al-Dossary HA, Alghamdi S (2020) Lipid-based nano delivery of tat-peptide conjugated drug or vaccine–promising therapeutic strategy for SARS-CoV-2 treatment. Expert Opin Drug Deliv 17:1–4

Ansari MA, Thiruvengadam M, Farooqui Z, Rajakumar G, Jamal QM, Alzohairy MA, Almatroudi A, Alomary MN, Chung IM, Al-Suhaimi EA (2021) Nanotechnology, in silico and endocrine-based strategy for delivering paclitaxel and miRNA: prospects for the therapeutic management of breast cancer. Sem Cancer Biol 69:109–128

Ansari MA, Thiruvengadam M, Venkidasamy B, Alomary MN, Salawi A, Chung IM, Shariati MA, Rebezov M (2022) Exosome-based nanomedicine for cancer treatment by targeting inflammatory pathways: current status and future perspectives. Semin Cancer Biol 86:678–696

Ansari MA, Shoaib S, Chauhan W, Gahtani RM, Hani U, Alomary MN, Alasiri G, Ahmad N, Jahan R, Yusuf N, Islam N (2024a) Nanozymes and carbon-dots based nanoplatforms for cancer imaging, diagnosis and therapeutics: current trends and challenges. Environ Res 241:117522

Ansari MA, Tripathi T, Venkidasamy B, Monziani A, Rajakumar G, Alomary MN, Alyahya SA, Onimus O, D'souza N, Barkat MA, Al-Suhaimi EA (2024b) Multifunctional nanocarriers for Alzheimer's disease: befriending the barriers. Mol Neurobiol 61(5):3042–3089

Araujo-Lima CF, Nunes RJM, Carpes RM, Aiub CAF, Felzenszwalb I (2017) Pharmacokinetic and toxicological evaluation of a zinc gluconate-based chemical sterilant using in vitro and in silico approaches. BioMed Res Int 2017:5746768

Bao P, Chen Z, Tai R-Z, Shen H-M, Martin FL, Zhu Y-G (2015) Selenite-induced toxicity in cancer cells is mediated by metabolic generation of endogenous selenium nanoparticles. J Proteome Res 14(2):1127–1136. https://doi.org/10.1021/pr501086e

Bhise K, Sau S, Rauf MA, Alzhrani R, Iyer A (2020) Hypoxia targeting nanoparticles for immunogenic cell death of triple negative breast cancer. FASEB J 34(S1):1. https://doi.org/10.1096/fasebj.2020.34.s1.09796

Bozzuto G, Molinari A (2015) Liposomes as nanomedical devices. Int J Nanomedicine 10:975

Bradbury MWB (1985) The blood-brain barrier. Transport across the cerebral endothelium. Circ Res 57(2):213–222

Brinker FJ (2006) Traditional medicines for modern times. Antidiabetic plants. Edited by A. Soumyanath (Oregon Health & Science University). CRC/Taylor & Francis, Boca Raton. 2006. xii + 527 pp. 7 × 10 in. $139.95. ISBN 0-415-33464-0. J Nat Prod 69(10):1528. https://doi.org/10.1021/np068233c

Bunggulawa EJ, Wang W, Yin T, Wang N, Durkan C, Wang Y, Wang G (2018) Recent advancements in the use of exosomes as drug delivery systems. J Nanobiotechnol 16(1):1–13

Cai X, Luo Y, Zhang W, Du D, Lin Y (2016) pH-sensitive ZnO quantum dots–doxorubicin nanoparticles for lung cancer targeted drug delivery. ACS Appl Mater Interfaces 8(34):22442–22450

Cetin H, Yalaz M, Akisu M, Hilmioglu S, Metin D, Kultursay N (2005) The efficacy of two different lipid-based amphotericin B in neonatal Candida septicemia. Pediatr Int 47(6):676–680

Chen H, Kim S, He W, Wang H, Low PS, Park K, Cheng J-X (2008) Fast release of lipophilic agents from circulating PEG-PDLLA micelles revealed by in vivo forster resonance energy transfer imaging. Langmuir 24(10):5213–5217

Chen F, Ehlerding EB, Cai W (2014) Theranostic nanoparticles. J Nucl Med 55(12):1919–1922. https://doi.org/10.2967/jnumed.114.146019

Chen G, Roy I, Yang C, Prasad PN (2016) Nanochemistry and nanomedicine for nanoparticle-based diagnostics and therapy. Chem Rev 116(5):2826–2885

Chen C-W, Syu W-J, Huang T-C, Lee Y-C, Hsiao J-K, Huang K-Y, Yu H-P, Liao M-Y, Lai P-S (2017) Encapsulation of Au/Fe3O4 nanoparticles into a polymer nanoarchitecture with combined near infrared-triggered chemo-photothermal therapy based on intracellular secondary protein understanding. J Mater Chem B 5(29):5774–5782. https://doi.org/10.1039/C7TB00944E

Croy SR, Kwon GS (2006) Polymeric micelles for drug delivery. Curr Pharm Des 12(36):4669–4684

Dahl JA, Maddux BLS, Hutchison JE (2007) Toward greener nanosynthesis. Chem Rev 107(6):2228–2269

Davis M, Georgia Chen Z, Shin D (2008) Nanoparticle therapeutics: an emerging treatment modality for cancer. Nat Rev Drug Discov 7:771–782. https://doi.org/10.1038/nrd2614

de Groot CJ, van Luyn MJA, van Dijk-Wolthuis WNE, Cadée JA, Plantinga JA, den Otter W, Hennink WE (2001) In vitro biocompatibility of biodegradable dextran-based hydrogels tested with human fibroblasts. Biomaterials 22(11):1197–1203

Dimov N, Kastner E, Hussain M, Perrie Y, Szita N (2017) Formation and purification of tailored liposomes for drug delivery using a module-based micro continuous-flow system. Sci Rep 7(1):1–13

Ding Z, Liu P, Hu D, Sheng Z, Yi H, Gao G, Wu Y, Zhang P, Ling S, Cai L (2017) Redox-responsive dextran based theranostic nanoparticles for near-infrared/magnetic resonance imaging and magnetically targeted photodynamic therapy. Biomater Sci 5(4):762–771

Duan Y, Yang BQ, Chang CC, Zhou J, Li HY, Xu ZH, Wang ZW, Li DY (2016) Preliminary study on assessment of lexiscan-induced blood-brain barrier opening and its level by CT perfusion imaging. Zhonghua Yi Xue Za Zhi 96(35):2825–2829

Duce JA, Bush AI (2010) Biological metals and Alzheimer's disease: implications for therapeutics and diagnostics. Prog Neurobiol 92(1):1–18. https://doi.org/10.1016/j.pneurobio.2010.04.003

Elbakry A, Zaky A, Liebl R, Rachel R, Goepferich A, Breunig M (2009) Layer-by-layer assembled gold nanoparticles for siRNA delivery. Nano Lett 9(5):2059–2064

El-Naggar ME, Othman SI, Allam AA, Morsy OM (2020) Synthesis, drying process and medical application of polysaccharide-based aerogels. Int J Biol Macromol 145:1115–1128

Fana M, Gallien J, Srinageshwar B, Dunbar GL, Rossignol J (2020) PAMAM dendrimer nanomolecules utilized as drug delivery systems for potential treatment of glioblastoma: a systematic review. Int J Nanomedicine 15:2789

Fang RH, Hu C-MJ, Luk BT, Gao W, Copp JA, Tai Y, O'Connor DE, Zhang L (2014) Cancer cell membrane-coated nanoparticles for anticancer vaccination and drug delivery. Nano Lett 14(4):2181–2188. https://doi.org/10.1021/nl500618u

Felnerova D, Viret J-F, Glück R, Moser C (2004) Liposomes and virosomes as delivery systems for antigens, nucleic acids and drugs. Curr Opin Biotechnol 15(6):518–529

Gacche RN, Shegokar HD, Gond DS, Yang Z, Jadhav AD (2011) Evaluation of selected flavonoids as antiangiogenic, anticancer, and radical scavenging agents: an experimental and in silico analysis. Cell Biochem Biophys 61(3):651–663

Gericke M, Pinches A (2006) Microbial production of gold nanoparticles. Gold Bull 39(1):22–28. https://doi.org/10.1007/BF03215529

Gref R, Lück M, Quellec P, Marchand M, Dellacherie E, Harnisch S, Blunk T, Müller RH (2000) 'Stealth'corona-core nanoparticles surface modified by polyethylene glycol (PEG): influences of the corona (PEG chain length and surface density) and of the core composition on phagocytic uptake and plasma protein adsorption. Colloids Surf B: Biointerfaces 18(3–4):301–313

Gunalan S, Sivaraj R, Rajendran V (2012) Green synthesized ZnO nanoparticles against bacterial and fungal pathogens. Prog Nat Sci 22(6):693–700. https://doi.org/10.1016/j.pnsc.2012.11.015

Han H-S, Niemeyer E, Huang Y, Kamoun WS, Martin JD, Bhaumik J, Chen Y, Roberge S, Cui J, Martin MR (2015) Quantum dot/antibody conjugates for in vivo cytometric imaging in mice. Proc Natl Acad Sci 112(5):1350–1355

Handa M, Beg S, Shukla R, Barkat MA, Choudhry H, Singh KK (2021) Recent advances in lipid-engineered multifunctional nanophytomedicines for cancer targeting. J Controll Rel 340:48–59. https://doi.org/10.1016/j.jconrel.2021.10.025

Hervé F, Ghinea N, Scherrmann JM (2008) CNS delivery via adsorptive transcytosis. AAPS J 10(3):455–472. https://doi.org/10.1208/s12248-008-9055-2

Hong S-P, Kang SH, Kim DK, Kang BS (2016) Paramagnetic nanoparticle-based targeting theranostic agent for c6 rat glioma cell. J Nanomater

Hsieh Y-K, Jiang P-S, Yang B-S, Sun T-Y, Peng H-H, Wang C-F (2011) Using laser ablation/inductively coupled plasma mass spectrometry to bioimage multiple elements in mouse tumors after hyperthermia. Anal Bioanal Chem 401(3):909–915

Islam T, Hasan M, Awal A, Nurunnabi M, Ahammad AJ (2020) Metal nanoparticles for electrochemical sensing: Progress and challenges in the clinical transition of point-of-care testing. Molecules 25(24):5787

Jahangir MA, Gilani SJ, Muheem A, Jafar M, Aslam M, Ansari MT, Barkat MA (2019) Quantum dots: next generation of smart nano-systems. Pharmac Nanotechnol 7(3):234–245. https://doi.org/10.2174/2211738507666190429113906

Jahromi EZ, Divsalar A, Saboury AA, Khaleghizadeh S, Mansouri-Torshizi H, Kostova I (2016) Palladium complexes: new candidates for anti-cancer drugs. J Iran Chem Soc 13(5):967–989. https://doi.org/10.1007/s13738-015-0804-8

Jain A, Jain SK (2016) Optimization of chitosan nanoparticles for colon tumors using experimental design methodology. Artif Cells Nanomed Biotechnol 44(8):1917–1926

Jain JP, Kumar N (2010) Development of amphotericin B loaded polymersomes based on (PEG) 3-PLA co-polymers: factors affecting size and in vitro evaluation. Eur J Pharm Sci 40(5):456–465

Jia X, Wang S, Zhou L, Sun L (2017) The potential liver, brain, and embryo toxicity of titanium dioxide nanoparticles on mice. Nanoscale Res Lett 12(1):478

Jong A, Wu C-H, Chen H-M, Luo F, Kwon-Chung KJ, Chang YC, LaMunyon CW, Plaas A, Huang S-H (2007) Identification and characterization of CPS1 as a hyaluronic acid synthase contributing to the pathogenesis of Cryptococcus neoformans infection. Eukaryot Cell 6(8):1486–1496

Juillerat-Jeanneret L (2008) The targeted delivery of cancer drugs across the blood–brain barrier: chemical modifications of drugs or drug-nanoparticles? Drug Discov Today 13(23–24):1099–1106

Kamaly N, Yameen B, Wu J, Farokhzad OC (2016) Degradable controlled-release polymers and polymeric nanoparticles: mechanisms of controlling drug release. Chem Rev 116(4):2602–2663

Kamat M, El-Boubbou K, Zhu DC, Lansdell T, Lu X, Li W, Huang X (2010) Hyaluronic acid immobilized magnetic nanoparticles for active targeting and imaging of macrophages. Bioconjug Chem 21(11):2128–2135. https://doi.org/10.1021/bc100354m

Kaur A, Jain K, Mehra NK, Jain NK (2017) Development and characterization of surface engineered PPI dendrimers for targeted drug delivery. Artif Cells Nanomed Biotechnol 45(3):414–425

Khan S, Ul-Islam M, Khattak WA, Ullah MW, Park JK (2015) Bacterial cellulose-titanium dioxide nanocomposites: nanostructural characteristics, antibacterial mechanism, and biocompatibility. Cellulose 22(1):565–579

Kharissova OV, Dias HVR, Kharisov BI, Pérez BO, Pérez VMJ (2016) The greener synthesis of nanoparticles. Trends Biotechnol 31(4):240–248. https://doi.org/10.1016/j.tibtech.2013.01.003

Kim ST, Chompoosor A, Yeh Y, Agasti SS, Solfiell DJ, Rotello VM (2012) Dendronized gold nanoparticles for siRNA delivery. Small 8(21):3253–3256

Kingston DGI (2011) Modern natural products drug discovery and its relevance to biodiversity conservation. J Nat Prod 74(3):496–511. https://doi.org/10.1021/np100550t

Klajnert B, Cangiotti M, Calici S, Majoral JP, Caminade AM, Cladera J, Bryszewska M, Ottaviani MF (2007) EPR study of the interactions between dendrimers and peptides involved in Alzheimer's and prion diseases. Macromol Biosci 7(8):1065–1074

Korthagen NM, Siebelt M, de Visser HM, Sandker M, Müller C, Weinans H (2019) Folate receptor positive macrophages in rat models for OA: phenotype switch in response to anti-inflammatory medication. Osteoarthr Cartil 27:S373–S373

Krishnamurthy S, Gnanasammandhan MK, Xie C, Huang K, Cui MY, Chan JM (2016) Monocyte cell membrane-derived nanoghosts for targeted cancer therapy. Nanoscale 8(13):6981–6985. https://doi.org/10.1039/C5NR07588B

Kumar Teli M, Mutalik S, Rajanikant GK (2010) Nanotechnology and nanomedicine: going small means aiming big. Curr Pharm Des 16(16):1882–1892

Lakhal S, Wood MJA (2011) Exosome nanotechnology: an emerging paradigm shift in drug delivery. BioEssays 33(10):737–741

Lalatsa A, Schatzlein AG, Uchegbu IF (2014) Strategies to deliver peptide drugs to the brain. Mol Pharm 11(4):1081–1093

Lapcık L Jr, Lapcık L, de Smedt S, Demeester J, Chabrecek P (1998) Hyaluronan: preparation, structure, properties, and applications. Chem Rev 98(8):2663–2684

Laroui H, Rakhya P, Xiao B, Viennois E, Merlin D (2013) Nanotechnology in diagnostics and therapeutics for gastrointestinal disorders. Dig Liver Dis 45(12):995–1002

Laurent S, Bridot J-L, Elst LV, Muller RN (2010) Magnetic iron oxide nanoparticles for biomedical applications. Future Med Chem 2(3):427–449

Lee H, Lee M-Y, Bhang SH, Kim B-S, Kim YS, Ju JH, Kim KS, Hahn SK (2014) Hyaluronate–gold nanoparticle/tocilizumab complex for the treatment of rheumatoid arthritis. ACS Nano 8(5):4790–4798

Lengke MF, Fleet ME, Southam G (2006) Morphology of gold nanoparticles synthesized by filamentous cyanobacteria from gold (I)– thiosulfate and gold (III)– chloride complexes. Langmuir 22(6):2780–2787

Leonor Pinzon-Daza M, Campia I, Kopecka J, Garzón R, Ghigo D, Rigant C (2013) Nanoparticle- and liposome-carried drugs: new strategies for active targeting and drug delivery across blood-brain barrier. Curr Drug Metab 14(6):625–640

Li Y, Li Y, Zhang X, Xu X, Zhang Z, Hu C, He Y, Gu Z (2016) Supramolecular PEGylated dendritic systems as pH/redox dual-responsive theranostic nanoplatforms for platinum drug delivery and NIR imaging. Theranostics 6(9):1293

Li X, Wang X, Zhao C, Shao L, Lu J, Tong Y, Chen L, Cui X, Sun H, Liu J, Li M, Deng X, Wu Y (2019) From one to all: self-assembled theranostic nanoparticles for tumor-targeted imaging and programmed photoactive therapy. J Nanobiotechnol 17(1):23. https://doi.org/10.1186/s12951-019-0450-x

Liong M, Lu J, Kovochich M, Xia T, Ruehm SG, Andre E, Tamanoi F, Zink JI (2008) Multifunctional inorganic nanoparticles for imaging, targeting, and drug delivery. ACS Nano 2(5):889–896. https://doi.org/10.1021/nn800072t.Multifunctional

Liu X, Zhou J, Yu T, Chen C, Cheng Q, Sengupta K, Huang Y, Li H, Liu C, Wang Y, Posocco P, Wang M, Cui Q, Giorgio S, Fermeglia M, Qu F, Pricl S, Shi Y, Liang Z et al (2014) Adaptive amphiphilic dendrimer-based nanoassemblies as robust and versatile siRNA delivery systems. Angew Chem Int Ed 53(44):11822–11827. https://doi.org/10.1002/anie.201406764

Loo C-Y, Rohanizadeh R, Young PM, Traini D, Cavaliere R, Whitchurch CB, Lee W-H (2016) Combination of silver nanoparticles and curcumin nanoparticles for enhanced anti-biofilm activities. J Agric Food Chem 64(12):2513–2522. https://doi.org/10.1021/acs.jafc.5b04559

Lu B, Lu F, Zou Y, Liu J, Rong B, Li Z, Dai F, Wu D, Lan G (2017) In situ reduction of silver nanoparticles by chitosan-l-glutamic acid/hyaluronic acid: enhancing antimicrobial and wound-healing activity. Carbohydr Polym 173:556–565

Luo D, Haverstick K, Belcheva N, Han E, Saltzman WM (2002) Poly (ethylene glycol)-conjugated PAMAM dendrimer for biocompatible, high-efficiency DNA delivery. Macromolecules 35(9):3456–3462

Maeda H, Wu J, Sawa T, Matsumura Y, Hori K (2000) Tumor vascular permeability and the EPR effect in macromolecular therapeutics: a review. J Control Release 65(1–2):271–284

Maeda H, Nakamura H, Fang J (2013) The EPR effect for macromolecular drug delivery to solid tumors: improvement of tumor uptake, lowering of systemic toxicity, and distinct tumor imaging in vivo. Adv Drug Deliv Rev 65(1):71–79

Martens TF, Remaut K, Deschout H, Engbersen JFJ, Hennink WE, van Steenbergen MJ, Demeester J, de Smedt SC, Braeckmans K (2015) Coating nanocarriers with hyaluronic acid facilitates intravitreal drug delivery for retinal gene therapy. J Control Release 202:83–92

Matthews L, Kanwar RK, Zhou S, Punj V, Kanwar JR (2010) Applications of nanomedicine in antibacterial medical therapeutics and diagnostics. Open Trop Med J 3(1):1–9

Mattos BD, Rojas OJ, Magalhães WLE (2017) Biogenic silica nanoparticles loaded with neem bark extract as green, slow-release biocide. J Clean Prod 142:4206–4213

Menjoge AR, Kannan RM, Tomalia DA (2010) Dendrimer-based drug and imaging conjugates: design considerations for nanomedical applications. Drug Discov Today 15(5–6):171–185

Miao T, Wang J, Zeng Y, Liu G, Chen X (2018) Polysaccharide-based controlled release systems for therapeutics delivery and tissue engineering: from bench to bedside. Advanced Science 5(4):1700513. https://doi.org/10.1002/advs.201700513

Miller LV, Hambidge KM, Naake VL, Hong Z, Westcott JL, Fennessey PV (1994) Size of the zinc pools that exchange rapidly with plasma zinc in humans: alternative techniques for measuring and relation to dietary zinc intake. J Nutr 124(2):268–276

Moreira JN, Santos A, Moura V, Pedroso, de L, Simoes S (2008) Non-viral lipid-based nanoparticles for targeted cancer systemic gene silencing. J Nanosci Nanotechnol 8(5):2187–2204

Mortensen L, Oberdörster G, Pentland AP, DeLouise LA (2008) In vivo skin penetration of quantum dot nanoparticles in the murine model: the effect of UVR. Nano Lett 8(9):2779–2787. https://doi.org/10.1021/nl801323y

Muhamad N, Plengsuriyakarn T, Na-Bangchang K (2018) Application of active targeting nanoparticle delivery system for chemotherapeutic drugs and traditional/herbal medicines in cancer therapy: a systematic review. Int J Nanomedicine 13:3921

Mun S-H, Kim S-B, Kong R, Choi J-G, Kim Y-C, Shin D-W, Kang O-H, Kwon D-Y (2014) Curcumin reverse methicillin resistance in Staphylococcus aureus. Molecules 19(11):18283–18295

Na HB, Song IC, Hyeon T (2009) Inorganic nanoparticles for MRI contrast agents. Adv Mater 21(21):2133–2148

Naeye B, Raemdonck K, Remaut K, Sproat B, Demeester J, de Smedt SC (2010) PEGylation of biodegradable dextran nanogels for siRNA delivery. Eur J Pharm Sci 40(4):342–351

Olerile LD, Liu Y, Zhang B, Wang T, Mu S, Zhang J, Selotlegeng L, Zhang N (2017) Near-infrared mediated quantum dots and paclitaxel co-loaded nanostructured lipid carriers for cancer theragnostic. Colloids Surf B: Biointerfaces 150:121–130

Onyeji CO, Nightingale CH, Marangos MN (1994) Enhanced killing of methicillin-resistant Staphylococcus aureus in human macrophages by liposome-entrapped vancomycin and teicoplanin. Infection 22(5):338–342

Oves M, Rauf MA, Hussain A, Qari HA, Khan AAP, Muhammad P, Rehman MT, Alajmi MF, Ismail IIM (2019) Antibacterial silver nanomaterial synthesis from Mesoflavibacter zeaxanthinifaciens and targeting biofilm formation. Front Pharmacol 10:801. https://doi.org/10.3389/fphar.2019.00801

Packiavathy IASV, Sasikumar P, Pandian SK, Ravi AV (2013) Prevention of quorum-sensing-mediated biofilm development and virulence factors production in vibrio spp. by curcumin. Appl Microbiol Biotechnol 97(23):10177–10187

Pan W, Banks WA, Fasold MB, Bluth J, Kastin AJ (1998) Transport of brain-derived neurotrophic factor across the blood–brain barrier. Neuropharmacology 37(12):1553–1561

Patil S, Sandberg A, Heckert E, Self W, Seal S (2007) Protein adsorption and cellular uptake of cerium oxide nanoparticles as a function of zeta potential. Biomaterials 28(31):4600–4607

Pelaz B, Alexiou C, Alvarez-Puebla RA, Alves F, Andrews AM, Ashraf S, Balogh LP, Ballerini L, Bestetti A, Brendel C (2017) Diverse applications of nanomedicine. ACS Nano 11(3):2313–2381

Peng BL, Dhar N, Liu HL, Tam KC (2011) Chemistry and applications of nanocrystalline cellulose and its derivatives: a nanotechnology perspective. Can J Chem Eng 89(5):1191–1206

Peng J, Weng J, Ren L, Sun L-P (2014) Interactions between gold nanoparticles and amyloid β 25–35 peptide. IET Nanobiotechnol 8(4):295–303

Pistone S, Goycoolea FM, Young A, Smistad G, Hiorth M (2017) Formulation of polysaccharide-based nanoparticles for local administration into the oral cavity. Eur J Pharm Sci 96:381–389

Prusty K, Swain SK (2018) Nano silver decorated polyacrylamide/dextran nanohydrogels hybrid composites for drug delivery applications. Mater Sci Eng C 85:130–141

Rauf MA, Zubair S, Ateeq H, Dabeer K, Pachauri S, Ajmal M, Owais M (2018) Synergistic effect of Diallyl sulphide with zinc oxide nanorods: a novel and effective approach for treatment of acute dermatitis in model animals. Front Microbiol 9:586

Rauf MA, Oves M, Ur Rehman F, Khan AR, Husain N (2019a) Bougainvillea flower extract mediated zinc oxide's nanomaterials for antimicrobial and anticancer activity. Biomed Pharmacother 116:108983. https://doi.org/10.1016/j.biopha.2019.108983

Rauf MA, Rehman FU, Zheng M, Shi B (2019b) In: Xue X (ed) The strategies of nanomaterials for traversing blood-brain barrier BT—nanomedicine in brain diseases: principles and application. Springer, Singapore, pp 29–57. https://doi.org/10.1007/978-981-13-8731-9_2

Ray PC, Yu H, Fu PP (2009) Toxicity and environmental risks of nanomaterials: challenges and future needs. J Environ Sci Health C Environ Carcinog Ecotoxicol Rev 27(1):1–35. https://doi.org/10.1080/10590500802708267

Rehman FU, Zhao C, Jiang H, Selke M, Wang X (2016) Protective effect of TiO2 nanowhiskers on tetra sulphonatophenyl porphyrin (TSPP) complexes induced oxidative stress during photodynamic therapy. Photodiagn Photodyn Ther 13:267–275

Rehman FU, Jiang H, Selke M, Wang X (2018) Mammalian cells: a unique scaffold for in situ biosynthesis of metallic nanomaterials and biomedical applications. J Mater Chem B 6(41):6501–6514

Rehman FU, Bao J, Muhammad P, He W, Hanif S, Rauf MA (2020) Blood-brain barrier amenable gold nanoparticles bio-fabrication in aged cell culture medium. Mater Today Bio 8:100072. https://doi.org/10.1016/j.mtbio.2020.100072

Ryu JH, Na JH, Ko HK, You DG, Park S, Jun E, Yeom HJ, Seo DH, Park JH, Jeong SY (2014) Non-invasive optical imaging of cathepsin B with activatable fluorogenic nanoprobes in various metastatic models. Biomaterials 35(7):2302–2311

Salvioni L, Rizzuto MA, Bertolini JA, Pandolfi L, Colombo M, Prosperi D (2019) Thirty years of cancer nanomedicine: success, frustration, and hope. Cancers 11(12):1855

Seo G, Lee G, Kim MJ, Baek S-H, Choi M, Ku KB, Lee C-S, Jun S, Park D, Kim HG (2020) Rapid detection of COVID-19 causative virus (SARS-CoV-2) in human nasopharyngeal swab specimens using field-effect transistor-based biosensor. ACS Nano 14(4):5135–5142

Sethi M, Sukumar R, Karve S, Werner ME, Wang EC, Moore DT, Kowalczyk SR, Zhang L, Wang AZ (2014) Effect of drug release kinetics on nanoparticle therapeutic efficacy and toxicity. Nanoscale 6(4):2321–2327

Shakir M, Jolly R, Khan MS, Rauf A, Kazmi S (2016) Nano-hydroxyapatite/β-CD/chitosan nanocomposite for potential applications in bone tissue engineering. Int J Biol Macromol 93:276–289. https://doi.org/10.1016/j.ijbiomac.2016.08.046

Shakir M, Jolly R, Khan AA, Ahmed SS, Alam S, Rauf MA, Owais M, Farooqi MA (2018) Resol based chitosan/nano-hydroxyapatite nanoensemble for effective bone tissue engineering. Carbohydr Polym 179(Supplement C):317–327. https://doi.org/10.1016/j.carbpol.2017.09.103

Shi Y, Pramanik A, Tchounwou C, Pedraza F, Crouch RA, Chavva SR, Vangara A, Sinha SS, Jones S, Sardar D (2015) Multifunctional biocompatible graphene oxide quantum dots decorated magnetic nanoplatform for efficient capture and two-photon imaging of rare tumor cells. ACS Appl Mater Interfaces 7(20):10935–10943

Shoeibi S, Mozdziak P, Golkar-Narenji A (2017) Biogenesis of selenium nanoparticles using green chemistry. Top Curr Chem 375(6):88. https://doi.org/10.1007/s41061-017-0176-x

Shukla AK, Iravani S (2016) Green synthesis and spectroscopic characterization of nanoparticles. In: Nanoscience in food and agriculture 1. Springer, pp 65–99

Silva MM, Calado R, Marto J, Bettencourt A, Almeida AJ, Gonçalves L (2017) Chitosan nanoparticles as a mucoadhesive drug delivery system for ocular administration. Mar Drugs 15(12):370

Sivanesan S, Rajeshkumar S (2019) Gold nanoparticles in diagnosis and treatment of Alzheimer's disease. In: Nanobiotechnology in neurodegenerative diseases. Springer, pp 289–306

Slocik JM, Knecht MR, Wright DW (2006) Biomimetic synthesis of nanoparticles. Encyclopedia of inorganic chemistry

Sokolova V, Epple M (2008) Inorganic nanoparticles as carriers of nucleic acids into cells. Angew Chem Int Ed 47(8):1382–1395

Stalmans S, Bracke N, Wynendaele E, Gevaert B, Peremans K, Burvenich C, Polis I, de Spiegeleer B (2015) Cell-penetrating peptides selectively cross the blood-brain barrier in vivo. PLoS One 10(10):e0139652–e0139652

Tandrup Schmidt S, Foged C, Smith Korsholm K, Rades T, Christensen D (2016) Liposome-based adjuvants for subunit vaccines: formulation strategies for subunit antigens and immunostimulators. Pharmaceutics 8(1):7

Tatiparti K, Rauf MA, Sau S, Alzhrani R, Iyer A (2020a) Ultra-small biomimetic nanoparticles for Alzheimer's disease targeted drug delivery. FASEB J 34(S1):1. https://doi.org/10.1096/fasebj.2020.34.s1.09774

Tatiparti K, Rauf MA, Sau S, Iyer AK (2020b) Carbonic anhydrase-IX guided albumin nanoparticles for hypoxia-mediated triple-negative breast cancer cell killing and imaging of patient-derived tumor. Molecules 25(10):2362

Tiwari SK, Agarwal S, Seth B, Yadav A, Nair S, Bhatnagar P, Karmakar M, Kumari M, Chauhan LKS, Patel DK, Srivastava V, Singh D, Gupta SK, Tripathi A, Chaturvedi RK, Gupta KC (2014) Curcumin-loaded nanoparticles potently induce adult neurogenesis and reverse cognitive deficits in Alzheimer's disease model via canonical Wnt/β-catenin pathway. ACS Nano 8(1):76–103. https://doi.org/10.1021/nn405077y

Tortorella S, Karagiannis TC (2014) Transferrin receptor-mediated endocytosis: a useful target for cancer therapy. J Membr Biol 247(4):291–307

Ulbrich K, Hola K, Subr V, Bakandritsos A, Tucek J, Zboril R (2016) Targeted drug delivery with polymers and magnetic nanoparticles: covalent and noncovalent approaches, release control, and clinical studies. Chem Rev 116(9):5338–5431

Velasco-Aguirre C, Morales F, Gallardo-Toledo E, Guerrero S, Giralt E, Araya E, Kogan MJ (2015) Peptides and proteins used to enhance gold nanoparticle delivery to the brain: preclinical approaches. Int J Nanomedicine 10:4919

Vyas TK, Shahiwala A, Marathe S, Misra A (2005) Intranasal drug delivery for brain targeting. Curr Drug Deliv 2(2):165–175

Wang J (2005) Nanomaterial-based electrochemical biosensors. Analyst 130(4):421–426

Wang C-H, Wang W-T, Hsiue G-H (2009) Development of polyion complex micelles for encapsulating and delivering amphotericin B. Biomaterials 30(19):3352–3358

Wang G, Gao S, Tian R, Miller-Kleinhenz J, Qin Z, Liu T, Li L, Zhang F, Ma Q, Zhu L (2018) Theranostic hyaluronic acid–iron micellar nanoparticles for magnetic-field-enhanced in vivo cancer chemotherapy. ChemMedChem 13(1):78–86

Wei X, Gao J, Zhan C, Xie C, Chai Z, Ran D, Ying M, Zheng P, Lu W (2015) Liposome-based glioma targeted drug delivery enabled by stable peptide ligands. J Control Release 218:13–21. https://doi.org/10.1016/j.jconrel.2015.09.059

Wen Y, Li J (2018) Ultrastable micelles boost chemotherapy. Nat Biomed Eng 2(5):273

Xia C-F, Boado RJ, Pardridge WM (2008) Antibody-mediated targeting of siRNA via the human insulin receptor using avidin– biotin technology. Mol Pharm 6(3):747–751

Xia W, Hilgenbrink AR, Matteson EL, Lockwood MB, Cheng J-X, Low PS (2009) A functional folate receptor is induced during macrophage activation and can be used to target drugs to activated macrophages. Blood 113(2):438–446

Yaghmur A, Tran BV, Moghimi SM (2019) Non-lamellar liquid crystalline nanocarriers for thymoquinone encapsulation. Molecules (Basel, Switzerland) 25(1):16. https://doi.org/10.3390/molecules25010016

Yang H (2010) Nanoparticle-mediated brain-specific drug delivery, imaging, and diagnosis. Pharm Res 27(9):1759–1771. https://doi.org/10.1007/s11095-010-0141-7

Yang S-J, Lin F-H, Tsai H-M, Lin C-F, Chin H-C, Wong J-M, Shieh M-J (2011) Alginate-folic acid-modified chitosan nanoparticles for photodynamic detection of intestinal neoplasms. Biomaterials 32(8):2174–2182

Zhang S, Fu R, Wu D, Xu W, Ye Q, Chen Z (2004) Preparation and characterization of antibacterial silver-dispersed activated carbon aerogels. Carbon 42(15):3209–3216

Zheng Y, Monty J, Linhardt RJ (2015) Polysaccharide-based nanocomposites and their applications. Carbohydr Res 405:23–32

Theranostic Application of Nanomedicine in Neurodegenerative Diseases: Current and Future Perspectives

9

Garima, Khalid Imtiyaz, Pooja, Prachi Pannu, Anjali Sharma, Smriti Raina, Sugandh Kumar, Syed Tauqeer Anwer, Moshahid Alam Rizvi, Shailendra Kumar Sinha, Khalid Umar Fakhri, and Anil Kumar Mavi

Abstract

Nanotechnology, operating at the nano-scale level (1–100 nm), offers innovative solutions in designing polymeric materials for molecular interactions with biological systems. These materials show potential in targeted drug delivery, minimizing adverse effects, and enhancing therapeutic efficacy, specifically in neurodegenerative disorders, i.e., Parkinson's disease (PD) and Alzheimer's disease (AD). However, delivering of drugs to the brain is one of the key challenging due to the blood-brain barrier's (BBB) restrictive nature. Strategies to overcome this barrier include using lipophilic analogues, prodrugs, and chemical drug delivery systems. Carrier-mediated and receptor/vector-mediated delivery

Garima · Pooja · P. Pannu · A. Sharma · S. Raina
Department of Biotechnology, Delhi Technological University (DTU), New Delhi, India

K. Imtiyaz · A. K. Mavi (✉)
Department of Botany and Life sciences, Shri Aurobindo college, University of Delhi, Delhi, India
e-mail: amavi_botany@aurobindo.du.ac.in

S. Kumar
School of Medicine, University of San Francisco California (UCSF), San Francisco, CA, USA

S. T. Anwer · M. A. Rizvi · K. U. Fakhri (✉)
Department of Biosciences, Jamia Millia Islamia, New Delhi, India

S. K. Sinha
Department of Zoology and Environmental Science and Disaster Management, BBMKU, Dhanbad, Jharkhand, India

M. A. Barkat et al. (eds.), *Nanotheranostics for Diagnosis and Therapy*, https://doi.org/10.1007/978-981-97-3115-2_9

exploit existing transport systems, while nanoparticle-mediated delivery efficiently transports drugs across the BBB. Nanotechnology holds vast potential in healthcare, bioscience, and medicine, with nanomedicine offering innovative solutions to conventional medicine's challenges. Utilizing multifunctional nanocarriers for targeted drug delivery and imaging, especially in treating neurodegenerative diseases, holds significant promise. The emergence of theranostic nanomedicines could revolutionize clinical trials in the future.

Keywords

Nanotechnology · Drug delivery · Blood-brain barrier's · Neurodegenerative disorders

9.1 Introduction

In order to develop devices characterized by minimal functional organization at the nanometer scale (1–100 nm), enabling molecular interactions with biological systems, nanotechnology offers opportunities to design polymeric materials. The nanoparticles may react, stimulate, and engage with target tissues and cells to minimize negative side effects and elicit desired physiological responses. Importantly, nanotechnology might provide methods for more selectively modifying intricate biological systems. This novel strategy overcomes traditional barriers and increases accessibility to target regions, addressing significant issues in medicine distribution. In addition to providing efficient solutions for difficult drug delivery issues, nanotechnology improves specificity and selectivity, which modifies diagnostic standards specifically. Neurodegenerative disorders (NDs) are a group of conditions that include gradual loss of neuronal subtypes over time and can be sporadic, familial, or both. The most well-known brain-related disease are Parkinson's disease (PD) and Alzheimer's disease (AD), that are two of the most crippling illnesses of the twenty-first century (Kabanov and Gendelman 2007).

Globally, about 24 million people experience dementia, with AD accounting for 60% of cases. AD affects 1% of people between the ages of 50 and 70, and its prevalence rises sharply to 50% in people over the age of 70 (Ferri et al. 2005). Simultaneously, Parkinson's disease (PD) is typified by the intermittent loss of midbrain nigrostriatal dopaminergic neurons, which lowers brain dopamine (DA) levels and produces the typical motor symptoms of stiffness, bradykinesia, and resting tremor (Singh et al. 2007). The brain transmitter known as DA is in charge of sending the electrical signals needed for typical physical movements. These aberrant movements stem from the DA deficit characteristic of Parkinson's disease. However the exact cause of PD and AD is not very well unknown yet and need to do more research to understand a complex chain of harmful variables is assumed to be its pathogenesis. A percentage of people over 65 who have PD is 1% of the population. While there are numerous biological reasons for the onset of PD, genetic factors contribute to slightly more than 5% of cases and are not considered prevalent.

Low molecular weight drugs, imaging contrast agents, nucleic acids, and proteins are among the biomacromolecules whose entry into the brain is severely restricted.

This represents a significant barrier to the development of safe and effective treatment and diagnostic modalities (Carradori et al. 2016). To deal with these barriers, specific physicochemical features are conferred by surface effects, macro quantum effects, and quantum dimension effects of nanomaterials. The advantage of using nanomaterials is that they can hold huge amounts of medication and shield it from enzymatic breakdown, which increases the medication's stability and effectiveness (Mura and Couvreur 2012; Lu et al. 2021). By altering surface molecules, drug-loaded nanomaterials can be guided to concentrate in lesioned cells and tissues, leading to improved target specificity and reduced off-target effects and side effects (Garbayo et al. 2014).

9.2 Structure and Role of BBB (Blood Brain Barrier)

The blood-brain barrier (BBB) serves as a highly specialized "semipermeable shield" (Rhaman et al. 2022), comprised of endothelial cells, safeguarding the central nervous system from indiscriminate substance passage between the bloodstream and the extracellular fluid where vital neurons reside. It's like a local security system that keeps out toxins and cells that might be floating around in the blood. The BBB acts like a strong shield, protecting the central nervous system from both inside and outside harms and stopping diseases from affecting it (Fig. 9.1). The BBB creates the perfect conditions for the brain to work well. It does this by keeping a balance in the brain, controlling the flow of fluids in and out, and defending against harmful substances. The blood-brain barrier (BBB) has distinctive characteristics of the small blood vessels within the central nervous system. Unlike other blood vessels, those in the CNS are continuous and lack openings. They possess specific features that enable precise control over the passage of molecules, ions, and cells between bloodstreams and the "CNS" (Teleanu et al. 2018). Moreover, the blood-brain barrier (BBB) is a dynamic structure that adjusts in response to normal physiological functions and pathological conditions. Factors such as inflammation, trauma, or specific diseases can modify BBB permeability, influencing the passage

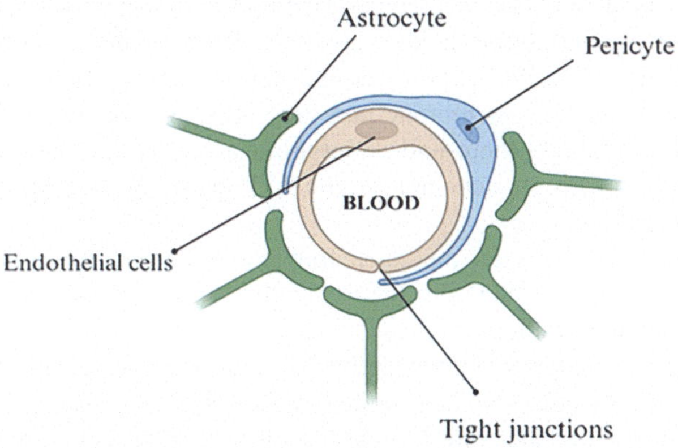

Fig. 9.1 Depicting blood brain barrier (BBB) with their components

of substances into the brain. Understanding these dynamic adaptations is imperative for the progression of therapeutic strategies and diagnostic methodologies.

9.2.1 Structure of BBB

This physiological border is an intricate assembly of tightly bound cerebral vascular components, including endothelial cells, hepatocytes, astrocytes, and basal membranes. The main components of BBB and their key functions are summarized in the Table 9.1.

1. The NVU "Neurovascular unit" includes the endothelial cells, astrocytes, pericytes, mural cells, and basement membranes. It is a dynamic and integrated system within the Central Nervous System (CNS) that involves the coordination of various cellular and structural elements to ensure the proper functioning of the brain (Lochhead et al. 2020).
2. The intercellular junctions in the BBB are the tight junctions which mainly control the property of low cellular permeability and is essential for maintaining CNS homeostasis.

9.2.2 Transport Mechanism of Molecules Across BBB

There are several para-cellular or transcellular "pathways" involved in the transport of macromolecules and small molecules across the BBB (Teleanu et al. 2018). Hence, the transport of molecules through the BBB is commonly categorized into various classes (as illustrated in Fig. 9.2), including:

(a) Passive diffusion: This pathway involves the passive diffusion of small lipophilic molecules, such as oxygen and carbon dioxide, across the lipid bilayer of endothelial cells, facilitated by their solubility in the lipid membrane (Teleanu et al. 2018; Barar et al. 2016).
(b) Facilitated diffusion: This mechanism enables the transport of larger molecules, such as glucose and amino acids, across the BBB, as they depend on specific carrier proteins to facilitate their passage. These carrier proteins are embedded in the endothelial cell membrane and they allow the passage of certain substances by binding to them and undergoing a conformational change. These solute carrier transports does not require ATP and is facilitated by the concentration gradients.
(c) Active transport or carrier-mediated transport via ABC transporters requires energy in the form of ATP and facilitates the movement of molecules against the concentration gradient.
(d) Transcytosis: whether receptor-mediated or adsorptive, entails the vesicular transport of substances across endothelial cells. This process involves the engulfment of molecules into vesicles on one side of the endothelial cell, their

Table 9.1 Major components of BBB and their respective functions

	BBB components	Functions	References
NVU (neurovascular unit)	1. Brain endothelial cells (ECs) 2. Mural cells (vascular smooth muscle cells and pericytes) 3. Astrocytes 4. Basement membrane	Form the BBB, creates a selective barrier by tightly regulating the entry of molecules, ions, and cells into the CNS. Essential for maintaining CNS homeostasis and protecting the brain from potentially harmful substances circulating in the bloodstream Vascular smooth muscle cells surround the large blood vessels which contribute to the regulation of blood flow to changing physiological demands Pericytes are situated on the abluminal surface of microvascular endothelial tubes and are embedded within the vascular basement membrane Major star shaped glial cells which are the fundamental components of NVU. The play a pivotal role in hemodynamic response, actively participating in the regulation of cerebral blood flow in response to neuronal activity Comprises of inner vascular basement membrane (containing ECM proteins such as laminin, type IV collagens) which provide a scaffold for formation of intercellular junction complexes and outer parenchymal basement membrane (contribute to structural support to NVU)	Persidsky et al. (2006) Persidsky et al. (2006), Attwell et al. (2016) Quintana (2017) Persidsky et al. (2006)
Intercellular junctions	1. Tight junctions (TJs) 2. Cadherins 3. Connexins	Controls the paracellular permeability, seal the intercellular space between adjacent brain microvascular endothelial cells, forming a tight barrier to regulate the passage of substances and essential for maintaining low permeability and restricting the entry of molecules and ions Transmembrane proteins which play a major role in cell adhesion and are integral components of the adherence junctions Protein subunits that form gap junction channels, thus enabling the exchange of small molecules between neighbouring cells	Greene and Campbell (2016) Persidsky et al. (2006) Persidsky et al. (2006)

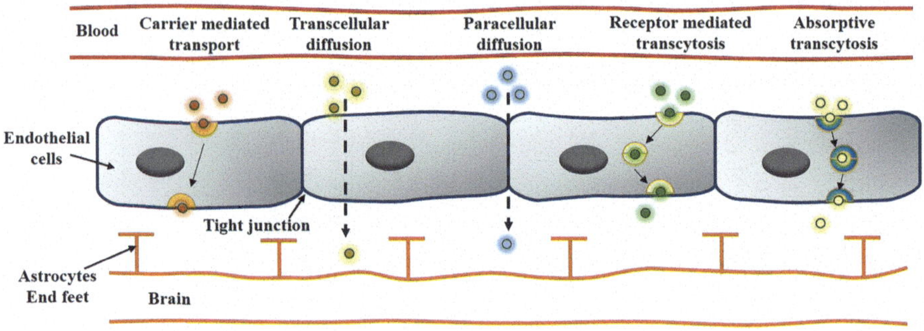

Fig. 9.2 Different methods of passing through blood brain barrier

transportation across the cell, and subsequent release on the other side. Large proteins, including protein and peptides, may utilize this transport mechanism (Barar et al. 2016; Bellettato and Scarpa 2018).

9.3 Strategies for Therapeutic Administration to the CNS

The brain, a delicate organ, is shielded by the microvascular endothelial cells of the BBB, which protect it from exogenous substances. However, addressing central nervous system diseases poses considerable challenges due to numerous formidable obstacles that hinder drug delivery effectively (Singh and Vellapandian 2023). Physiological barriers, including the BBB and blood-cerebrospinal fluid barrier, alongside a range of efflux transporter proteins, serve a crucial role in protecting the central nervous system (CNS) from toxic and infectious agents. These barriers also help maintain the ionic and volumetric balance essential for the apposite functioning of the nervous system (Furtado et al. 2018). However, these barriers also pose significant challenges to effective systemic drug delivery to the central nervous system (Rueda and Cruz 2017). Hence, the BBB stands as one of the most formidable barriers in the body, primarily due to the distinctive organization of blood vessels within the brain. These protective barriers effectively limit the entry of macromolecules into the brain, which can hinder the delivery of therapeutically valuable substances for treating CNS problems such as brain tumors, epilepsy, and cerebrovascular and neurodegenerative diseases (Puri et al. 2023). The formidable nature of these barriers poses significant challenges for drug delivery to the brain, as conventional methods have proven ineffective in selectively targeting drug molecules to the CNS (Chen and Liu 2012). To address this issue, current research endeavors are focused on the development of novel strategies and methodologies aimed at enhancing the delivery of drug compounds to the central nervous system. This section provides a concise overview of various approaches utilized to augment drug delivery to the CNS (Ou et al. 2021). The mechanism by which drugs are administered to the central nervous system can be categorized into two main groups: (1) invasive approaches and (2) non-invasive techniques.

9.3.1 Invasive Techniques

The use of invasive techniques, like osmotic pumps, for delivering drugs to the CNS is linked with several complications, including inflammatory reactions and neuronal damage. On the other hand, intracerebral implantation of therapeutic agents offers an alternative method, albeit one associated with significant pain. Despite the discomfort, this approach has been utilized in clinical practice (Pathan et al. 2009). Hence, there is a growing preference for drug delivery through non-invasive techniques to mitigate such complications. These also have many advantages which include the delivery of a better therapeutic drug concentration, which can be achieved by modifications like prodrugs, nano drug delivery, etc. (Witt et al. 2001). In general, it is observed that only a limited number of peptides and nutrients can effectively traverse the BBB to attain a sufficient concentration within the brain tissue, whether administered intravenously or orally (Carradori et al. 2016). Delivering adequate quantities of highly potent drugs to the CNS through systemic routes is essential. However, it is crucial to prevent these drugs from dispersing into other tissues, which could lead to severe toxic side effects (Chen et al. 2019a). Hence sometimes, it is necessary to either administer drugs directly into the brain tissue or to disrupt the blood brain barrier (Furtado et al. 2018).

9.3.2 Non-invasive Techniques

These techniques offer valuable means for delivering drugs to the CNS without causing patient discomfort or compromising the integrity of the BBB. Some non-invasive approaches that could be employed for drug delivery include (Achar et al. 2021):

1. Prodrug approach.
2. Changing the drug solubility.
3. Chimeric peptide.
4. Trojan approach.
5. Nano drug delivery.

9.3.3 Strategies for Enhanced CNS Drug Delivery

Different methods are being studied to avoid the multitude of barriers that inhibit the penetration of therapeutic drugs into the brain illustrated in Fig. 9.3 (Su and Sinko 2006). These generally come under one of the following three broad categories which include the manipulation of drugs, disruption of the BBB, and exploration of alternate pathways for the delivery of drug namely, manipulating drugs, disrupting the BBB, and exploiting alternative routes for the delivery of drug (Mittal et al. 2022).

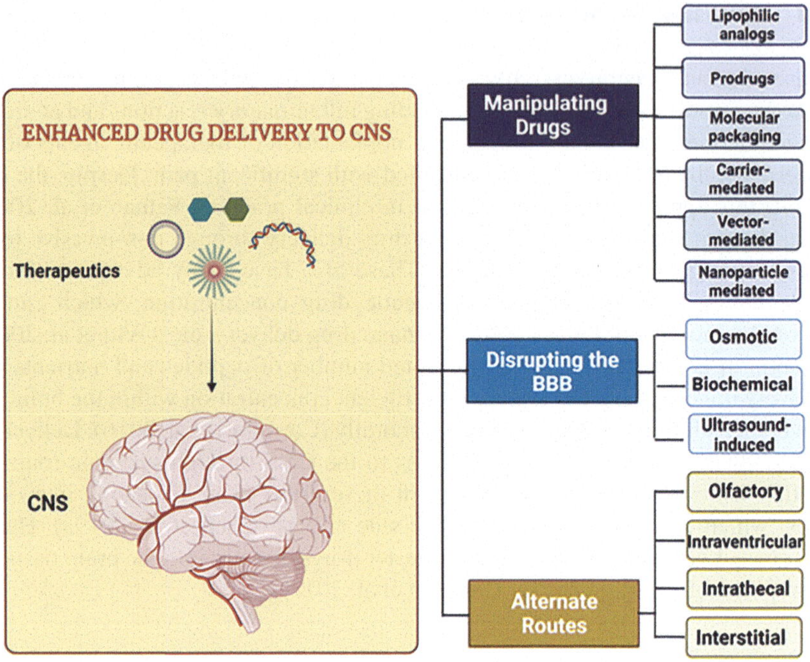

Fig. 9.3 Strategies for enhanced CNS drug delivery system

Under physiological conditions, the entry of circulating molecules into the brain occurs predominantly through a transcellular route across capillary endothelial cells, facilitated by passive or active transport mechanisms, or a combination of both (Yazdani et al. 2019). Free diffusion serves as the primary passive pathway, permitting the passage of lipid-soluble molecules with a molecular mass below 400–600 Da (Mamun et al. 2022). In contrast, a range of chemicals have the ability to pass through endothelial cells via distinct endogenous transport systems, predominantly employing carrier-mediated, receptor-mediated, or absorptive-mediated transport pathways. The transportation of vital nutrients, including amino acids, carboxylic acids, glucose, and nucleosides, to the brain is a critical function performed by these systems (Kumar et al. 2021). Absorptive-mediated transport refers to a form of endocytosis that is initiated by molecules possessing either a cationized ligand or a peptide component, such as albumin (Lu 2012). Receptor-mediated transport processes are the primary means by which macromolecules conjugated with surface receptor ligands, such as transferrin, insulin, and lactoferrin, are transported (Pardridge 1988; Alhodieb et al. 2023).

9.3.4 Drug Manipulation for Enhancing BBB Permeability

9.3.4.1 Prodrugs

Drug's pharmacokinetic properties can be enhanced by using prodrug which basically is a drug that is covalently attached to an unrelated chemical moiety. The enzymatic or hydrolytic processes in the body help activate the initially inactive prodrug, by cleaving the chemical moiety that is attached to it (Vale et al. 2018). In order to increase the drugs penetration into the brain, the prodrugs can also be made by attaching certain chemical groups that will increase their lipophilicity (Pavan et al. 2008).

9.3.4.2 Lipophilic Analogs

The BBB efficiently obstructs more than 98% of small-molecule medicines and all macromolecular therapies from reaching the brain (Mikitsh and Chacko 2014). Passive diffusion is restricted to lipid-soluble medicines with a molecular weight below 400–600 Da due to the tight connections of the BBB. Improving the lipophilicity of medicinal drugs is a potentially effective approach for enhancing the permeability of the blood-brain barrier (Wong et al. 2020). The degree of lipophilicity exhibited by a medication is generally associated with its capacity to effectively penetrate the cerebrovascular system. The augmentation of lipophilicity serves to improve the ability of a medication to traverse the blood-brain barrier. As a result, hydrophobic derivatives of tiny hydrophilic medications exhibit enhanced ability to traverse the blood-brain barrier (Waterhouse 2003). An instance of an effective anti-cancer treatment is Crizotinib, which is an oral selective small-molecule tyrosine kinase inhibitor. Nevertheless, its efficacy against brain tumor metastases is constrained due to its inadequate ability to traverse the BBB (Lo and li. 2003).

9.3.4.3 Chemical Drug Delivery Systems (CDS)

In this delivery system, we use the lipophilic precursors of the compounds that we wish to transport to the brain via the BBB (Pavan et al. 2008). These inactive chemical derivatives of a drug are obtained through various chemical modifications. They play a crucial role in facilitating site-specific delivery to the brain by undergoing multi-step enzymatic or chemical transformations (Garg et al. 2015). These derivatives primarily consist of two types of removable moieties. The first is the targeter, responsible for targeting and achieving site-specificity. The second is the modifier, which acts as a lyophilizer and protects certain functions of the brain (Nagappa and Dwibhashyam 2008).

9.3.4.4 Molecular Packaging

The transport of peptides like TRH (thyrotropin-releasing hormone) through the BBB is even more difficult as the ubiquitous peptidases can easily inactivate them (Bodor and Buchwald 2002). Hence to enhance the penetration of molecules through the BBB, we need to increase the passive transport by enhancing the lipophilic content, prevent the premature degradation of the peptides by ensuring enzymatic stability, and discover the lock-in mechanism to help target specific part of the brain (Ghosh et al. 2019). The molecular packaging strategy involves incorporating the peptide unit into a bulky molecule, primarily comprised of groups that facilitate blood-brain barrier (BBB) penetration while also preventing recognition by peptidases (Ghorai et al. 2023). Typically, a brain-targeted packaged peptide delivery system includes several components:

1. Redox targeter: A functional group responsible for targeting and directing the peptide across the BBB.
2. Spacer function: Comprising selected amino acids, this function ensures the removal of the charged targeter from the peptide.
3. Peptide: The active component of the molecule, responsible for its therapeutic action.
4. The bulky lipophilic moiety: Connected to the molecule either through an ester link or, in some cases, through a C-terminal adjuster. This moiety serves to increase the solubility of lipids and mask the peptide structure of the molecule, thereby facilitating its transportation across the BBB (Nikolova et al. 2022; Bodor and Buchwald 1999).

9.3.4.5 Carrier-Mediated Drug Delivery

Active transport systems are present for transporting various substances such as monosaccharides, vitamins, hormones, and different amino acids. The carrier mediated method uses the facilitative transport which is found in the brain endothelium (Cura and Carruthers 2012). Usually the facilitative carrier systems are found mainly in the cerebro vascular membranes which are used to transport both glucose and large neutral amino acids (LNAAs) (Zaragozá 2020). These carrier systems however are very specific for the transport of monosaccharides and hence cannot be used to deliver drugs. However, many endogenous as well as exogenous LNAAs are transported efficiently with the help of LNAA carrier system (Nałęcz 2017). Hence CNS drug delivery via this system has been an attractive strategy. The LNAA carrier system has been used to deliver levodopa, a precursor to dopamine, to the brain (Bodor et al. 1979; Mann et al. 2003).

9.3.4.6 Receptor/Vector-Mediated Drug Delivery

The extravasation of various essential nutrients and signaling molecules, which cannot diffuse directly through the cerebro-microvasculature, occurs through specific transcytosis systems (Archie et al. 2021). These systems encompass the transport of insulin, transferrin, and insulin-like growth factor. Among them, chimeric peptide delivery stands as a strategy where receptor-mediated transcytosis facilitates the transportation of a non-transportable peptide. This peptide is attached to a transportable peptide or protein, aiding in its passage across the barrier (Pardridge 1986; Wang et al. 2009).

Similarly, the process of endocytosis can be initiated through the interaction between the vector and its receptor located on the luminal surface of capillary endothelial cells in the brain (Smith and Gumbleton 2006). The active moiety of the chimeric peptide is released through exocytosis at the abluminal plasma membrane subsequent to enzymatic cleavage of the linkage between the drug and the vector. Following this, the peptide medication is released into the interstitial space of the brain, facilitating its interaction with the specific target receptors located on the cells of the brain (Pardridge 1992; Bickel et al. 2001). Similarly, immunoliposomes consist of liposomes that are coated with the inert polymer PEG and specifically linked to the OX-26 antitransferrin-receptor monoclonal antibody. In order to augment the intracellular permeability of membrane-permeable proteins, proteins can serve as carriers (Nagappa and Dwibhashyam 2008; Pardridge et al. 1995).

9.3.4.7 Nanoparticle Mediated Drug Delivery

This method is now widely used for the transport of drugs or peptides. One of the most effective biodegradable polymeric nanoparticle include the poly (lactic-co-glycolic acid) PLGA nanoparticles (Alvarado et al. 2019). Nanoparticles have facilitated the successful transport of drugs across the BBB, including tubocurarine, loperamide, and doxorubicin. The main mechanisms by which these nanoparticles are transported include endocytosis by the endothelial cells of blood capillaries or transcytosis across the endothelial cells (Ye et al. 2013). Applying polysorbate to nanoparticles enhances the process of apolipoprotein E (apoE) and other plasma components being taken up by the cells. These components then engage with the LDL receptors on the brain endothelial cells. The efflux pump may be inhibited, leading to the effective transport of nanoparticles coated with polysorbates (Madrid et al. 1991; Hartl et al. 2021).

9.4 Disrupting the BBB

Systemic administration of drugs combined with transient blood-brain barrier (BBB) disruption aids in enhanced drug delivery to the central nervous system (CNS). The successful disruption of the BBB relies on the creation of transient pores large enough to permit the entry of drug molecules without causing damage (Bors and Erdő 2019). Some techniques for transiently disrupting the BBB include:

9.4.1 Osmotic BBB Disruption

One of the most commonly utilized methods for osmotic blood-brain barrier (BBB) disruption is the intracranial infusion of a hyperosmolar solution of mannitol (Fink 2012). This procedure involves the direct infusion of the hyperosmolar solution into an artery that reaches the target area of the brain (Himmelseher 2007). The acute dehydration of endothelial cells induced by the hyperosmolar solution causes the cells to shrink, resulting in the widening of tight junctions connecting adjacent membranes. Complete restoration of the blood-brain barrier occurs after subsequent rehydration in the presence of normal plasma, typically about 4 h after treatment (Abbott and Romero 1996).

9.4.2 Biochemical BBB Disruption

Biochemical disruption techniques offer a safer approach to disrupting the blood-brain barrier (BBB). Various molecules, such as histamine and vasoactive peptides, which contribute to inflammatory reactions, also temporarily increase vascular permeability and leakage (Khosa et al. 2019). This heightened permeability facilitates an increased uptake of drug molecules by the cells. For instance, vascular permeability is enhanced through the use of bradykinin, a vasodilator that acts on the B2 receptors (van Nieuw Amerongen and van Hinsbergh 2002).

9.4.3 Ultrasound-Induced Disruption

Recently, there have been trials undertaken on the enhancement of medication transport to the central nervous system using hyperthermia. Ultrasound can generate mild hyperthermia in a restricted area of the tissue (Seynhaeve et al. 2020). The utilization of focused ultrasound allows for the reversible rupture of the BBB, occurring only within the designated area. At first, air that is contained in a lipid carrier is injected into the veins (McMahon et al. 2021). Following this, low-frequency ultrasound is administered through the transcranial pathway, resulting in the formation of oscillations in the injected air bubbles. The interaction between cerebral capillaries and the BBB results in a reversible disruption (Qiu et al. 2021).

9.5 Alternative Routes for CNS Drug Delivery

There are several alternative routes proposed to enhance drug penetration into the brain while bypassing the cardiovascular system. These include:

1. Intranasal delivery: Direct administration of drugs through the nasal mucosa, allowing them to bypass the blood-brain barrier and enter the CNS.
2. Intrathecal administration: Injection of drugs directly into the cerebrospinal fluid (CSF) through the spinal canal, providing direct access to the CNS.
3. Intracerebroventricular administration: Delivery of drugs into the cerebral ventricles, allowing them to diffuse throughout the CNS.
4. Intracerebral administration: Direct injection of drugs into specific brain regions or lesions.
5. Intracerebral implantation: Placement of drug-releasing devices or implants directly into the brain tissue.

9.5.1 Olfactory and Trigeminal Pathways to the CNS

The neural connections linking the nasal mucosa to the brain offer an effective route for delivering therapeutic substances to the central nervous system (CNS) without invasive procedures (Pardeshi and Belgamwar 2013). The olfactory pathway provides two routes within and outside neurons to access the brain. Within neurons, drugs travel via axonal transport, taking hours to days to reach various brain regions. The extraneuronal route involves bulk transport through perineural channels, enabling direct delivery of drugs to brain tissue or the cerebrospinal fluid (CSF) within minutes (Miyake and Bleier 2015). Intranasal delivery allows substances to reach the CSF easily since the CSF typically drains alongside olfactory axons as they pass through the skull's cribriform plate and reach the nasal cavity's roof, where the CSF enters nasal lymphatics (Gänger and Schindowski 2018). Characteristics like small molecular size and high lipid solubility favor rapid absorption of small molecules into the CNS through intranasal uptake. However, limitations exist for intranasal delivery as a method to bypass the BBB (Bahadur and Pathak 2012). Concentration limitations may vary in different brain and spinal cord regions, potentially decreasing as the drug's molecular weight increases. Moreover, some therapeutic agents may undergo partial degradation in the nasal mucosa or induce irritation to the mucosal tissues (Alhodieb et al. 2023).

9.5.2 Intraventricular/Intrathecal/Interstitial Delivery

An Ommaya reservoir, a type of plastic reservoir, is surgically placed under the skin of the scalp. It connects to the brain's ventricles through a catheter, enabling the direct infusion of medications into these ventricles (Almadrones et al. 1995). By manually pressing the reservoir through the scalp, drug solutions injected under the skin can reach the ventricles. Using this method offers several benefits over delivering drugs through the bloodstream. This approach circumvents the BBB, leading to higher concentrations of drugs in the cerebrospinal fluid (CSF) (Patel and Patel 2017). Smaller doses suffice, significantly reducing the risk of systemic toxicity. However, there are drawbacks to this technique, including slow drug distribution within the CSF and potential complications such as increased intracranial pressure, which can result in hemorrhages, CSF leaks, and CNS infections when injecting or infusing fluids into small ventricular spaces (Shen et al. 2004).

Another alternative method, known as the interstitial route, bypasses the blood-brain barrier (BBB) as well, allowing for the achievement of high drug concentrations within the brain with minimal systemic exposure and toxicity (Pandey et al. 2016). Sustained concentrations of drugs within the brain can be crucial for treating various neurodegenerative disorders and optimizing the effectiveness of certain chemotherapy drugs against tumours (Alam et al. 2010).

9.5.3 Biotechnological Approaches

Exploring biotechnological methods for interstitial drug delivery via biological tissues holds promise (Som et al. 2022). This typically involves transplanting dopamine-releasing neurons from embryos due to their robust post-transplantation survival and growth properties. However, a major challenge arises from insufficient neovascularization, which limits the success of this approach (Zhuang et al. 2022). In the absence of rapid formation of new blood vessels, transplanted solid grafts suffer irreversible ischemic damage, leading to cell death (Muthu et al. 2014). Another avenue in drug delivery to the CNS involves gene therapy. Cells undergo genetic modification before implantation to produce and release specific therapeutic substances. This technique has been investigated for its potential in treating brain tumors (An et al. 2015). Furthermore, there is ongoing exploration of antisense drug delivery, a recent technology in CNS drug delivery (Kabay et al. 2022). This method utilizes peptide nucleic acids (PNAs), which are antisense oligonucleotides featuring a polypeptide backbone. Receptor-mediated transcytosis has also been investigated as a means to facilitate PNA delivery to the CNS (Patra et al. 2018).

9.6 Neurodegenerative Diseases

The persistent and gradual degradation of the structure and function of the central nervous system is the main characteristic shared by a diverse range of conditions known as neurodegenerative diseases (Alhodieb et al. 2023; Vadakkan 2016). Clinical symptoms, such as behavioral and cognitive impairments, are used to classify the different illness types; the most prevalent forms are extrapyramidal and pyramidal movement disorders. Oxidative stress, neuroinflammation, programmed cell death, and proteotoxic stress cause progressive neuronal malfunction and death. Abnormalities in the ubiquitin-proteasomal and lysosomal/autophagosomal systems have also been observed (Dugger and Dickson 2017). These neurodegenerative illnesses presently only have palliative treatment available; there is no known cure. Western medicine has helped individuals with these conditions feel better. For example, dopaminergic treatment for Parkinson's disease movement abnormalities and antipsychotic medication are administered for dementia/AD to improve psychological and behavioral symptoms (Wei et al. 2023).

Two different types of lesions are seen in AD neuropathology namely: β-amyloid (Aβ) neurofibrillary tangles (NFT) and senile plaques (Castellani et al. 2010). Plaques and NFT are commonly associated with AD pathology, but they are insufficient alone to account for the substantial and profound neuronal loss of the disease. Similar to AD, PD symptoms are distinct, subtype-specific, and develop progressively in the affected brain sub-region. These include tremors, inflexibility/stiffness of the trunk and limbs, walking difficulty, and bradykinesia (slowness of movement) (Lee et al. 2022). The pathological hallmark of Parkinson's disease (PD) is the death of dopaminergic neurons, or neurons that produce dopamine, in the substantia nigra pars compacta (SNpc). This is subsequently accompanied by the reduction of striatal dopamine levels, impacting the caudate-putamen (Dauer and Przedborski 2003). Damage to neurons is caused by neurotoxic substances, which are inflammatory reactions elicited by activated immunological glia. This harm adds to the BBB vulnerability and is linked to cognitive and motor decline (Zeevi et al. 2010). As a result, the BBB disruption speeds up a neuroinflammatory cascade by facilitating leukocyte entrance into the brain.

Lou Gehrig's disease, also known as ALS, is another prevalent neurodegenerative disorder impacting the cognitive function of numerous adult's brain (Newell et al. 2022). This condition affects 60% of the population and is characterized by symptoms such as muscle weakness and twitching, eventually leading to complete neuromuscular failure and death. Common symptoms include slurred speech, tripping, dropping objects, atypical arm and/or leg weariness, and cramping (often associated with impaired respiratory function), typically appearing within 3–5 years (Bella and Chad 1998). Worldwide, 5–7 per 100,000 persons are affected with ALS, with 95% of cases occurring sporadically. Neuronal shrinkage and degeneration almost exclusively affecting the upper and lower motor neurons are hallmarks of ALS disease (Muddasir Qureshi et al. 2006). Activated microglia in neurodegenerative diseases influence neuronal injury by secreting pro-inflammatory factors, quinolinic acid, glutamate, and reactive oxygen species, among other things. Additionally, they activate cell chemotaxis and adaptive immune responses, which prolong neuronal injury by facilitating trans endothelial immune cell migration across the blood-brain barrier (Marchetti and Engelhardt 2020). Inflammatory secretions interact with surrounding glial cells, including astrocyte and endothelial cells, as the situation develops. This results in a damaging loop of inflammation amplification, both autocrine and paracrine, which seriously damages tissue and impairs neurogenesis (Gusev and Zhuravleva 2022).

9.6.1 Amyloidoses and Tauopathies

Fibrous proteins with a β-sheet-rich structure, known as amyloids, have certain structural characteristics that enable the binding of specific dyes like Congo red. A subgroup of prevalent neurodegenerative disorders exhibits significant extracellular protein aggregates, despite the fact that all of them have amyloid features (Hernández and Avila 2007). Amyloid-like filamentous aggregates in the cytoplasm of neurons and glia develop in these disorders, while extracellular deposits show up as plaques or in the walls of blood vessels. These plaques are subdivided into diffusive "primitive," dense cored "burnt-out," "classical" and cotton wool plaques (Dickson 1997). Depending on the kind, illness stage, and neuroanatomical location, different amyloid plaques can exist. Alzheimer's disease (AD) is commonly associated with β-amyloid (Aβ), however, amyloid deposits can also be found in other illnesses, particularly in people who have apolipoprotein E4, a major genetic risk factor for AD. Other amyloid forms, such as prion protein in Creutzfeldt-Jakob disease and ABri in family dementias in Britain and Denmark, are seen in less prevalent illnesses (Liu et al. 2013).

Tauopathies, however, are associated with pathological tau protein accumulation in neurons and glia and can be classified into primary and secondary forms. As in the cases of familial British dementia, AD, and GSS, the main driver in secondary tauopathies is not tau itself (Kovacs et al. 2022). In degenerative tauopathies, aberrant tau accumulation is coupled with posttranslational alterations of tau, a microtubule-associated phosphoprotein that is widely distributed in axons. Tau's significance in neurodegeneration was highlighted by the discovery of tau protein-encoding mutations in frontotemporal dementia and parkinsonism (FTDP-17T) (Hutton et al. 1998). Six isoforms of the tau protein are generated by alternative mRNA splicing (Andreadis 2005). Critical for microtubule binding, exon 10 splicing results in the production of 4R tau with four repeats and 3R tau with three repeats. The proportions of 3R and 4R tau differ depending on the type of cell and the condition; 4R tau predominates in glial cells. Neurodegenerative tauopathies comprise 3R/4R tauopathies (AD, CTE, tangle-dominated dementia, Guam Parkinson dementia complex, some FTDP-17T forms), 4R tauopathies (PSP, CBD, AGD, some FTDP-17T forms), and 3R/4R tauopathies (Pick's disease, some FTDP-17T forms) (Okamoto et al. 2019; Rösler et al. 2019).

9.6.2 Parkinson's Disease (PD) and Implicated Gene

A preliminary illustration of heritable connections in PD has been found, and the earliest known inheritable deformity or mutation was found on the SNCA gene on chromosome 4, which codes for α-synuclein. Subsequent analysis turned up a duplication or triplication of SNCA, which raised the possibility that PD and toxicity could result from overexpressing α-synuclein (García-González et al. 2019). Leucine-Rich Repeat Kinase 2 (LRRK2), DJ-1, ubiquitin carboxyl-terminal esterase L1 (UCHL 1), phosphatase and tensin homolog (PTEN), and Parkin are among the other genes that have been associated with this condition (Feng and Maguire-Zeiss 2010). When instances with autosomal-dominant PD present, mutations in the Gly-2019-Ser gene of the LRRK2 gene are the most frequently detected monogenic mutations affecting the whole globe, this mutant type leads to 1 case of sporadic PD's and 4 cases of familial PD's. Other mutations were observed in Arg-1441-His, Arg-1441-Cys, Arg-1441-Gly, ile-2020-Thr, Tyr-1699-Cys, and Asn-1437-His in addition to the Gly-2019-Ser alterations. This specific gene possesses many domains for protein–protein interactions in addition to an enzymatic core responsible for GTPase and serine–threonine kinase activity. The Gly-2019-Ser mutation in the kinase domain of the gene, positioned adjacent to the Ras-of-complex (Roc) GTPase domain, leads to an undesirable elevation in LRRK2 kinase activity (Tolosa et al. 2020). Because of this, current gene therapy approaches employ LRRK2 kinase inhibitors to target this domain.

9.6.3 Alzheimer's Disease (AD) and Implicated Genes

The amyloid precursor (APP), presenilin 1 (PSEN1), and presenilin 2 (PSEN2) genes are more prevalent in the autosomal dominant types of Alzheimer's disease (AD), whereas in sporadic AD, apolipoprotein E (ApoE) is the predominant gene (Cortini et al. 2018). The conversion of amyloid precursor protein (APP) into amyloid-β peptide (Aβ) by the β-secretase and γ-secretase complexes is a significant mechanism implicated in the etiology of Alzheimer's disease (AD). These complexes are widely recognized as important targets for therapeutic intervention. The potential approach involves the utilization of inhibitors that specifically target enzymes responsible for cleaving APP, including caspases, meprin-β, rhomboid-like protein-4 (RHBDL4), membrane-type matrix metalloproteinases (MT-MMPs/β-secretases), and legumain (δ-secretase) (García-González et al. 2019).

Mutations in the PSEN1 gene are identified as the principal etiological factor contributing to familial Alzheimer's disease (FAD). The gene encoding Presenilin-1 (PS1) is responsible for the enzymatic activity of γ—secretase, which facilitates the cleavage of specific type 1 transmembrane proteins, including Notch and APP (Ansari et al. 2023). Prior to this cleavage, β-secretase commences the process of cleavage to generate Aβ. As a consequence, Aβ40 and Aβ42 are formed, with Aβ becoming hydrophobic and causing the aggregation of Aβ in the brain. This leads to the accumulation of amyloid plaques, which is the documented pathogenic mechanism responsible for FAD. Presenilin is an inherent factor in memory, cognition, and the preservation of neurons as individual's age. Nevertheless, a mutation in the PSEN1 gene results in the impairment of presenilin function, leading to the development of dementia and neurodegeneration in individuals with familial Alzheimer's disease (Harilal et al. 2019). The gene ApoE, which plays a crucial role in lipid transport and injury repair inside the brain, has been linked to an increased vulnerability to Alzheimer's disease as a result of genetic variations (Fazil et al. 2012).

9.7 Nanotheranostics

The development of targeted drug delivery system is the major challenge in the treatment of neurodegenerative diseases. Various novel methods have emerged for the delivery of the drugs to the central nervous system (CNS). However, the transport of the therapeutic moieties to the target site in the CNS has been constrained due to the presence of BBB (Saha et al. 2023). The advancement in nanotechnology shows the promising potential toward the treatment of neurodegenerative disorder because nanoparticles can cross the BBB (Balachandran and Yuana 2019). Extensive research has been conducted on NP based therapies to identify the starting point for the initiation of theranostic application (Bragazzi 2019). Theranostic nanomedicine is an emerging field which utilize nanoparticles that has multifunctional abilities combining both therapeutic and diagnostic capabilities (Fig. 9.4) (Ho et al. 2012). Such nanoparticles that perform both functions can overcome the drawbacks of conventional nanoparticle system leading to more efficient therapeutic outcome (Ramanathan et al. 2018).

Nanotheranostics is designed by integrating nanotechnology with theranostic. It involves administering a nanotheranostic agent via different types of nanoparticles (Xu et al. 2023). Nanoparticles can be functionalized to target specific cell surface receptor and after reaching the target site in the body, the shell surrounding the medicine (theranostic agent) break down leading to the release of therapeutics to the target site (Kumar et al. 2020). Nanotheranostics offer advantage of targeted delivery of drugs, ability to cross BBB due to high surface area to volume ratio and also prevents the damage to the neighbouring healthy tissues (Aggarwal et al. 2022). New technical inventions include wide range of nanoplatforms, nanocarriers, and nanodelivery agents such as nanotubes, nanowires, nanospheres, liposomes, microemulsions, solid lipid NPs for accurate diagnosis and treatment. The application of theranostic nanoparticles for sensing, imaging, and medication delivery is currently

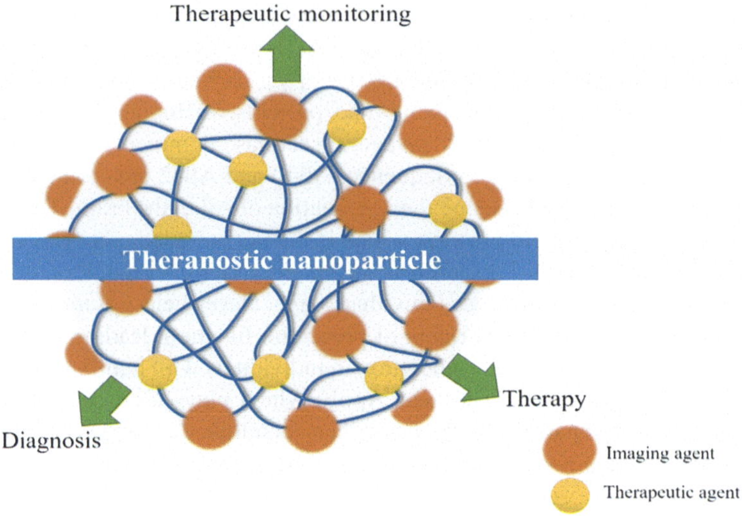

Fig. 9.4 Multifunctional ability of theranostic nanoparticle

a topic of significant interest (Chen et al. 2019b; Kumar et al. 2021). However, several potential challenges could hinder the success of theranostic nanoparticles (TNP). These include the discovery and targeting of new biomarkers, the inherent toxicity of nanoparticle elements, formulation stability, production costs, and control of intellectual property (Muthu et al. 2014).

9.7.1 Types of NPs Used in the Theranostic Application of Nanomedicines

Nanotechnology serves as a crucial bridge between the realms of biology and physics, leveraging minuscule structures and particles across various research domains. Among these, nanomedicine stands out as an essential component (Patra et al. 2018). In various nano based drug delivery systems, these nanomaterials ranging from size 1 to 100 nanometres are of utmost importance. Different kinds of magnetic, biological, mechanical, structural, and chemical properties are exhibited by these nanoparticles which eventually contribute to the development of nanomedicine (Zhu et al. 2021). Hence, the type of nanoparticles (NPs) generally used are discussed below:

(a) **Liposomes:** The initial wave of innovative colloidal nanocarriers, referred to as "Liposomes" (Rhaman et al. 2022), was created and experimented with as a drug delivery system during the 1970s. These are microscopic spheres made up of phospholipids comprising of 1 or more concentric lipid bilayers. It is often believed that conjugating the liposomes with a molecule called as PEG (polyethylene glycol) which is a biocompatible hydrophilic molecule increases the biocompatibility of the NP and decreases its immunogenicity. Liposomes have several advantages of liposomes such as they are suitable for delivery of hydrophobic, hydrophobic, and amphipathic drugs and agents, they are more biocompatible, have localized action and controlled release (Karthivashan et al. 2020).

(b) **Polymeric NPs:** have unique modifiable structure where the core structure is formed by the matrix of polymers such as natural polymers (alginate, chitosan) or synthetic polymers (PLGA, PEG) which encapsulates the therapeutic agent such as drugs. Size ranges from 10 nm to 100 nm. These "Polymeric NPs" (Teleanu et al. 2018; Xie et al. 2019) have some inherent properties which makes them distinct from others are—non toxicity, enhanced biocompatibility, and biodegradability and high surface area (drug adsorbed on the surface). PEGylation often increases the shelf life of nanoparticles useful in the systemic circulation. Stimuli-responsive polymers can release therapeutic payloads in response to changes in pH, temperature, or enzyme activity characteristic of the disease microenvironment, enhancing treatment precision.

(c) **Quantum dots (QDs):** are semiconducting materials which consists of semiconductor core (of CdTe) coated by shell (mainly ZnS) to improve optical properties and a cap for solubility in aqueous buffer. The diameter ranges from 1 nm to 10 nm. Quantum dots have size-dependent band gap which allows the precise control over the emitted light colour. So, this emission makes it suitable for various scientific applications such as in biological imaging and real time monitoring of intracellular processes (Valizadeh et al. 2012). These have several advantages such as remarkable photostability, bioconjugation, and biocompatibility, efficient electron transport properties. Various applications of QDs include MRI (magnetic resonance imaging), in vitro and in vivo detection, and gene therapy (World Molecular Imaging Congress 2022 2022).

(d) **Carbon based NPs:** carbon based nanoparticles mainly include fullerenes and carbon nanotubes which serve as most effective nanocarriers of therapeutic agents. Carbon nanotubes (CNTs) (Zhu et al. 2021; Rhaman et al. 2022) are cylindrical structures which are composed of rolled up graphene sheets which can be either single walled CNTs or multi walled CNTs with varying layers of graphene sheets. They have high electrical conductivity, better thermal stability, and extraordinary strength. These unique properties enable them to penetrate the cell membrane and deliver the therapeutic agents efficiently.

On the other hand, fullerenes are spherical carbon molecules composed of 60 carbon atoms (C_{60}). They have a unique ability to act as antioxidants which makes them valuable in overcoming oxidative stress conditions. Fullerenes exhibit inherent fluorescence properties used in bioimaging. Due to their hollow structure, drugs can be encapsulated inside the fullerene cage protecting them from degradation and enhance targeted delivery. Fullerenes have shown promise in various therapeutic applications, including neuroprotection, anti-inflammatory effects, and potential antiviral activities (Dhiman et al. 2022).

9.7.2 Imaging and Early Diagnosis

Imaging methods gather information about drug delivery to the specific tissue or facilitate controlled drug release by exploiting the physical, chemical, and optical properties of nanomaterials (Fig. 9.5) (Siafaka et al. 2021). Multifunctional nanoparticles that are conjugated with dyes, contrast agents, and biomarkers are to study pathological changes that occur within the brain (Shim et al. 2010). Magnetic resonance imaging (MRI) and positron-emission tomography can be used to monitor theranostics and clinical nuclear imaging systems after they have been successfully delivered (Li et al. 2016).

9.7.3 Positron Emission Tomography (PET)

PET imaging offers a method for observing aggregated Aβ and tau proteins in the brain, providing insights into pathological changes occurring in Alzheimer's disease (AD) patients. PPET radiotracer technology enables visualization of amyloid accumulation in Alzheimer's disease (AD), mild cognitive impairment, and aging. However, its broader application is limited by challenges such as low spatial resolution, radiation exposure, and high costs in AD. On the other hand, poly(D, L-lactide-co-glycolide) (PLGA) nanoparticles offer biocompatible properties. By employing fluorescently labeled PLGA, amyloid-β (Aβ) aggregates can be detected in AD. Among the clinically approved PET tracers, three include 18 F-Florbetaben, 18 F-Flutemetamol, and 18 F-Florbetapir (Govindarajan and Kar 2023).

9.7.4 Magnetic Resonance Imaging (MRI)

A new revolutionary approach involves nanoparticles that is used in Magnetic Resonance Imaging (MRI). MRI relies on the interaction between radiofrequency pulses and the magnetic properties of hydrogen atom. A high static magnetic field is required to achieve this interaction. As compared to PET, Magnetic Resonance

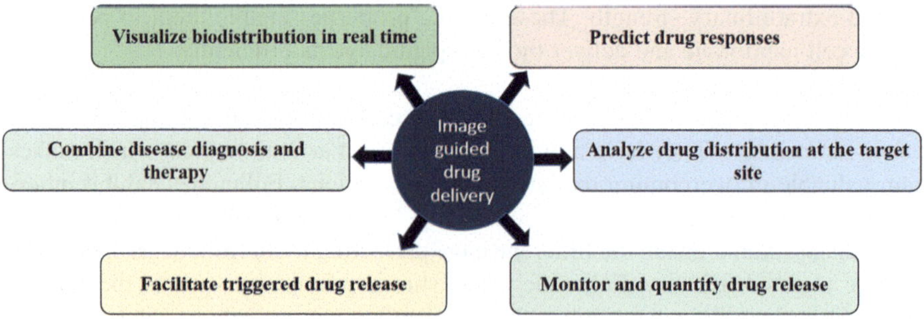

Fig. 9.5 Significance of image guided drug delivery system

Imaging has much higher resolution. Different types of magnetic nanoparticles have been employed, with a special attention directed towards iron oxide containing nanoparticle called as superparamagnetic iron oxide nanoparticles. This nanomaterial has excellent magnetic properties. Another paramagnetic nanoparticle involves gadolinium and manganese (Bruno et al. 2022).

Lesions in the CNS caused by multiple sclerosis are detected using gadolinium enhanced MRI scans for blood brain barrier breakdown (Gauberti et al. 2018).

9.8 Challenges and Future Directions

Numerous novel strategies have emerged for the transport of drugs to the CNS. However, there are few potential downsides that needs to be focussed first that involves lack of adequate equipment for detailed and scalable synthesis of nanomaterials determining the efficacy and safety of nanomaterials, the presence of blood brain barrier that has constrained the transport of drug material across it. Advancements in the field of nanotechnology have sparked a revolution in effectively transporting drugs across the BBB. Neurotoxicity poses a significant concern linked with nanotheranostic agents, largely attributed to the initiation of oxidative stress. This oxidative stress is influenced by several factors, such as the morphology, size, surface area, solubility, concentration, and duration, as well as the mode of administration of nanotherapeutic agents. Despite these difficulties, the unique physiochemical properties of nanoparticles as noted earlier make them attractive candidate in the nanomedicine (Jagaran and Singh 2021). Few points should be considered while manufacturing of nanoparticle in the future: minimum cost and high efficacy, biocompatibility, should have ability to penetrate across various biological barriers, encapsulated molecules should be released freely at the target site and possess potential therapeutic efficacy (Rahman et al. 2022). In vitro studies suggest that drugs encapsulated nanoparticles are good candidate for the treatment of neurodegenerative diseases. However, in vivo studies and clinical trials are necessary to effectively trace the causes of disease and enhance strategy for delivery of drugs across the BBB.

9.9 Conclusion

The emergence of cutting-edge nanotechnologies holds great potential to drive innovation and progress across a spectrum of disciplines, with profound implications for key sectors like healthcare, bioscience engineering, and medical applications. Nanomedicine is becoming increasingly effective to overcome the barriers that are still challenge for conventional medicine. Multi-functional nanocarriers used for targeted drug delivery and in vivo imaging hold significant promise for treatment of neurodegenerative diseases. Prospects for the future include the potential emergence of theranostic nanomedicines undergoing clinical trials. However, it is imperative to address present challenges and delve into innovating new nano tools and devices for diagnosing and treating neurodegenerative diseases.

Acknowledgements KUF thanks the Indian Council of Medical Research-Senior Research Fellow, Govt. of India for providing financial assistance (File No.: 45/75/2020-Nan-BMS). We acknowledge the use of BioRender.com for creating the figures in this work.

References

Abbott NJ, Romero IA (1996) Transporting therapeutics across the blood-brain barrier. Mol Med Today 2(3):106–113

Achar A, Myers R, Ghosh C (2021) Drug Delivery Challenges in Brain Disorders across the Blood–Brain Barrier: Novel Methods and Future Considerations for Improved Therapy. Biomedicines 9(12):1834

Aggarwal N, Choudhury S, Chibh S, Panda JJ (2022) Aptamer-nanoconjugates as emerging theranostic systems in neurodegenerative disorders. Colloid Interface Sci Commun 46:100554

Alam MI, Beg S, Samad A, Baboota S, Kohli K, Ali J et al (2010) Strategy for effective brain drug delivery. Eur J Pharm Sci 40(5):385–403

Alhodieb FS, Rahman MA, Barkat MA, Alanezi AA, Barkat HA, Hadi HA, Harwansh RK, Mittal V (2023) Nanomedicine-driven therapeutic interventions of autophagy and stem cells in the management of Alzheimer's disease. Nanomedicine (Lond) 18(2):145–168. https://doi.org/10.2217/nnm-2022-0108. Epub 2023 Mar 20

Almadrones L, Campana P, Dantis EC (1995) Arterial, peritoneal, and intraventricular access devices. Semin Oncol Nurs 11(3):194–202

Alvarado Y, Muro C, Illescas J, Riera F (2019) Polymer nanoparticles for the release of complex molecules. Mater Biomed Eng:135–163

An SS, Lee S, Kim J, Bark CW, Lee B, Ju H et al (2015) Spotlight on nano-theranostics in South Korea: applications in diagnostics and treatment of diseases. Int J Nanomedicine 10(Spec Iss):3–8

Andreadis A (2005) Tau gene alternative splicing: expression patterns, regulation and modulation of function in normal brain and neurodegenerative diseases. Biochim Biophys Acta (BBA) - Mol Basis Dis 1739(2–3):91–103

Ansari MA, Tripathi T, Venkidasamy B, Monziani A, Rajakumar G, Alomary MN, Alyahya SA, Onimus O, D'souza N, Barkat MA, Al-Suhaimi EA, Samynathan R, Thiruvengadam M (2023) Multifunctional nanocarriers for Alzheimer's disease: befriending the barriers. Mol Neurobiol 61:3042–3089. https://doi.org/10.1007/s12035-023-03730-z

Archie SR, Al Shoyaib A, Cucullo L (2021) Blood-brain barrier dysfunction in CNS disorders and putative therapeutic targets: An overview. Pharmaceutics 13(11):1779

Attwell D, Mishra A, Hall CN, O'Farrell FM, Dalkara T (2016) What is a pericyte? J Cereb Blood Flow Metab 36(2):451–455

Bahadur S, Pathak K (2012) Physicochemical and physiological considerations for efficient nose-to-brain targeting. Expert Opin Drug Deliv 9(1):19–31

Balachandran B, Yuana Y (2019) Extracellular vesicles-based drug delivery system for cancer treatment. Schumacher U, editor. Cogent Med 6(1):1635806

Barar J, Rafi MA, Pourseif MM, Omidi Y (2016) Blood-brain barrier transport machineries and targeted therapy of brain diseases. Bioimpacts 6(4):225–248

Bella I, Chad DA (1998) Neuromuscular disorders and acute respiratory failure. Neurol Clin 16(2):391–417

Bellettato CM, Scarpa M (2018) Possible strategies to cross the blood–brain barrier. Ital J Pediatr 44(S2):131

Bickel U, Yoshikawa T, Pardridge WM (2001) Delivery of peptides and proteins through the blood–brain barrier. Adv Drug Deliv Rev 46(1–3):247–279

Bodor N, Buchwald P (1999) Recent advances in the brain targeting of neuropharmaceuticals by chemical delivery systems. Adv Drug Deliv Rev 36(2–3):229–254

Bodor N, Buchwald P (2002) Barriers to remember: brain-targeting chemical delivery systems and Alzheimer's disease. Drug Discov Today 7(14):766–774

Bodor N, Prokai L, Wu WM, Farag H, Jonalagadda S, Kawamura M et al (1992) A strategy for delivering peptides into the central nervous system by sequential metabolism. Science (1979) 257(5077):1698–1700

Bors L, Erdő F (2019) Overcoming the blood–brain barrier. Challenges and tricks for CNS drug delivery. Sci Pharm 87(1):6

Bragazzi NL (2019) Nanomedicine: insights from a bibliometrics-based analysis of emerging publishing and research trends. Medicina (B Aires) 55(12):785

Bruno F, Granata V, Cobianchi Bellisari F, Sgalambro F, Tommasino E, Palumbo P et al (2022) Advanced magnetic resonance imaging (MRI) techniques: technical principles and applications in nanomedicine. Cancers (Basel) 14(7):1626

Carradori D, Gaudin A, Brambilla D, Andrieux K (2016) Application of nanomedicine to the CNS diseases. Int Rev Neurobiol 130:73–113

Castellani RJ, Rolston RK, Smith MA (2010) Alzheimer disease. Dis Mon 56(9):484–546

Chen Y, Liu L (2012) Modern methods for delivery of drugs across the blood–brain barrier. Adv Drug Deliv Rev 64(7):640–665

Chen JC, Li LM, Gao JQ (2019a) Biomaterials for local drug delivery in central nervous system. Int J Pharm 560:92–100

Chen X, Zheng G, Cheng J, Yang YY (2019b) Supramolecular nanotheranostics. Theranostics 9(11):3014–3016

Cortini F, Cantoni C, Villa C (2018) Epileptic seizures in autosomal dominant forms of Alzheimer's disease. Seizure 61:4–7

Cura AJ, Carruthers A (2012) Role of monosaccharide transport proteins in carbohydrate assimilation, distribution, metabolism, and homeostasis. Compr Physiol 2(2):863–914

Dauer W, Przedborski S (2003) Parkinson's disease. Neuron 39(6):889–909

Dhiman S, Kaur A, Sharma M (2022) Fullerenes for anticancer drug targeting: teaching an old dog a new trick. Mini Rev Med Chem 22(22):2864–2880. https://doi.org/10.2174/1389557522666220317145544

Dickson DW (1997) The pathogenesis of senile plaques. J Neuropathol Exp Neurol 56(4):321–339

Dugger BN, Dickson DW (2017) Pathology of neurodegenerative diseases. Cold Spring Harb Perspect Biol 9(7):a028035

Fazil M, Shadab, Baboota S, Sahni JK, Ali J (2012) Nanotherapeutics for Alzheimer's disease (AD): past, present and future. J Drug Target 20(2):97–113

Feng LR, Maguire-Zeiss KA (2010) Gene therapy in Parkinson's disease. CNS Drugs 24(3):177–192

Ferri CP, Prince M, Brayne C, Brodaty H, Fratiglioni L, Ganguli M et al (2005) Global prevalence of dementia: a Delphi consensus study. Lancet 366(9503):2112–2117

Fink ME (2012) Osmotherapy for intracranial hypertension: mannitol versus hypertonic saline. Continuum 18:640–654

Furtado D, Björnmalm M, Ayton S, Bush AI, Kempe K, Caruso F (2018) Overcoming the blood–brain barrier: the role of nanomaterials in treating neurological diseases. Adv Mater 30(46):e1801362

Gänger S, Schindowski K (2018) Tailoring formulations for intranasal nose-to-brain delivery: a review on architecture, physico-chemical characteristics and mucociliary clearance of the nasal olfactory mucosa. Pharmaceutics. 10(3):116

Garbayo E, Mendoza A, Blanco-Prieto M (2014) Diagnostic and therapeutic uses of nanomaterials in the brain. Curr Med Chem 21(36):4100–4131

García-González L, Pilat D, Baranger K, Rivera S (2019) Emerging alternative proteinases in APP metabolism and Alzheimer's disease pathogenesis: a focus on MT1-MMP and MT5-MMP. Front Aging Neurosci 11:244

Garg T, Bhandari S, Rath G, Goyal AK (2015) Current strategies for targeted delivery of bio-active drug molecules in the treatment of brain tumor. J Drug Target 23(10):865–887

Gauberti M, Fournier AP, Docagne F, Vivien D, Martinez de Lizarrondo S (2018) Molecular magnetic resonance imaging of endothelial activation in the central nervous system. Theranostics 8(5):1195–1212

Ghorai SM, Deep A, Magoo D, Gupta C, Gupta N (2023) Cell-penetrating and targeted peptides delivery systems as potential pharmaceutical carriers for enhanced delivery across the blood–brain barrier (BBB). Pharmaceutics. 15(7):1999

Ghosh S, Lalani R, Patel V, Bhowmick S, Misra A (2019) Surface engineered liposomal delivery of therapeutics across the blood brain barrier: recent advances, challenges and opportunities. Expert Opin Drug Deliv 16(12):1287–1311

Govindarajan K, Kar S (2023) Detection of β-amyloid aggregates/plaques in 5xFAD mice by labelled native PLGA nanoparticles: implication in the diagnosis of Alzheimer's disease. J Nanobiotechnol 21(1):216

Greene C, Campbell M (2016) Tight junction modulation of the blood brain barrier: CNS delivery of small molecules. Tissue Barriers. 4(1):e1138017

Gusev E, Zhuravleva Y (2022) Inflammation: a new look at an old problem. Int J Mol Sci 23(9):4596

Harilal S, Jose J, Parambi DGT, Kumar R, Mathew GE, Uddin MS et al (2019) Advancements in nanotherapeutics for Alzheimer's disease: current perspectives. J Pharm Pharmacol 71(9):1370–1383

Hartl N, Adams F, Merkel OM (2021) From adsorption to covalent bonding: apolipoprotein e functionalization of polymeric nanoparticles for drug delivery across the blood–brain barrier. Adv Ther (Weinh) 4(1):2000092

Hernández F, Avila J (2007) Tauopathies. Cell Mol Life Sci 64(17):2219–2233

Himmelseher S (2007) Hypertonic saline solutions for treatment of intracranial hypertension. Curr Opin Anaesthesiol 20(5):414–426

Ho JA, Wang LS, Chuang MC (2012) Nanotheranostics: a review of recent publications. Int J Nanomedicine 7:4679–4695

Hutton M, Lendon CL, Rizzu P, Baker M, Froelich S, Houlden H et al (1998) Association of missense and 5′-splice-site mutations in tau with the inherited dementia FTDP-17. Nature 393(6686):702–705

Jagaran K, Singh M (2021) Nanomedicine for neurodegenerative disorders: focus on Alzheimer's and Parkinson's diseases. Int J Mol Sci 22(16):9082

Kabanov AV, Gendelman HE (2007) Nanomedicine in the diagnosis and therapy of neurodegenerative disorders. Prog Polym Sci 32(8–9):1054–1082

Kabay G, Manz A, Dincer C (2022) Microfluidic roadmap for translational nanotheranostics. Small Methods 6(2):e2101217

Karthivashan G, Ganesan P, Park SY, Lee HW, Choi DK (2020) Lipid-based nanodelivery approaches for dopamine-replacement therapies in Parkinson's disease: from preclinical to translational studies. Biomaterials 232:119704

Khosa A, Saha RN, Singhvi G (2019) Drug delivery to the brain. Nanomaterials for drug delivery and therapy. Elsevier, pp 461–514

Kovacs GG, Ghetti B, Goedert M (2022) Classification of diseases with accumulation of Tau protein. Neuropathol Appl Neurobiol 48(3):e12792

Kumar A, Chaudhary RK, Singh R, Singh SP, Wang SY, Hoe ZY et al (2020) Nanotheranostic applications for detection and targeting neurodegenerative diseases. Front Neurosci 14:305

Kumar R, Chhikara BS, Gulia K, Chhillar M (2021) Review of nanotheranostics for molecular mechanisms underlying psychiatric disorders and commensurate nanotherapeutics for neuropsychiatry: the mind knockout. Nano 5(3):288–308

Lee SH, Park SM, Yeo SS, Kwon O, Lee MK, Yoo H et al (2022) Parkinson's disease subtyping using clinical features and biomarkers: literature review and preliminary study of subtype clustering. Diagnostics 12(1):112

Li X, Zhang XN, Li XD, Chang J, Li X, Zhang XN et al (2016) Multimodality imaging in nanomedicine and nanotheranostics. Cancer Biol Med 13(3):339–348

Liu CC, Kanekiyo T, Xu H, Bu G (2013) Apolipoprotein E and Alzheimer disease: risk, mechanisms and therapy. Nat Rev Neurol 9(2):106–118

Lo Y, li. (2003) Relationships between the hydrophilic–lipophilic balance values of pharmaceutical excipients and their multidrug resistance modulating effect in Caco-2 cells and rat intestines. J Control Release 90(1):37–48

Lochhead JJ, Yang J, Ronaldson PT, Davis TP (2020) Structure, function, and regulation of the blood-brain barrier tight junction in central nervous system disorders. Front Physiol 11:914

Lu W (2012) Adsorptive-mediated brain delivery systems. Curr Pharm Biotechnol 13(12):2340–2348

Lu W, Yao J, Zhu X, Qi Y (2021) Nanomedicines: redefining traditional medicine. Biomed Pharmacother 134:111103

Madrid Y, Langer LF, Brem H, Langer R (1991) New directions in the delivery of drugs and other substances to the central nervous system. Adv Pharmacol 22:299–324

Mamun AA, Uddin MS, Perveen A, Jha NK, Alghamdi BS, Jeandet P et al (2022) Inflammation-targeted nanomedicine against brain cancer: from design strategies to future developments. Semin Cancer Biol 86:101–116

Mann GE, Yudilevich DL, Sobrevia L (2003) Regulation of amino acid and glucose transporters in endothelial and smooth muscle cells. Physiol Rev 83(1):183–252

Marchetti L, Engelhardt B (2020) Immune cell trafficking across the blood-brain barrier in the absence and presence of neuroinflammation. Vasc Biol 2(1):H1–H18

McMahon D, O'Reilly MA, Hynynen K (2021) Therapeutic agent delivery across the blood–brain barrier using focused ultrasound. Annu Rev Biomed Eng 23(1):89–113

Mikitsh JL, Chacko AM (2014) Pathways for small molecule delivery to the central nervous system across the blood-brain barrier. Perspect Medicin Chem 6:PMC.S13384

Mittal KR, Pharasi N, Sarna B, Singh M, Rachana HS et al (2022) Nanotechnology-based drug delivery for the treatment of CNS disorders. Transl Neurosci 13(1):527–546

Miyake MM, Bleier BS (2015) The blood-brain barrier and nasal drug delivery to the central nervous system. Am J Rhinol Allergy 29(2):124–127

Muddasir Qureshi M, Hayden D, Urbinelli L, Ferrante K, Newhall K, Myers D et al (2006) Analysis of factors that modify susceptibility and rate of progression in amyotrophic lateral sclerosis (ALS). Amyotroph Lateral Scler 7(3):173–182

Mura S, Couvreur P (2012) Nanotheranostics for personalized medicine. Adv Drug Deliv Rev 64(13):1394–1416

Muthu MS, Leong DT, Mei L, Feng SS (2014) Nanotheranostics - application and further development of nanomedicine strategies for advanced theranostics. Theranostics 4(6):660–677

Nagappa A, Dwibhashyam VSNM (2008) Strategies for enhanced drug delivery to the central nervous system. Indian J Pharm Sci 70(2):145

Nałęcz KA (2017) Solute Carriers in the Blood–Brain Barier: Safety in Abundance. Neurochem Res 42(3):795–809

Newell ME, Adhikari S, Halden RU (2022) Systematic and state-of the science review of the role of environmental factors in amyotrophic lateral sclerosis (ALS) or Lou Gehrig's disease. Sci Total Environ 817:152504

Nikolova MP, Kumar EM, Chavali MS (2022) Updates on responsive drug delivery based on liposome vehicles for cancer treatment. Pharmaceutics. 14(10):2195

Okamoto K, Amari M, Fukuda T, Suzuki K, Takatama M (2019) Astrocytic tau pathologies in aged human brain. Neuropathology 39(3):187–193

Ou LC, Zhong S, Ou J, Tian J (2021) Application of targeted therapy strategies with nanomedicine delivery for atherosclerosis. Acta Pharmacol Sin 42(1):10–17

Pandey PK, Sharma AK, Gupta U (2016) Blood brain barrier: An overview on strategies in drug delivery, realistic in vitro modeling and in vivo live tracking. Tissue Barriers 4(1):e1129476

Pardeshi CV, Belgamwar VS (2013) Direct nose to brain drug delivery via integrated nerve pathways bypassing the blood–brain barrier: an excellent platform for brain targeting. Expert Opin Drug Deliv 10(7):957–972

Pardridge WM (1986) Receptor-mediated peptide transport through the blood-brain barrier. Endocr Rev 7(3):314–330

Pardridge WM (1988) Recent advances in blood-brain barrier transport. Annu Rev Pharmacol Toxicol 28(1):25–39

Pardridge WM (1992) Recent developments in peptide drug delivery to the brain. Pharmacol Toxicol 71(1):3–10

Pardridge WM, Boado RJ, Kang YS (1995) Vector-mediated delivery of a polyamide ('peptide') nucleic acid analogue through the blood-brain barrier in vivo. Proc Natl Acad Sci 92(12):5592–5596

Patel MM, Patel BM (2017) Crossing the blood–brain barrier: recent advances in drug delivery to the brain. CNS Drugs 31(2):109–133

Pathan S, Iqbal Z, Zaidi S, Talegaonkar S, Vohra D, Jain G et al (2009) CNS drug delivery systems: novel approaches. Recent Pat Drug Deliv Formul 3(1):71–89

Patra JK, Das G, Fraceto LF, Campos EVR, Rodriguez-Torres MP, Acosta-Torres LS et al (2018) Nano based drug delivery systems: recent developments and future prospects. J Nanobiotechnol 16(1):71

Pavan B, Dalpiaz A, Ciliberti N, Biondi C, Manfredini S, Vertuani S (2008) Progress in drug delivery to the central nervous system by the prodrug approach. Molecules 13(5):1035–1065

Persidsky Y, Ramirez SH, Haorah J, Kanmogne GD (2006) Blood–brain barrier: structural components and function under physiologic and pathologic conditions. J Neuroimmune Pharmacol 1(3):223–236

Puri S, Mazza M, Roy G, England RM, Zhou L, Nourian S et al (2023) Evolution of nanomedicine formulations for targeted delivery and controlled release. Adv Drug Deliv Rev 200:114962

Qiu W, Bouakaz A, Konofagou EE, Zheng H (2021) Ultrasound for the brain: a review of physical and engineering principles, and clinical applications. IEEE Trans Ultrason Ferroelectr Freq Control 68(1):6–20

Quintana FJ (2017) Astrocytes to the rescue! Glia limitans astrocytic endfeet control CNS inflammation. J Clin Invest 127(8):2897–2899

Rahman MM, Islam MR, Akash S, Harun-Or-Rashid M, Ray TK, Rahaman MS et al (2022) Recent advancements of nanoparticles application in cancer and neurodegenerative disorders: at a glance. Biomed Pharmacother 153:113305

Ramanathan S, Archunan G, Sivakumar M, Tamil Selvan S, Fred AL, Kumar S et al (2018) Theranostic applications of nanoparticles in neurodegenerative disorders. Int J Nanomedicine 13:5561–5576

Rhaman MM, Islam MR, Akash S, Mim M, Alam MN, Nepovimova E et al (2022) Exploring the role of nanomedicines for the therapeutic approach of central nervous system dysfunction: at a glance. Front Cell Dev Biol 10:989471

Rösler TW, Tayaranian Marvian A, Brendel M, Nykänen NP, Höllerhage M, Schwarz SC et al (2019) Four-repeat tauopathies. Prog Neurobiol 180:101644

Rueda F, Cruz LJ (2017) Targeting the brain with nanomedicine. Curr Pharm Des 23(13):1879–1896

Saha P, Kathuria H, Pandey MM (2023) Intranasal nanotherapeutics for brain targeting and clinical studies in Parkinson's disease. J Control Release 358:293–318

Seynhaeve ALB, Amin M, Haemmerich D, van Rhoon GC, ten Hagen TLM (2020) Hyperthermia and smart drug delivery systems for solid tumor therapy. Adv Drug Deliv Rev 163–164:125–144

Shen D, Artru A, Adkison K (2004) Principles and applicability of CSF sampling for the assessment of CNS drug delivery and pharmacodynamics. Adv Drug Deliv Rev 56(12):1825–1857

Shim MS, Kim CS, Ahn YC, Chen Z, Kwon YJ (2010) Combined multimodal optical imaging and targeted gene silencing using stimuli-transforming nanotheragnostics. J Am Chem Soc 132(24):8316–8324

Siafaka PI, Okur NÜ, Karantas ID, Okur ME, Gündoğdu EA (2021) Current update on nanoplatforms as therapeutic and diagnostic tools: a review for the materials used as nanotheranostics and imaging modalities. Asian J Pharm Sci 16(1):24–46

Singh SA, Vellapandian C (2023) Structure of the blood brain barrier and its role in the transporters for the movement of substrates across the barriers. Curr Drug Metab 24(4):250–269

Singh N, Pillay V, Choonara YE (2007) Advances in the treatment of Parkinson's disease. Prog Neurobiol 81(1):29–44

Smith MW, Gumbleton M (2006) Endocytosis at the blood–brain barrier: from basic understanding to drug delivery strategies. J Drug Target 14(4):191–214

Som A, Rosenboom JG, Chandler A, Sheth RA, Wehrenberg-Klee E (2022) Image-guided intratumoral immunotherapy: developing a clinically practical technology. Adv Drug Deliv Rev 189:114505

Su Y, Sinko PJ (2006) Drug delivery across the blood–brain barrier: why is it difficult? How to measure and improve it? Expert Opin Drug Deliv 3(3):419–435

Teleanu D, Chircov C, Grumezescu A, Volceanov A, Teleanu R (2018) Blood-brain delivery methods using nanotechnology. Pharmaceutics 10(4):269

Tolosa E, Vila M, Klein C, Rascol O (2020) LRRK2 in Parkinson disease: challenges of clinical trials. Nat Rev Neurol 16(2):97–107

Vadakkan KI (2016) Neurodegenerative disorders share common features of "loss of function" states of a proposed mechanism of nervous system functions. Biomed Pharmacother 83:412–430

Vale N, Ferreira A, Matos J, Fresco P, Gouveia M (2018) Amino acids in the development of prodrugs. Molecules 23(9):2318

Valizadeh A, Mikaeili H, Samiei M, Farkhani SM, Zarghami N, Kouhi M, Akbarzadeh A, Davaran S (2012) Quantum dots: synthesis, bioapplications, and toxicity. Nanoscale Res Lett 7(1):480. https://doi.org/10.1186/1556-276X-7-480

van Nieuw Amerongen GP, van Hinsbergh VWM (2002) Targets for pharmacological intervention of endothelial hyperpermeability and barrier function. Vasc Pharmacol 39(4–5):257–272

Wang YY, Lui PC, Li JY (2009) Receptor-mediated therapeutic transport across the blood–brain barrier. Immunotherapy 1(6):983–993

Waterhouse R (2003) Determination of lipophilicity and its use as a predictor of blood–brain barrier penetration of molecular imaging agents. Mol Imaging Biol 5(6):376–389

Wei M, Yang Z, Li S, Le W (2023) Nanotherapeutic and stem cell therapeutic strategies in neurodegenerative diseases: a promising therapeutic approach. Int J Nanomedicine 18:611–626

Witt KA, Gillespie TJ, Huber JD, Egleton RD, Davis TP (2001) Peptide drug modifications to enhance bioavailability and blood-brain barrier permeability. Peptides (NY) 22(12):2329–2343

Wong XY, Sena-Torralba A, Álvarez-Diduk R, Muthoosamy K, Merkoçi A (2020) Nanomaterials for nanotheranostics: tuning their properties according to disease needs. ACS Nano 14(3):2585–2627

World Molecular Imaging Congress 2022 (2022) Mol Imaging Biol 24(Suppl 2):63–480. https://doi.org/10.1007/s11307-022-01794-2

Xie J, Shen Z, Anraku Y, Kataoka K, Chen X (2019) Nanomaterial-based blood-brain-barrier (BBB) crossing strategies. Biomaterials 224:119491

Xu X, Liu A, Liu S, Ma Y, Zhang X, Zhang M et al (2023) Application of molecular dynamics simulation in self-assembled cancer nanomedicine. Biomater Res 27(1):39

Yazdani S, Jaldin-Fincati JR, Pereira RVS, Klip A (2019) Endothelial cell barriers: transport of molecules between blood and tissues. Traffic 20(6):390–403

Ye D, Raghnaill MN, Bramini M, Mahon E, Åberg C, Salvati A et al (2013) Nanoparticle accumulation and transcytosis in brain endothelial cell layers. Nanoscale 5(22):11153

Zaragozá R (2020) Transport of amino acids across the blood-brain barrier. Front Physiol 11:11

Zeevi N, Pachter J, McCullough LD, Wolfson L, Kuchel GA (2010) The blood–brain barrier: geriatric relevance of a critical brain–body interface. J Am Geriatr Soc 58(9):1749–1757

Zhu W, Wei Z, Han C, Weng X (2021) Nanomaterials as promising theranostic tools in nanomedicine and their applications in clinical disease diagnosis and treatment. Nano 11(12):3346

Zhuang J, Zhang X, Liu Q, Zhu M, Huang X (2022) Targeted delivery of nanomedicines for promoting vascular regeneration in ischemic diseases. Theranostics 12(14):6223–6241

Nanotheranostics: Clinical Status, Toxicity, Regulatory Consideration, and Future Prospects

10

Rutik Hirulkar, Nishtha Chaurawal, Fahad Saad Alhodieb, Harshita Barkat, Simran Preet, and Kaisar Raza

Abstract

This chapter highlights the overview and basic idea about nanotheranostics and their use in medical research. Nanotechnology is a multidisciplinary approach that incorporates physical, chemical, and beneficial effects on living matter at the microscopic level. Recently, nano-medicine has become an emerging field of science that combines nano-science, nanoengineering, and nano-technology with life sciences. Theranostics is the combination of therapy and diagnostics. Nanotheranostics are carrier-based strategies of sub-micron and nano sizes. In the present chapter, you will encounter novel theranostics, such as liposomes, micelles, quantum dots, and carbon nanotubes (CNTs). This chapter also helps you better understand their clinical status and toxicological effects. It also provides an overview of the regulatory considerations required for marketing approval and their future perspective for more efficient treatment strategies and smooth operationalization of medical research.

R. Hirulkar · N. Chaurawal · K. Raza (✉)
Department of Pharmacy, School of Chemical Sciences and Pharmacy, Central University of Rajasthan, Ajmer, Rajasthan, India
e-mail: drkaisar@curaj.ac.in

F. S. Alhodieb
Department of Clinical Nutrition, College of Applied Health Sciences in Ar Rass, Qassim University, Ar Rass, Saudi Arabia
e-mail: f.alhodieb@qu.edu.sa

H. Barkat
Department of Pharmaceutics, College of Pharmacy, University of Hafr Al-Batin, Hafr Al-Batin, Saudi Arabia

S. Preet
Department of Biophysics, Panjab University, Chandigarh, India
e-mail: simranpreet@pu.ac.in

M. A. Barkat et al. (eds.), *Nanotheranostics for Diagnosis and Therapy*,
https://doi.org/10.1007/978-981-97-3115-2_10

Keywords

Nanotheranostics · Nanomedicine · Liposomes · Dendrimers · Gold nanoparticles · Carbon nanotubes · Quantum dots

10.1 Introduction

Nanotechnology is a multidisciplinary approach that incorporates the physical, chemical, and positive impact on living matter at the nanoscale (Maheswaran et al. 2018). Nanomedicine is an emerging field of science that combines nano-science, nano-engineering, and nano-technology along with life sciences, which reveals appealing results for the medical community and society (Siafaka et al. 2021). Nanomedicine uses information and facilities nano-technology in the domain of medicine. They enable the preliminary stage diagnosis, prevention, treatment and follow-up of diseases.

Theranostic is a new blooming branch of medicine and is a blend of two (or more) words into one new word. Theranostics is the coalition of therapy and diagnostics. Theranostics was coined to bring a guild about their diagnostic and therapeutic applicability within solitary agent (Maheswaran et al. 2018). Theranostics nanomedicine comprises the behoof of nano-sized theranostics with multiple functionalities such as target-oriented delivery, control/prolonged release, and better transport proficiency. Nanotheranostics are the carrier-based strategies of submicron or nano sizes (Siafaka et al. 2021). Theranostics help in the governance of several ailments and cancer therapeutics. Chemotherapy mainly relies upon orthodox therapy and radiotherapy. The shortcomings of traditional therapies are that healthy cells are also influenced by the therapeutic effects of the anticancer drugs, leading to several side effects (Maheswaran et al. 2018). An ideal nano theranostics should travel and retain for a more extended period of time, should present better release behavior, and should have greater tissue target specificity and penetration ability (Siafaka et al. 2021). Mainly for medical usage: theranostic nanomedicines are colloidal nanoparticles (NPs) which vary from 10 to 1000 nm (1 μm). They especially merge in macro-molecular polymers/materials in which theranostic agents are either adsorbed on, conjugated with, entrapped or encased for the diagnosis and its treatment simultaneously at cell and molecular levels.

Nanotheranostics is the implementation and development of nanomedicine strategies, i.e., designing and building the various nanocarriers let's say as liposomes, dendrimeric nanocarriers, micellar NPs, polymeric conjugates, metals or gold NPs (AuNPs), carbon nano-tubes (CNTs), for controlled, extended and target specific joint delivery of diagnostic and also the therapeutic products for refined theranostic effects with lesser secondary consequences. The objective points to identify and then cure the disease at the preliminary level itself, when the disease is curable or at least treatable (Muthu et al. 2014).

10.1.1 Nanoparticle Nanotheranostics

10.1.1.1 Liposomes

Liposomes were fabricated by Alec D. Bangham at the Babraham Institute at the University of Cambridge in 1960 (Maheswaran et al. 2018). Liposomes are small bilayer vesicular structures which enclose aqueous compartments that are enclosed along with the incorporation of amphiphilic phospholipids and cholesterol (Muthu et al. 2014). The primitive structures were comprised only of natural lipids; presently, they affixed natural and synthetic lipids and surface-active agents. They have the aptitude to encase both lipophilic plus hydrophilic agents within the lipid membrane and in the aquatic core, respectively (Maheswaran et al. 2018). The liposome structure with an aquatic core surrounded by a lipidic bilayer is shown in Fig. 10.1. These preparations are spheric and diversify in size; most are 400 nm or below. Most commonly, liposomes with smaller diameters and zero or negative charge are employed in the case of theranostics. Three chief methods for the arrangement of liposomes are mechanical dispersion, solvent dispersion, and detergent removal.

Nevertheless, batch-to-batch variance in production, inadequate stability, and low drug loading efficiency are possible disadvantages of liposomes. Theranostic liposomes have the ability to entrap nanosized diagnostic substances like gold, iron oxide, and quantum dots (QDs), while the therapeutic agent can be incorporated in the lipophilic bilayer shell or enclosed in the core. In order to prevent immune system opsonization and faster removal from blood circulation, PEG-coated liposomes,

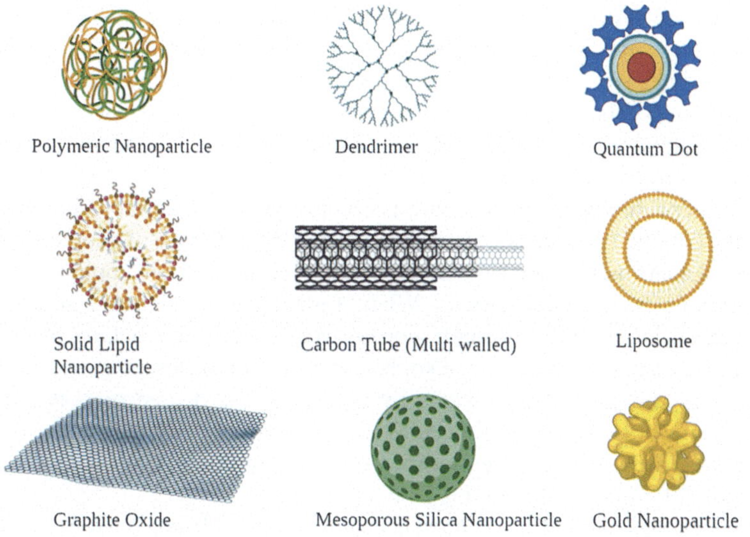

Fig. 10.1 Schematic representation of various nanocarriers utilized for theranostics purposes. (Reproduced under Creative Commons Attribution 4.0 International License with Springer Nature, Gavas et al. 2021)

which are also referred to as stealth liposomes, were designed with permanence and an extended half-life inside the blood (Muthu et al. 2014).

Liposomes are lipid carriers which are structurally biocompatible and biodegradable. In the majority of instances, liposomes are customized or laminated with the aid of various active molecules, including PEG and vitamins, which can encourage their bio-compatibility. For instance, vitamin-E TPGS-coated liposomes were extensively fabricated. Long-circulating and pH-susceptive liposomes, which are PEG-wrapped and folate-PEG-enveloped, are loaded with ^{159}Gd, and poly-L-lysine were examined as anti-cancer nano-theranostics. These fabricated liposomes demonstrated improved animal survival and increased tumor uptake. A fresh liposomal system containing NIR carbon dots and cinobufagin was developed and identified as potential anti-cancer nano-theranostics. The produced system then demonstrated significantly higher bio-imaging, and liposomes were grasped by the cells and then transported to the specific tumor site. Moreover, a prolonged release behavior and improved antineoplastic activity were noted (Siafaka et al. 2021).

Instances when Wen et al. designed apomorphine, embodied liposomes and QDs to abolish intake by the liver in order to enhance brain-specific targeting. The prepared medication was assessed further in vitro for brain endothelial cell uptake and in vivo for bio-imaging. In vitro cell intake study also ensured the dispersion of liposomes across receptor-arbitrated endocytosis routes. In the case of theranostic liposomes, a compelling circulation was noted in contrast to free QDs. Moment when the QD's were encased into the liposomes the brain uptake was noted to be increased 2.4 folds than the previous one (Fang et al. 2012).

10.1.1.2 Dendrimers

Dendrimers are synthetic nanomedicines which are comprised of highly branched spherical polymers, Fig. 10.1. Dendrimers that are used as theranostics are 10–100 nm in range. They can be prepared by following two approaches, namely "Divergent synthesis" and "Convergent synthesis" (Muthu et al. 2014). "Exclusive branching points" are available in their structures for conjugation to numerous entities (Maheswaran et al. 2018).

Dendrimers' fundamental structure is composed mainly of recurring branching around the core that produces a three-dimensional design made of geometric shapes, Fig. 10.2. Dendrimer's positive surface charge caused them to rupture cell membranes, demonstrating their toxicity. When a medication is encased in dendrimers for targeted delivery, it occasionally releases quickly before reaching the intended location. Dendrimeric NPs can also be prepared in different sizes, molecular weights, and chemical arrangements by regulating the degree of polymerization (Muthu et al. 2014). The drug molecules can be encapsulated in the dendrimer interior; the drug molecules can be modified by covalent interactions with the peripheral functional groups and complex formation.

Different types of dendrimers have been employed to deliver a diversity of therapeutic agents in an effort to maximize therapeutic effectiveness in the treatment of cancer. Doxorubicin, paclitaxel (TAX), cisplatin, methotrexate (MTX), and camptothecin are a few of the most researched anti-neoplastic medications, which are

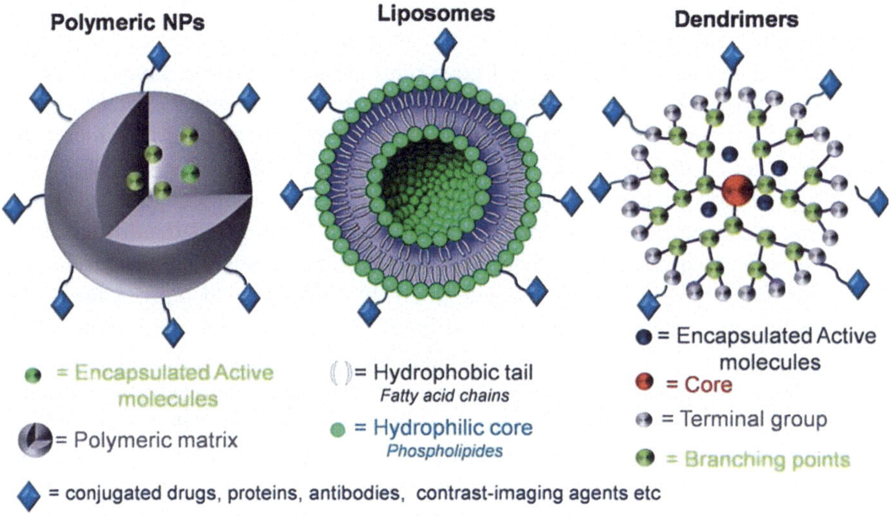

Fig. 10.2 Various structures of nanotheranostics based on polymeric NPs, liposomes, and dendrimers. (Reproduced under Creative Commons Attribution 4.0 International License with Elsevier, Siafaka et al. 2021)

transformed into dendrimer-based drug delivery systems. The dendrimers are also employed in gene therapeutics, antiviral treatments, vaccines, and other diagnostic tools. Dendrimer-based nano-devices can be made more biocompatible and tumor-specific by selecting the appropriate chemical structures and appropriate alterations at boundaries.

Dendrimeric preparations can perform as mono-molecular NPs with adaptable sizes and chemical networks. These characteristics are of interest for reproducibility both chemically and biologically. At the nano-level, they are identified to be the most functional integrally and structurally controlled devices for nano-medicines. In general, dendrimer nanodevices with 10–100 nm in size are employed as theranostic platforms.

When Tomalia et al. published the first study on commercially available polyamidoamine (PAMAM) dendrimers. Numerous other dendrimers, including poly(L-glutamic acid) and poly(propylene-imine) (PPI), were investigated for their role in medicine delivery. The glycoprotein-reliant dendrimeric preparations have also been employed in anti-cancer or anti-fungal prophylaxis or therapy responses (Niederhafner et al. 2008; Nanjwade et al. 2009).

10.1.1.2.1 Theranostics Dendrimer for Fluorescence Imaging

Numerous uses for fluorescence sensing technology exist in the biological sciences and in diagnostics (Giepmans et al. 2006). While many artificial organic fluorophores possessing sensing properties have been fabricated, there are significant obstacles to in vivo medicinal applications, including poor solubility, unpredictable intracellular localization, and lack of ability to aim for desired cells or tissues.

Fabrication of a dendrimer-based imaging agent is the best feasible alternative to overcome their shortcomings.

A research group of Majoros and coworkers designed diverse kinds of multi-operational PAMAM-linked dendrimers as theranostic anti-cancer devices. These PAMAM allied dendrimers hold MTX, fluorescent dye fluorescein isothiocyanate (FI), and targeting ligand folic acid (FA) in its structures (Majoros et al. 2005). The group further approached the study of saccharide-terminated G3 PAMAM dendrimer as a transporter for theranostic applications. Conjugated dendrimers then demonstrated enhanced endocytosis and cytotoxicity versus folic acid receptor (FR)-expressing KB cell lines (Zhang et al. 2011). This group also proclaimed that the multifunctional PAMAM-linked dendrimeric conjugates were built using click chemistry to obtain a target-specificity in theranostic anti-neoplastic (Huang et al. 2012). The same team then developed a dendrimer-based conjugate with the anti-cancer drug paclitaxel (TAX) and a fluorescent dye (Cy5), which demonstrated the mechanism of action of TAX over microtubules. With the help of a single microtubule imaging technique, his panel of scientists successfully demonstrated that the TAX-dendrimer conjugate affixes to microtubules (Majoros et al. 2005; Zhang et al. 2011; Huang et al. 2012; Cline et al. 2013).

10.1.1.2.2 Dendrimer Cored Theranostics for MRI

MRI is a notable clinical diagnostic method. It is possible to obtain high-resolution, non-invasive three-dimensional images that mirror the state of water molecules. Contrast agents impacting the relaxation time of protons of water are suggested to be beneficial for MRI. In T1-weighted MRI, ion-coordinated chelates of gadolinium (Gd(III)) are the most suitable choice for these types of contrast agents. But, presently, Gd(III) based marketable agents showed incompetent divergence enhancement capabilities and restricted target-oriented action at the target site.

Instance when, Wiener and his team developed the maiden gadolinium encased PAMAM linked dendrimer for MRI imaging alone PAMAM linked dendrimer with many Gd chelates raised the relativity, which corresponded to the MR signal intensity. As the relaxivity of the Gd(III) chelates was responsible for affecting the water exchange rate, the relaxivity of macro-molecular Gd(III) chelates is greatly enhanced upon lowering their molecular tumbling.

Various dendrimer-reliant macromolecular contrast agents are designed by the use of conjugating Gd (III) chelates towards the peripheries of the dendrimer.

Superparamagnetic iron oxide nanoparticles (SPIO NPs) are an appropriate contrasting agent for T2-loaded MRI. Chang et al. (2011) created a hybrid material with SPIO NPs and pH-susceptible PEGylated dendrimer (G2.5)/DOX conjugation. By use of a pH-separable hydrazone linkage, the dendrimer was affixed with DOX and subsequently complexed with SPIO NPs. At pH 5.2, the produced PEG-G2.5-DOX@SPIO NPs released more DOX than at pH 7.4. The utilization of mPEG-G2.5-DOX@SPIO NPs as an MRI contrasting agent was investigated by the authors for in vivo cancer detection. From 1 to 24 h following the injection, tumors were evident as hyper-intense regions on T2-weighted MRI, suggesting that the complexes can be utilized as MR probes (Morishita et al. 2005; Thorek et al. 2006; Lee et al. 2006, 2009; Chang et al. 2011).

10.1.1.2.3 Dendrimer Cored Theranostic for CT Scan

As a potent diagnostic imaging technique, X-ray CT is used in clinics at a reasonable cost. Contrasting agents find their applicability in CT imaging of specified tissues and organs in the body. Presently, iodinated components are utilized as divergence agents in clinical applications. However, iodinated compounds reveal the general issues associated with short span of circulation, potential renal noxiousness, and lack of targeting specificity. In addition, MRI is a more sensitive imaging method than X-ray CT, and iodine compounds are needed in higher concentrations. Dendrimers have been coupled with ionized substances to increase their biocompatibility and prolong the span of their blood circulation.

A dendrimer-reliant CT imaging probes have been reported by Fu et al. The dendrimer consists of a PEG core, which is then conjugated with responsive tri-iodo phthalimide moieties. Dendrimer-stabilized Bi2 S3, NPs are useful for CT imaging because they can lessen X-ray penetrability much more efficiently than iodinated compounds at a similar molar percentage of the active element (Bi versus I). This nano-device can be applicable as a flexible platform in various biological systems for targeted CT imaging and therapy (Fu et al. 2006; Fang et al. 2013). A simple method for creating PAMAM-linked dendrimer-stabilized gold NPs (Au DSNPs) connected with folic acid (FA) was described by Liu et al. The KB cells were selectively targeted in an induced mouse tumor model and enhanced by this nanodevice (Liu et al. 2013).

10.1.1.2.4 Dendrimer Cored for PET and SPECT Imaging

Nuclear medicine imaging is highly sensitive and is useful for quantitative full-body examination. Nuclear imaging methods like PET and SPECT can provide tomographic as well as quantitative functional data inside a lively subject. Dendrimer nano-devices are greatly employed in PET and SPECT studies.

Grünwald et al. developed PAMAM-linked dendrimer-covered viral vectors with the hNIS gene (theranostic sodium iodide symporter genes). Dendrimer-coated adenovirus vectors containing the hNIS gene (dcAd5-CMV/NIS) on i.v. administration demonstrated a significant reduction in hepatic accumulation and liver damage. On the other hand, dcAd5-CMV/NIS accumulated and transduced more efficiently at the tumor tissue. This delivery system permits imaging and radio-virotherapy of non-thyroidal malignancies. For further enhancement of the targeting activity, a PAMAM dendrimer having an epidermal growth factor receptor (EGFR)-specific ligand peptide (GE11) was developed by the same group of authors, with which the adenovirus vectors holding the hNIS gene were wrapped. High mass in the tumor tissues was observed, and the EGFR specificity was assured by a decrease in tumoral aggregations after pre-treatment with the EGFR-specific antibody cetuximab (Grünwald et al. 2013a, b). A brief outline of the dendrimer-derived theranostics is showcased in Table 10.1.

Table 10.1 Brief on dendrimer-cored theranostics and its applications (Sk and Kojima 2015)

Sr. no.	Types of dendrimers	API	Diagnostic agent	Target/disease	Imaging technique	References
1.	G3 PAMAM linked dendrimer	Methotrexate (MTX)	Fluorine (Fl)	In vitro cytotoxicity within "FR-expressing KB cell line"	Fluorescence	Zhang et al. (2011)
2.	G5 PAMAM linked dendrimer	Paclitaxel (TAX)	Cy5	Microtubules	Fluorescence	Cline et al. (2013)
3.	PEGylated-G2.5 dendrimer	Doxorubicin (DOX)	SPIO NPs	In vivo cancer identification in mice models	Magnetic resonance imaging (MRI)	Chang et al. (2011)
4.	G4 PAMAM dendrimer's glycidol hydroxyl-terminated	N/A	Bi	CT-image of Mouse and Rabbits	Computed tomography	–
5.	PAMAM dendrimer	AuNPs	AuNPs	In vitro and in vivo targeted CT-imaging of human liver cell carcinomas	Computed topography	Liu et al. (2014)
6.	G5 PAMAM-dendrimer	hNIS gene	Radio-active Iodine (^{123}I)	^{123}I scintigraphy and 'NIS-mediated therapy' in a 'liver cell carcinoma' in xenograft mouse models	Positron emission tomography	Grünwald et al. (2013a)

10.1.2 Polymeric Nanoparticles

10.1.2.1 Micelles

Micelles are self-constructed colloidal structures with hydrophobic cores encapsulated with hydrophilic shells. Micelles are the additional secure option for parenteral delivery of weakly water-soluble drugs. The stability of micelles mainly relies upon the cohesive forces among the drug and the core and also on the cross-linking between the core and shell. They are, in general, prepared using the organic solvent method and direct dissolution method. Therapeutic agents are sealed within the aquaphobic core of micelles and the outer hydrophilic layer along with the targeting agent, which is then delivered by intravenous route. Theranostic micelles with diameters less than 50 nm prevent renal exclusion and reticuloendothelial systems and offer improved endothelial cell permeation inside solid tumors.

Theranostic micelles are effective dais for cancer medications. One such product is Genexol-PMTM, a micellar formulation containing paclitaxel that is made from poly(ethylene glycol)-block-poly(D,L-lactide) (Kim et al. 2001). Genexol-PMTM has previously undergone a few clinical trials to assess its efficacy and safety. TPGS is utilized as an alternate material for construction of drug/diagnostic material loaded targeted/non-targeted micelles (Muthu et al. 2012; Kutty and Feng 2013; Zhao et al. 2013). Superparamagnetic iron oxide-containing thermodynamic TPGS micelles were created for cancer treatment and diagnosis. Enhanced thermal as well as magnetic properties, cytotoxicity, in vitro cellular intake, and in vivo imaging effects were shown in contrast to Pluronic F127 and Resovist® micelles for commercial use. It was observed that the produced TPGS micelles were extremely monodisperse, stable, and of suitable size range. The cytotoxicity and cellular uptake were studied using a breast cancer cell line, MCF-7, in vitro for a full day. With the usage of micelles, cell viability was observed to be reduced after hyperthermia treatment. By establishing amide bonds in the midst of the carboxylic (C=O) groups of hyaluronic acid and the amine (NH_2) groups of DOX, Kim and co-workers developed the conjugated hyaluronic acid-DOX preparation (Kim et al. 2012). These conjugates got self-assembled to form NPs that resemble micelles.

Additionally, micelles were enveloped with gold half-shell by the process of thermal vapor deposition, potentially leading to utilizing the gold surface to extend the residence time. The cytotoxicity investigation conducted on cancer cells validated the anticancer properties of the hyaluronic acid-DOX conjugate micelle. The conjugate half-shell micelles of gold-coated hyaluronic acid-DOX are an effective option for potential uses of cancer theranostics (Kim et al. 2012). Because of hydrophobic encapsulation, stability, simplicity of preparation, and effectiveness in preclinical and clinical investigations, micelles are gaining popularity as mechanisms for theranostic delivery.

Hyaluronic acid-DOX conjugate micelle's anticancer activity was validated by a cytotoxicity investigation conducted on cancer cells (Muthu et al. 2014).

10.1.3 Metallic Nanoparticles

10.1.3.1 Gold Nanoparticles

Gold nanoparticles (AuNPs) are composed of gold cores and are alternative carrier systems that provide desirable values of theranostic systems (Muthu et al. 2014). A diagrammatic representation of gold NPs is demonstrated in Fig. 10.1. They have directed the arena of theranostics due to their rare physico-chemical properties (Pedrosa et al. 2015). AuNPs are distinct in the field of cancer theranostics due to their numerous properties. AuNPs exhibit biocompatibility and are easily functionalized through the attachment of diverse molecules onto their surfaces. Due to their versatility, AuNPs attach to the drug delivery system with ease (Maheswaran et al. 2018).

They are built with a core diameter ranging from 1.5 to 10 nm, which offers a bigger surface area for effective drug-ligand conjugates (Muthu et al. 2014). They can be crafted in varied shapes such as round, rod-shaped, spheric, prismatic, diamond, cages or in single solid bodies (Pedrosa et al. 2015). AuNPs are mainly prepared by chemical treatment of hydrogen tetra-chloroacetate. Advanced therapeutics that precisely identify the target receptor for active targeting can be created by conjugating AuNPs with a drug and a targeting ligand. AuNPs possess some integral characteristics such as diagnostic properties, adjustable core size, mono-dispersity, lesser toxicity, surface plasmon absorption, large surface-to-volume ratio, capability to adhere biological molecules through Au–S bonds, light-scattering properties and facilitated fabrication (Muthu et al. 2014).

10.1.3.1.1 Gold Nanosphere

One of the first nanotheranostics systems to be assessed in vivo were hollow gold nanospheres ranging about 44 nm in diameter, demonstrating comparable photothermal capacity at concentrations 100 times lesser than that of superparamagnetic silica-coated iron oxide (FeO) equivalents. Furthermore, these empty NPs and nanocages have the benefit of functioning as encasing vectors at the same time, enabling an elevated agent load per particle. In gold nano-shells, a thin gold nanoscale shell surrounds a spherical dielectric core particle. They are typically made with silica, although several cores have been reported that utilize the advantages of other nanomaterials, such as magnetic (FeO) and high scattering (Ag).

Huang and his associates recently created bellflower-shaped gold nanoparticles (AuNPs; GBFs) using a liquid-liquid-gas tri-phase interface system that created vacuum bubbles in a two-phase liquid-liquid system by the use of ultrasound. GBFs have been demonstrated to exhibit a robust plasmon band at around 800 nm, rendering them suitable photo-thermal antenna candidates for nanotheranostics usage. Additionally, their intravenous administration has been demonstrated to either postpone the growth of tumor mass or facilitate total tumor regression after a couple of weeks with the modest laser power (0.5 W/cm^2), while permitting photo-acoustic imaging. Even though the in vivo and in vitro toxicity was minimal (control mice did not lose weight), more biodistribution and toxicity research is required before these GFB can be used in clinical settings (Huang et al. 2014).

AuNPs have recently demonstrated the potential to treat multidrug resistant (MDR) tumors using targeted photo-thermal therapy in conjunction with a chemotherapeutic agent. Theranostic AuNPs were created, implying that a peptide substrate was used to conjugate DOX to AuNPs via an Au–S link. Ala-Gly-Gly-Leu-Pro-Gly-Gly (CPLGLAGG), which is cleaved specifically by the enzyme protease. Studies conducted on animals revealed that following the functionalized AuNPs injection given to tumor-bearing mice, Intracellular glutathione and the overexpressed protease in tumor tissue caused the functionalized AuNPs to rapidly release DOX (Muthu et al. 2014).

10.1.3.2 Silver Nanoparticles (AgNPs)

AgNPs have been widely recognized as a paradigm-shifting advancement for cancer detection and therapy due to their remarkable properties. AgNPs have the highest "plasmon excitation efficiency" and may be tuned to any visible wavelength through their plasmon resonance. This special quality has been used to reduce detection limits and monitor the progression of diseases. The top-down and bottom-up methods are typically employed to synthesize different AgNPs (Table 10.2), and both provide them with the desired photophysical properties and usefulness in in vitro and in vivo applications (Li et al. 2018; Barkat et al. 2018; Abada et al. 2024).

10.1.3.2.1 Top-Down Approaches (Physical Techniques)

In the top-down approach, the appropriate bulk material is broken down into smaller species with the help of size reduction techniques, namely milling, grinding,

Table 10.2 Methods of synthesis of AgNPs and their corresponding merits and demerits (Li et al. 2018)

Synthesis methods	Implementation methods	Merits	Demerits
Physical methods	• Thermal decomposition • Mechanical grinding • Etching • Cutting • Sputtering • Arc-discharge • Energy ball • Milling • Physical vapor condensation	Lack of chemical reagents and by-products Size uniformity	Utilizes high energy Tedious for thermal stability Structural defects
Chemical methods	Chemical reduction Electrochemical Irradiation-aided chemical Irradiation-assisted pyrolysis	Controlled silver content Narrow size distribution	Use of highly deleterious organic solvents
Biological methods	Biological agents Microorganisms (bacteria, fungi, yeasts, algal, plant extracts, etc.)	Effortless and mild Large scale Environment friendly Cost-effective Biocompatible	Ambiguous mechanism action potential bacterial contamination and transfection

sputtering, thermal/laser ablation, and other physical processes. Bulk metals are grounded mechanically and then stabilized using colloidal protecting agents. Typically, AgNPs are synthesized by evaporation deposition techniques in a tube furnace at atmospheric pressure. One of the top-down physical techniques for creating AgNPs with distinct surface characteristics and using various solvents is the laser ablation technique. It is possible to produce clean and uncontaminated metal NPs for use in biomedical applications without compelling chemical reagents in solutions (Li et al. 2018; Barkat et al. 2018; Abada et al. 2024).

10.1.3.2.2 Bottom-Up Approaches (Chemical and Biological Methods)

Owing to the bottom-up method, metal atoms grow into NPs by using chemical or biological processes to self-assemble into new nuclei. The uniformity of products and the ability to produce NPs in huge quantities in a short span of time are the exclusive advantages. However, there are still issues that need to be addressed because they are not well understood (Barkat et al. 2018; Abada et al. 2024).

10.1.3.2.3 Silver Nanoparticles Prepared by Chemical Methods

Chemical techniques are based on chemical reductions, and the bottom-up AgNPs preparation procedures involve metal precursors/silver salts, capping agents/stabilizers, and organic and inorganic reducing agents, irradiation-assisted chemical transformation, electrochemical methods, and pyrolysis. Silver ions (Ag^+) get reduced to metallic silver (Ag^0) with the aid of reducing agents succeeded by aggregation into oligomeric clusters inside the solution to form AgNPs. The stabilizing agents shield them from sedimentation and help them retain their surface characteristics (Li et al. 2018; Barkat et al. 2018; Abada et al. 2024).

Frequently utilized reducing agents are hydrogen gas, borohydrides, citrates, ascorbic acid, polyols, Tollen's reagent, etc. Among these, borohydrite is more beneficial due to its efficient reducing and stabilizing properties to withstand the aggregation of AgNPs during crystal growth and decay. Prior research has shown that a powerful reductant rapidly provides a smaller and sharper distribution of NPs, but it requires more cautious control (Li et al. 2018; Barkat et al. 2018; Abada et al. 2024).

10.1.3.2.4 Silver Nanoparticles Prepared by Biological Methods

Biological processes are a green and eco-friendly alternative to chemical and physical methods. These techniques are rapidly improving fields where NPs are being developed using biological entities such as plant extracts, biomasses, and microorganisms. The glaring benefits of these techniques are eco-friendliness, moderate reaction conditions, and easy mass production. Plant extracts are the desirable preparatory medium owing to their ease of improvement, their lessened biohazards, and the scarcity of any requirement for incubation processes. A number of researchers have spotlighted the plant-assisted synthesis of AgNPs with specific morphology and size for targeted theranostic applications (Li et al. 2018; Barkat et al. 2018; Abada et al. 2024).

10.1.3.2.5 Silver Nanoparticles in Cancer Theranostics

Depending upon size and shape, differential optical, electrical, and magnetic properties of AgNPs are of greater interest and are utilized in medical imaging and antineoplastic therapy applications (Table 10.3).

Srinivasan et al. used an amide and an acid-cleavable hydrazone link to decorate 70 nm AgNPs with a folic acid (FA) targeting ligand and the chemotherapy medication doxorubicin (DOX) through polyethylene glycol (PEG) spacer. When NPs enter the cells preferentially, which overexpresses the folic acid (FA), silver NPs show pH-triggered drug release and photo-dynamic activity (PDA) (De Matteis et al. 2018).

Raju Prasad Mandal's group developed AgNPs in situ inside sodium deoxycholate micelles. The developed AgNPs showed a green fluorescence and had a unique ability to interfere with the components of gall stones/pigmented stones, which was established by an interaction study between the endobiotic pigment bilirubin (BR) and bile salt (NaDC). Bile salts encourage the dimerization of BR, enhancing its solubilization, pigment stone production, and red fluorescence emission, all of which are significant for cancer diagnosis and the study of biological tissue metabolism, as well as having implications for future diagnostic applications. Additionally, these AgNPs exhibit considerable cytotoxicity toward breast cancer cells (MCF-7), which is crucial for therapeutic purposes. This original work opened up new avenues for AgNPs in the disruption of gall stones/pigment components, authorizing them with greater physiological importance as potential theranostic material (Mandal et al. 2017).

For the first time, Sudip Mukherjee et al. presented a 4-in-1 multifunctional system. They biosynthesized AgNPs (b-AgNPs) by reducing an $AgNO_3$ solution with an Olax scandens leaf extract. In comparison to chemically synthesized AgNP, biosynthesized one exhibited increased antibacterial and anticancer activity, drug delivery, and diagnostic effectivity. The intense red fluorescence allowed drug localization within the cells to be detected. For next-generation biomedical applications, the 4-in-1 system could be used as an antibacterial, anticancer, drug delivery vehicle, and imaging agent (Mukherjee et al. 2014).

Dan Shao et al. developed ingenious Janus silver-mesoporous silica nanoparticles (Ag-MSNs) in an alcohol-free reaction system using the anisotropic island nucleation and growth mechanism. Because of the innovative Janus structure, Ag-MSN nanocomposites were advocated as an effective drug delivery method capable of endocytosis and pH-responsive drug distribution, followed by SERS imaging. Furthermore, the hydrophobic anticancer medication doxorubicin (DOX) was loaded into the mesopores to produce DOX-Ag-MSNs, which reduced the proliferation of cancer cells versus healthy human cells. These inspiring results brought the potential of Janus Ag-MSNs to the limelight as an attainable platform for effective anti-neoplastic theranostics (Shao et al. 2016).

Zhang and colleagues deposited AgNPs as the capping agent on mesoporous silica-coated gold nanorods (AuNR@MS) in situ to tackle early release and adverse effects of exogenous stimulus issues within the mesoporous silica-coated metal nanoparticle system and developed a multifunctional glutathione (GSH)-responsive

Table 10.3 In vitro applications of AgNPs in cancer theranostics (Li et al. 2018)

Group of scientist	Method of synthesis	Size (nm)	Coatings	Concentration	Cell lines	Diagnostic modes	Therapeutic regimen	References
Supriya Srinivasan et al.	Chemical	70	PEG, FA, and DOX	Ag: 32.5 ppb DOX: 4 μmol/L	MES-SA, MES-SA/Dx5, SKOV-3: human	SERS	PDT Chemotherapy	De Matteis et al. (2018)
Ranju Prasad Mandal's group	Biological	6–12	Bile salt	1–2 μmol/L	MCF-7: human	FL	Chemotherapy	Mandal et al. (2017)
Sudip Mukherjee et al.	Biological	~100	Olax scandens leaf extract	3–30 μmol/L	A549: human FL Oxidative stress MCF-7: human B16: mouse H9C2: rat HUVEC: Human CHO: hamster	FL	Oxidative stress	Mukherjee et al. (2014)
Dan Shao et al.	Chemical	80	CTAB	0–10 μmol/L	HepG2, A549, MCF-7: human	SERS	Oxidative stress	Shao et al. (2016)
Zhang and coworkers	Biological	–	AuNR@MS and TMPyP	1.3 μmol/L (IC_{50} for Hela); 4.0 μmol/L (IC_{50} for LC02)	Hela; L02: human	SPR	PDT	Zhang et al. (2015)

nanocarrier, AuNR@MS@AgNPs. The inner AuNR core served as a hyperthermia agent. To further provide the AuNR@MS@AgNPs with phototherapy and particular recognition capabilities, the photodynamic therapy (PDT) reagent TMPyP4 was introduced within the mesoporous structure, and the aptamer AS1411 was decorated on the AgNPs. TMPyP4 was produced after targeted internalization and stimulated cell death, which was aided by the photothermal impact of the AuNPs subjected to NIR irradiation and caused the local temperature to rise. AuNR@MS@AgNPs are multifunctional and synergistic theranostic systems (Zhang et al. 2015).

Sriram et al. biosynthesized AgNPs and demonstrated in vitro and in vivo cytotoxicity against dalton's lymphoma ascites (DLA) cell lines (Table 10.4). Their potent pharmacological activity emerged from the activation of the caspase 3 enzyme, successive to induction of apoptosis with no adverse effects. AgNPs enhanced the survival time of tumor-bearing mice by 50%, lowered ascitic fluid by 65%, hoisted ascites fluid white blood cell and platelet counts, decreased malignant cell count, and restored mouse body weight to normal. All of the findings indicated that AgNPs have outstanding cell activity, tumor inhibitory effects, and disease-controlling ability without damaging the normal tissue, making them a cost-effective choice in cancer theranostics (Sangiliyandi 2010).

Activatable aptamer probes (AAPs) were self-assembled on the surface of Au@Ag/AuNPs, yielding an activatable theranostic nanoprobe (ATNP) with good fluorescence quenching and near-infrared (IR) photo-thermal characteristics. Under 980 nm illumination, the Au@Ag/AuNPs absorbed from 400 to 1100 nm, resulting in a 4.5-fold greater potential for hyperthermia than Au NRs. ATNP was made capable of signal activation by a target binding-induced conformational shift in the AAP and providing on-demand treatment via image-guided irradiation using a thiolated S6 aptamer and fluorophore-labeled complementary DNA. In vitro and in vivo experiments were performed on A549 lung carcinoma. This eventually confirmed that the ATNP significantly increased imaging contrast and targeted demolition, implying its promise as a versatile theranostic method for patient-centered cancer treatment (Shi et al. 2014).

Kim and colleagues used galvanic displacement to arrange AgNPs (size 13 nm) within and on porous silicon nanoparticles (pSiNPs), and the intrinsic photoluminescence (PL) of the porous silicon matrix was quenched by the AgNPs. When exposed to an aqueous oxidant, Ag^+ was released, which killed Gram-negative bacteria and Gram-positive bacteria, and PL was recovered. In vitro, a transactivator of transcription (TAT) peptide conjugated pSi-Ag nanocomposite targeted to *S. aureus* bacteria accumulated in the liver, spleen, and lungs of mice and released Ag^+ with antibacterial activity, particularly in liver tissue and restored the PL of pSi in vivo. The findings support the use of pSi-AgNPs as a targeted self-reporting and nontoxic nanocarrier for imaging-triggered disease treatment (Kim et al. 2016).

Farrag et al. radiolabeled three different types of AgNPs by ^{125}I isotope adsorption, achieving maximum labelling yields of >90% and no optical quality problems. In vivo, tissue uptake and biodistribution studies using ^{125}I-AgNPs demonstrated that solid tumor-bearing mice absorbed the most DOX/PVP/AgNPs. They displayed prolonged high affinity to the tumor site, resulting in a synergistic imaging and

Table 10.4 In vivo applications of AgNPs in cancer theranostics (Li et al. 2018)

Research group	Method of synthesis	Size (nm)	Dose (per day)	Tumor model	Way of delivery	Diagnostic mode	Tumor inhibition rate	References
Sriram et al.	Biological	50	500 nmol/L	Dalton's lymphoma ascites (DLA)	IP, 15 days	–	65%	Sriram et al. (2010)
Kim and coworkers	Chemical	13	2.5–50 mol/L for *P. aeruginosa*; 15–300 nmol/L for *S. aureus*	*P. aeruginosa* (Gram-negative bacteria); *S. aureus* (Gram-positive bacteria)	IV, 1 h	PL	>90%	Kim et al. (2016)
Farrag et al.	Chemical	10.0 ± 1.2 (AgNPs); 19.8 ± 3.4 (PVP/AgNPs); 26.1 ± 5.3 (DOX/PVP/AgNPs)	Bolus dose in 0.2 mL fresh filtered DIW	Ehrlich ascites carcinoma	IV	SPR	Uptake rate: 35.3% (DOX/PVP/AgNPs); 28.4% (PVP/AgNPs); 22.3% (AgNPs)	Farrag et al. (2017)

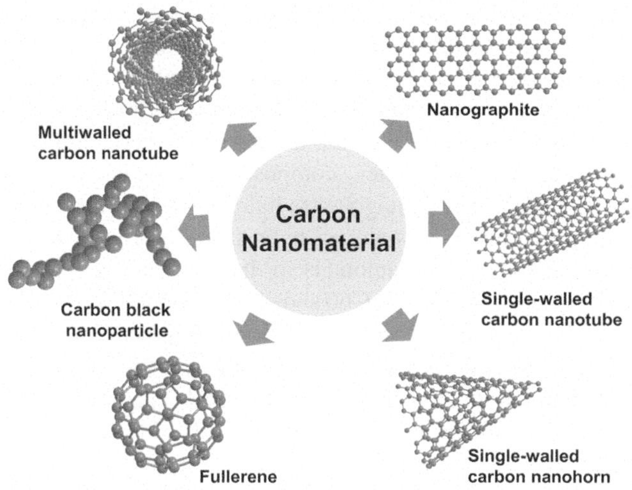

Fig. 10.3 Various carbon-based nanoparticles. (Reproduced under Creative Commons Attribution 4.0 International License with Springer Nature, Yuan et al. 2019)

anticancer impact of DOX and the AgNPs. All of these findings suggest that radiolabeled AgNPs can be used as tumor-specific radiotheranostic agents for cancer detection and treatment (Farrag et al. 2017).

10.1.4 Carbon-Based Nanoparticles

Scientists are exploring carbon nanomaterials for theranostic objectives due to their exclusive physical and chemical characteristics. They are categorized into sp^2 carbon nano-materials such as zero-dimensional fullerenes, one-dimensional (1D) carbon nanotubes (CNTs), two-dimensional (2D) graphenes, and carbon dots. The larger surface area, their capacity to encase the diagnostic agents, and their aptness for surface modifications make them a good choice for theranostic applications (Muthu et al. 2014). Various carbon-based NPs are shown in Fig. 10.3.

10.1.4.1 Fullerene

In 1985, Harold Kroto, along with James R. Heath, Sean O'Brien, Robert Curl, and Richard Smalley, invented fullerenes in soot, which were fabricated by evaporating the carbon under the influence of helium atmosphere which resulted in the observation of discrete peaks when checked in the mass spectrum of the product corresponding to molecules with the exact mass of 60/70 carbon atoms, namely C60 and C70. Fullerene is an acronym for buckminsterfullerene, given after the American architect who propagated this kind of structure (Kościk et al. 2022).

Fullerenes constructed are also referred to as buckyballs and come in hollow or ellipsoid forms, Fig. 10.3. It is possible for fullerenes to trigger the host immunity and produce fullerene-specific antibodies (Maheswaran et al. 2018). Their distinct electron-deficient nature and cage-like shape give rise to captivating characteristics that have made them a desirable subject for research in a variety of domains. With a van der Waals diameter of 10.18 Å, the most prevalent fullerene, C-60, is firmly situated in the nanotechnology domain. The distinctive hybridization of the strained carbon–carbon bonds is the source of the fullerene's important characteristics and complex chemistry. The fullerenes compose a network of both hexagons (1,3,5-cyclo-hexatriene1) and pentagons (radialene) inside them, whereas the carbon atoms in graphene and CNTs are sp^2-hybridized to provide hexagonal lattice, and the tetrahedral network of a diamond is sp^3-hybridized, Fig. 10.3. Firstly, fullerenes are good acceptors of electrons; C60 shows six reversible reductions. Secondly, an extensive array of chemical reactions can be employed to embellish the outside of the cages. Lastly, the relative inertness of the cage interiors makes it easier to catch and isolate atomic, ionic, and even tiny molecule species. The special chemical properties of the fullerenes are a result of both orbital rehybridization and topological considerations.

10.1.4.1.1 Fullerene in Theranostics

C60, which was examined in several sectors of medicine, exhibited no cytotoxicity or extremely low acute toxicity under in vitro and in vivo cytotoxic investigations in human as well as animal cells.

Encapsulating a Gd (gadolinium) core into a fullerene is one method of utilizing fullerenes in theranostic medicine. Gd^{3+} ion is housed inside a fullerene cage; this structure protects the metallic characteristics, prevents leaking, and hence safeguards its dissociation in vivo. Li et al. designed an enveloped gado-fullerene before functionalizing it with the cytokine interleukin-13, which was overexpressed in human glioblastoma cell lines. To improve water solubility and tumor uptake, the molecule was further functionalized with amine groups, which may sustain a positive charge at physiological pH. However, no cytotoxicity data have been published by this group of researchers, and initial tumor intake and MRI investigations appeared to be promising enough (Li et al. 2015). Fullerenes' photo-chemical capabilities also suggest that they could be used directly in medicine for local photodynamic treatment.

Chiang et al. utilized hydrophilic nano-spheres produced from hexa(sulfo-*n*-butyl)-C60 (FC4S) to conduct an initial in vivo photodynamic therapy (PDT) investigation. This research was carried out subcutaneously in ICR mice with sarcomas. FC4S was given intraperitoneally or intravenously before being exposed to an argon ion laser beam or an argon-pumped dye laser. It was determined that inhibiting tumor growth using a low wavelength, i.e., a 515 nm laser, which was better absorbed than a 633 nm laser, was more successful. This sort of delivery proved to be marginally more effective in terms of inhibitory effects, demonstrating the prospective usage of fullerenes as PS for cancer PDT (Liu et al. 2007).

Tabata et al. reported tumor elimination of sarcoma in mice after treatment of C-60 conjugated with polyethene glycol (PEG) and irradiation. Fullerenes were associated with metals belonging to the "lanthanide group" (lanthanoid-metallo-fullerenes) and were used as radioactive tracers for the purpose of diagnosis as well as therapy medications (Tabata and Ikada 1999).

10.1.4.2 Carbon Nanotubes

Many accounts place their discovery in 1991 and credit Sumio Iijima with it. Morinobu Endo, who saw hollow tubes formed from coiled graphite created using the chemical vapor growth technique in 1976, is responsible for the mass manufacturing method utilized today, which is referred to as the Endo process (Kośćik et al. 2022). Nanotubes are structures that resemble tubes. Different types of nanotubes exist, such as membrane, carbon, inorganic, and DNA nanotubes. They are also employed in gene therapy for carrying DNA across cell membranes (Maheswaran et al. 2018). CNTs usually have cylindrical structures, and based on the number of carbon sheets, they are further classified into Single-walled carbon nanotubes (SWCNTs) and multi-walled carbon nanotubes (MWCNTs) (Fig. 10.3). CNTs are synthesized using different methods such as the discharge method, laser method, ball milling method, and chemical vapor deposition method (Muthu et al. 2014).

10.1.4.2.1 CNTs in Theranostics

CNTs have been prepared as drug-delivering systems (DDS) and therapeutic carriers due to their ability to penetrate plasma membranes easily. Heister and his associates designed a drug-delivering system by conjugating Doxorubicin with monoclonal antibodies and fluorescein, all of which were bonded to the peripheral walls of oxidized SWCNTs. The monoclonal antibody (mAb) detected carcino-embryonic antigen, a tumor marker, and allowed DOX to bind to antigenic target sites present on cancer cells. When SWCNT-DOX complexes were delivered to WiDr colon cancer cells, they penetrated the cells and released DOX to the nucleus, while SWCNTs remained in the cytoplasm. The approach is speedy and inexpensive, allowing scientists to conduct numerous trials (Sheikhpour et al. 2020). Golubewa's team revealed that glioma cells can efficiently accumulate SWCNTs in the year 2020, resulting in comparatively large agglomerations, allowing photo-induced mortality of cancer cells via pico-second laser irradiation method. They also used the CARS imaging technique and demonstrated that it can be used to efficiently visualize a cancer cell as well as its photo-induced SWCNT-conditioned death, providing solid evidence that this technology is a powerful tool for nanotheranostics. The benefit of utilizing stable SWCNT suspensions in place of traditional NPs in photo-acoustic therapy is the cells which agglomerate are demolished, while free nanotubes in the surroundings are photo-acoustically inactive, so they are not hazardous to nearby tissues and fluids (Golubewa et al. 2020). CNTs are employed as a diverse agent in the magnetic resonance imaging (MRI) approach in diagnosis. The most prevalent study involved gadolinium chelates, which can be integrated into CNTs (Hartman et al. 2008). Because of their fluorescence in the NIR region, carbon nanotubes can also be used in cell imaging. CNTs offer the advantage of not affecting the viability of the diagnosed cells (Choi et al. 2007).

10.1.4.3 Graphenes

Konstantin Novoselelov and Andre Geim both successfully extracted graphene at the University of Manchester in the year 2004. Graphene layers were successfully extracted from graphite using a simple adhesive tape in a procedure known as micromechanical cleavage or Scotch tape technique. However, the first signs of this nano-material occurred in 1947. P.R. Wallace was the first to investigate graphene theory as a foundation for exploring the electrical characteristics of graphite. Graphene is extensively researched for drug and gene delivery because of its larger surface area, which is four magnitudes greater than the surface area of any other nanomaterial studied for the purpose of nano-medicine and ease of functionalization.

10.1.4.3.1 Graphene in Theranostics

In 2020, Rosa Garriga and associates investigated the toxic activity of four carbon nano-materials and their pertinency as delivery systems for drugs in colon cancer (Caco-2) and breast cancer (MCF-7) cell lines in in vitro research. Nanodots, graphene oxide, CNTs, and reduced graphene oxide were conjugated with two diverse medicines (camptothecin, CPT, and doxorubicin, DOX). Carbon-drug complexes excelled the free drug in terms of anticancer efficacy. When paired with CPT, graphene oxide produces extraordinary outcomes. DOX is more effective when used with R-GO (Garriga et al. 2020).

Zhang et al. also used Cisplatin (CPT) and Doxorubicin (DOX) in combination with graphene. They used the simplest approach for creating such a type of complexes: sorption. The selection of both treatments was not by chance, as studies suggest that when administered in conjunction, they exerted a better anti-cancer effect. Unluckily, such systems have a significant disadvantage: quick release of drug molecules bound over the surface. As a result, this kind of carrier cannot be employed for long-term therapy (Zhang et al. 2010).

In the year 2018, Wang et al. carried out an experiment on an optical sensing system setup for live colon cancer cells. A graphene-based optical biosensor (GOB) was employed. An ultrafast RI shift caused by faint ultrasonic vibrations was observed by this biosensor. They used the GOB for the first time to demonstrate ultra-sensitive and real-time detection of unlabeled cancer cell responses to paclitaxel (PTX) during primitive delivery of drugs. Colon cancer cells responded more intensely and quickly to the least-concentration paclitaxel treatment than to high-concentration paclitaxel treatment. This unanticipated reaction was confirmed by two colon cancer cell lines, namely, LoVo and HCT-116 (Wang et al. 2018).

10.1.4.4 Quantum Dots

Carbon quantum nano-dots (CQNDs), also referred to as carbon dots (CQDs) or C-dots, are smaller carbon NPs (less than 10 nm in size) encountered accidentally while purifying single-walled carbon nanotubes (Kościk et al. 2022). CQDs are the emerging class of fluorescent NPs which are under lime lights due to their biocompatibility and versatility for cancer theragnostic applications. QDs are nanosized semiconductor crystals. Their smaller size helps them act either alone or in combination with other theranostics to impart additional functional capacities which are beneficial for their therapeutic applications (Truskewycz et al. 2022). Conventional types of QDs are mainly comprised of two parts: Heavy metal core and wide bandgap semiconductor shells, as demonstrated in Fig. 10.1 (Kargozar et al. 2020).

10.1.4.4.1 Quantum Dot in Theranostics

Carbon nanodots can be utilized as photosensitizers (PSs) in photodynamic therapy (PDT). With the growing application for improved cancer detection and therapy, much research is focusing on the usage of QDs in photodynamic therapy (PDT), which has proven to be quite effective, and the approach used allows us to limit side effects (Kościk et al. 2022).

Zheng and coworkers attempted the synthesis of a complex based on QDs in 2014 with the aim of developing a bioimaging agent. The researchers developed a CQD-Oxa complex via a condensation process among the amine (NH_2) groups of oxaliplatin. It combines the optical features of carbon nano-dots with oxaliplatin's tumor-fighting action. Due to the reduction in the environment of cancer cells, collectively, they operate as drug carriers and permit regulated drug release. The vigor of the CQD-Oxa fluorescence allows us to track the movement and distribution of medication molecules. This technique was used to study tumor decrease in mice (H22 liver cancer). According to this research, tumor size shrinks monotonously with the course of CQD-Oxa treatment. After 6 days, the tumor volume dropped from 535.2 to 46.6 mm^3, indicating that 91.3% of the tumors were healed (Zheng et al. 2014). In the year 2017, the Ko group developed Herceptin (HER)-labelled QDs in order to come up with a theranostic drug that targets HER2 overexpression tumors. The QDs were also meant to be used as a bioimaging agent platform in this system. These compounds were created through a facile coupling process and can produce more advanced complexes with medications like doxorubicin (DOX). This complex's blue fluorescence (from QD) allows for quick diagnosis, and HER allows for efficient, targeted accumulation of these cells. Changes in pH and temperature can easily influence DOX release. Such a complex provides diversity in cancer treatment and enables rapid diagnosis, which is typically crucial in the case of breast cancer (Ko et al. 2017).

10.2 Clinical Trials

Clinical trials are carried out in order to foster novel medications, medical devices, procedures, and treatments to treat, in some cases, cure diseases/disorders. Clinical trials compare advanced devices, medications, or procedures to ones that are already used. For instance, in a clinical study, a novel drug for hypertension treatment might be compared to an already-approved drug in a particular patient or subject group. In certain clinical trials, patients might be randomized (as in a coin flip) to receive either the approved treatment or the investigational or experimental treatment. Every study looking into a novel pharmaceutical, therapy, or technology needs FDA approval. When a brand-new treatment or medication has been performed in a laboratory setting and safety information has been collected, as well as outputs have been assured, human trials may begin. A protocol must be submitted before a human clinical trial may begin to the FDA with information containing the following:

- Study
- List of investigators
- Treatment information
- Literature review
- Study population
- Study design and methodology
- Anticipated benefits and risks
- Informed consent process

An Institutional Review Board (IRB) at the facility should approve the protocol where the research will be conducted further. For volunteers to take part in a study, their consent must be given freely. Anonymity and the privacy of all subjects must be protected at all costs. The procedure of obtaining written informed permission from the patient is an essential component of the research process. In some cases, though, informed consent may be gained from a family member in a study involving critically sick patients.

10.2.1 Different Phases of Clinical Trials

Once FDA approval has been granted, the research study is all set to go ahead with human trials. Each clinical trial process mainly consists of four main phases, as discussed below:

10.2.1.1 Phase I

A tiny population of patients (typically not more than "80–100," although it could be as little as 5–10) receiving the new medication or treatment makeup Phase I. These participants can be in good health or, based on the study, have the illness process under investigation. A Phase I trial's objective is to assess the emerging treatment's safety as well as any negative consequences or adverse effects and to assist in establishing a safe margin of dose. Details regarding the pharmacological effects and metabolism of new drugs are learned in Phase I of a clinical trial. Once the new medication has passed Phase I without any significant setbacks, the Phase II trial is good to go.

10.2.1.2 Phase II

Phase II of the trial involves a treatment, drug, or device being applied to a greater number of volunteers—typically between "100 and 300"—who have the illness under investigation. This stage is carried out to assess the effectiveness of the new treatment and to obtain further data regarding safety concerns in a timely regulated environment. Risks and side effects are typically determined during this stage. This stage is very closely managed and attentively observed.

10.2.1.3 Phase III

Phase III of the novel medication is frequently carried out with many more individuals, usually thousands in number. Phase III's objective is to deduce more details about the effectiveness, secondary consequences, benefit-to-risk ratio, and safety. "Phase III" trials typically compare a novel treatment to an established one. Information acquired in Phase III is stretched to the generalized population. If the novel treatment proves to be unsafe, ineffective, or both, or it has adverse effects that could be dangerous, the FDA or the research sponsor may, at any point, terminate the trial. On the other hand, the trial might end sooner than anticipated if it shows that the novel treatment is safer or more successful than the current one.

10.2.1.4 Phase IV

Phase IV of a clinical trial is the last step in the drug approval process. This phase is also known as "Post-Marketing Phase Trials." Phase IV trials are conducted following the FDA's approval of the novel medicine and its current availability in the marketplace. The objectives of these experiments are intended to collect long-term information regarding the therapy and evaluate the impact of the therapy on various demographics. The duration of time before novel drugs, equipment, or therapies are made accessible to the general public is quite lengthy. Starting from laboratory testing, human clinical trials are the next step in the process. Occasionally, the procedure is quite expensive and might take years (Sims and Miracle 2002).

10.3 Clinical Status of Different Nano-theranostics

Extensive exploration of lipidic carriers has resulted in the development of numerous innovative liposomal formulations for the treatment of a variety of illnesses. A variety of products are undergoing clinical trial items, as well as their current progressing phase and indication. Various other formulations present in clinical trials are summarized in Table 10.5. Different nanoparticle nanomaterials undergoing clinical trials are mentioned in Table 10.6. Whereas nanotheranostics under preclinical studies are mentioned in Table 10.7, those undergoing clinical trials under Table 10.8 and those being clinically approved for cancer are described in Table 10.9, respectively.

10.3.1 Drugs Under Phase III Trials

10.3.1.1 Stimuvax

Stimuvax®, which was previously referred to as the BLP25 liposome vaccine, is a therapeutic vaccine that indicates specific types of carcinomas that express tumor-specific antigens (TSA). Tecemotide (TCM), an antigenic lipopeptide, is integrated

Table 10.5 Glimpse of formulations present in clinical trials (Bulbake et al. 2017)

S. no.	Clinical trials	Product	API	Route	Indications	Manufacturer
1.	Phase III	Stimuvax	Tecemotide	Sub-cutaneous	Lung cancer	Oncothyreon Inc.
2.		ThermoDox	Doxorubicin	Intra-venous	Liver cell carcinoma and concurrent locally advanced breast cancer	Celsion
3.		Lipoplatin	Cisplatin	Intra-venous	Lung carcinoma	Regulon Inc.
4.	Phase II	Aroplatin	Platinum analogue	Intra-venous	Metastatic colon cancer	Agenus Inc.
5.		Annamycin Liposomes	Semi-synthetic Doxorubicin analogue Annamycin	Intra-venous	Descend acute myeloid leukemia	Aronex Pharmaceuticals
6.		OSI-211	Lurtotecan	Intra-venous	Ovarian, neck and head cancer	OSI Pharmaceuticals
7.		SPI-077	Cisplatin	Intra-venous	Lung, head and neck carcinoma	Alza Corporation
8.		S-CKD602	Potent topoisomerase inhibitor	Intra-venous	Cancer	Alza Corporation
9.		LEP-ETU	Paclitaxel	Intra-venous	Malignant tumors	Neopharm Labs Ltd.
10.		LE-SN38	Irinotecan	Intra-venous	Advanced colon carcinoma	Neopharm Labs Ltd.

(continued)

Table 10.5 (continued)

S. no.	Clinical trials	Product	API	Route	Indications	Manufacturer
11.	Phase I	LEM-ETU	Mitoxantrone	Intra-venous	Various cancer	NeoPharm Labs Ltd.
12.		INX-0125	Vinorelbine tartarate	Intra-venous	Advanced carcinoma	Inex Pharmaceuticals
13.		INX-0076	Topotecan	Intra-venous	Advanced carcinoma	Inex Pharmaceuticals
14.		2B3-101	Doxorubicin	Intra-venous	Carcinoma	2-BBB therapeutic
15.		MTL-CEBPA	CEBPA siRNA	Intra-venous	Hepatocellular carcinoma	MiNA Therapeutics
16.		ATI-123	Docetaxel	Intra-venous	Carcinoma	Azaya therapeutic
17.		LiPlaCis	Cisplatin	Intra-venous	Advanced carcinoma	Oncology Venture
18.		MCC-465	Doxorubicin	Intra-venous	Gastric cancer	Mitsubishi Tanabe Pharma Corporation
19.		Alocrest	Vinorelbine	Intra-venous	Breast and lung cancer	Spectrum Pharmaceuticals

Table 10.6 Nanoparticle nanomedicine (NNM's) under clinical trials (Hua et al. 2018)

Type	Product	API	Disease	Trial status
Lipid NNM's	LE-DT	Docetaxel (DTX)	Carcinoma, ductal cancer	Phase I/II trials
	EndoTAG-1	Paclitaxel (PCT)	Breast carcinoma, stomach carcinoma	Phase II trials
	TLI	Topotecan	Ovary cancer, carcinomas	Phase I trials
	MM-302	Doxorubicin	Breast cancer	Phase I trials
	Patisiran	siRNA	TTR-mediated amyloidosis	Phase I/II/III trials
Polymer NNM's	Paclical	Paclitaxel	Ovarian carcinoma	Phase III concluded
	NK 105	Paclitaxel	Stomach cancer	Phase III concluded
	BIND-014	Docetaxel	Carcinomas	Phase II concluded
	CALAA-01	RRM2 siRNA	Carcinomas	Phase II concluded
	CRLX 101	Camptothecin	NSCLC	Phase II concluded

by this cancer vaccine in a liposomal delivery. TCM aims mucin 1 (MUC1), which is overexpressed in diverse cancer cells, such as prostate, non-small cell lung cancer (NSCLC), breast, and colonic cancer. MUC1 undergoes abnormal glycosylation when a cell turns carcinogenic in its behaviour. Even though Stimuvax is referred to

Table 10.7 Elected nanotheranostics in preclinical study of breast and prostate cancer (Akpa et al. 2023)

Nanoparticles	Cancer type	Therapeutic agent	Diagnostic agent	References
Polystyrenes	Breast	Doxorubicin	Cyanin dye	Cano-Cortes et al. (2020)
Gold (Au) nanoclusters	Breast	Paclitaxel	Indocyanine dye	Liu et al. (2019)
Graphene-based quantum dots	Breast	Doxorubicin	–	Ko et al. (2017)
PEG-liposomes	Breast	Doxorubicin	Gadoteridol	Rizzitelli et al. (2015, 2016)
Micelle	Breast	Docetaxel	NIR probe DiR	Ma et al. (2012)
Iron oxide	Breast	Curcumin	–	Yallapu et al. (2012)
Gold	Prostate	Aptamer/doxorubicin	–	Bagalkot et al. (2007)
Quantum dots (QD's)	Prostate	Aptamer/doxorubicin	–	Kim et al. (2010)

PEG polyethylene glycol, *DiR* lipophilic fluorochrome (1,1-dioctadecyl-3,3,3,3-tetramethylindotricarbocyanine iodide)

Table 10.8 Elected nano theranostics in clinical studies for cancer (Akpa et al. 2023)

Nanoparticle	Drug	Clinical phase	Applications
Liposomes	Daunorubicin	1	Breast cancer
Magnetic NP's	–	Early Phase 1	Prostate cancer
Quantum dots	Veldoreotide	1	Breast cancer
Polymeric	BIND-014	2	Prostate cancer
Polymeric	DOX, cyclo-phosphamide and filgrastim, followed by PTX	2	Breast cancer
Polymeric	PTX and cyclo-phosphamide	2	Breast cancer
Hafnium oxide's	–	1 and 2	Breast cancer
Magnetic NP's	Superparamagnetic iron	1 and 2	Breast cancer

DOX doxorubicin, *PTX* paclitaxel

as an anti-neoplastic drug, it doesn't directly prohibit malignancies but assists in improving the life span of cancer patients. This was an inaugural preparation designed by Cancer Research UK and was taken forward by Oncothyreon, Inc. (previously known as Biomira, Inc.). Product rights were attained to onchothyreon by Merck KgaA.

In phase II trials, Stimuvax exhibited a decisive response and showed significant improvement in life expectancy from 13.3 to 30.6 months. As an unexpected

Table 10.9 Elected nanotheranostics clinically approved for breast and prostate cancer (Akpa et al. 2023)

Product™	Nanoparticle	API	Disease (cancer)	Company	Approved (year)
Doxil	Liposome	Doxorubicin	Breast	Ortho Biotech (USA)	FDA (1995)
Caelyx	Liposome	Doxorubicin	Breast	Schering-Plough (USA)	EMA (1996)
Myocet	Liposome	Doxorubicin	Breast	Teva UK	EMA (2000)
Eligard	Polymer	Leuprorelin acetate	Prostate	Recordati Industria Chimicae Farmaceutica (Italy)	FDA (2002)
Nano Therm	Metallic	Fe_2O_3	Prostate	Magforce (Germany)	EMA (2013)
Pazenir	Albumin	Paclitaxel	Breast	Ratiopharm GmbH (Germany)	EMA (2019)

FDA Food and Drug Administration, *EMA* European Medicines Agency

revelation in the world of cancer research, Stimuvax failed to meet the study's primary and secondary end-points, which resulted in its discontinuation.

10.3.1.2 ThermoDox

ThermoDox is a liposomal formulation that underwent clinical trials and is the thermo-sensitive medication of the endorsed drug Doxorubicin (DOX). A leading oncology firm named "Celsion Corporation" developed ThermoDox, which focused on the evolution of novel anti-cancer medications. It is mainly specified in primary liver cancer as well as in recurrent "chest wall breast cancer." These liposomes comprise DPPC, Myristoylstearoyl phosphatidylcholine (MSPC) and 1,2-distearoyl-sn-glycero-3-phosphoethanolamine-*N*-[amino(polyethylene glycol)-2000] (DSPE-PEG-2000). These triple components are conjugated altogether to acquire the definite functions and material properties which are affiliated with lipids. This explicit combination results in the development of proprietary thermal-activated DOX liposomal medication.

ThermoDox formulation is administered via intravenous route conjugated with Radio-frequency ablation (RFA) and can also be utilized along with microwave hyperthermia and high-intensity focused ultrasound. ThermoDox attained a 25-fold increased drug concentration in comparison to standard DOX liposomes. Grain boundary permeabilization at phase transition temperature from liposomes has improved DOX release.

This formulation has acknowledged the status of orphan drug given by the European Commission, USA orphan drug status as well as the FDA fast-track title for the treatment of hepatocellular carcinoma. At present, Phase III trials (HEAT study) are under evaluation in patients with primary hepatic cancers, which are unable to be removed with surgery. Heat study then succeeded in getting regulatory agency support from 11 countries all around the globe. Alternative treatments with this product are OPTIMA (for liver cancer along with ThermoDox) and TARDOX (targeted chemotherapy with the use of "focused ultrasound" for Liver carcinomas).

10.3.1.3 Lipoplatin

Lipoplatin is a patented liposomal medication prepared from Cisplatin (CPT) and is endorsed by the FDA as an economic cytotoxic agent. It is enumerated under DNA cross-linked agents or DNA synthesis cytotoxic agents which has been suggested by the name Lipoplatin for the treatment of pancreatic cancer and as Nanoplatin for lung cancer. Lipoplatin is mainly composed of various lipids, lipid conjugates, and cholesterol.

Lipoplatin is utilized in conjunction with gemcitabine in case of pancreatic malignancies. This formulation is under Phase I trials for malignant pleural effusion and under Phase II trials for its application in breast cancer and gastric cancer, Phase II/III stage for malignancies of the pancreas. Lipoplatin was able to decrease adverse effects of CPT involving renal toxicity, peripheral neuropathy, ototoxicity, etc. For pancreatic cancer treatment, European Medicine Agency gave the status of an orphan drug.

10.3.2 Drugs Under Phase II Trials

10.3.2.1 Aroplatin

Aroplatin is a chemotherapy-induced platinum analogue cis-(trans-*R*,*R*-1,2-diamino cyclohexane)bis(neodecanoato)platinum(II) (NDDP) encased liposomal preparation. This product is a leading liposomal platinum medication which underwent clinical trials, and its loaded analogues are similar in structures to Eloxatin. Platinum analogues show cytotoxicity of tumors by the formation of inter- and intra-strand cross-links of DNA, inhibiting its synthesis. The investigational new drug application of Aroplatin was owned by "Aronex Pharmaceuticals" and received the designation of an orphan drug by the US FDA. A monophasic trial was concluded for the formulation in 2002, which was suggested for Metastatic colorectal cancer. Phase I/II trial of Aroplatin was ended for treatment of advanced sarcomas. As an escalation trial of the product after Phase I for maximum tolerated dose, the trial was terminated, and no internal progress of the product was pursued further.

10.3.2.2 Liposomal Annamycin

Liposomal annamycin (L-Annamycin, Annamycin-LF, S-ANNA) is a liposomal medication which encloses the partially-synthetic doxorubicin analogue Annamycin. Vesicular structures constitute DMPC and DMPG in the (molar ratio of 7:3). Liposomes help in higher entrapment of drugs and also in flexibility for selection of formulation, which transport drugs to the tumor site in higher proportion. It acts by intercalating DNA and inhibiting topoisomerase II, resulting in the inhibition of DNA replication and protein synthesis. L-Annamycin is designed to overcome the cardiac toxicity associated with ready-for-use anthracyclines and was detected to be less toxic along with enhanced anti-cancer activity.

In 2015, Aronex Pharmaceuticals developed L-Annamycin, and an agreement was signed by Moleculin Biotech, Inc. to attain the permit for clinical trials of L-Annamycin. The product was able to complete Phase I and II trials within patients with refractory acute lymphocytic leukemia (Pui and Jeha 2007). Lesser dose-related toxicities were observed in comparison to DOX (Wetzler et al. 2013). "Moleculin Biotech Inc." has a plan to accelerate the approval process for myeloid leukemia and also has intentions to apply for its orphan drug status in the USA for the treatment of acute myeloid leukemia (Bulbake et al. 2017).

10.3.3 Drugs Under Phase I Trials

Besides of liposome-based formulations under Phase III and II trials, few liposomal preparations like LEM-ETU, INX-0125, INX-0076 etc., are under Phase I clinical trials as well and are mentioned as follows;

10.3.3.1 LEM-ETU

LEM-ETU is liposome-based preparation encapsulated with Mitoxantrone, comprised of DOPC, cholesterol, and cardioplipin in a molar ratio (90:5:5) respectively (Wicki et al. 2015). This product is developed by NeoPharm's NeoLipid liposome technology for effective management of different cancers (Ahmad et al. 2005). Cardiolipin has an important role in drug entrapment and due to its electrically negative charge, electrostatic interactions are observed with loaded moiety and in turns shows increased drug loading as compared to other liposomal preparations (Bulbake et al. 2017).

10.3.3.2 INX-0125 and INX-0076

INX-0125 (sphingosome-entrapped vinorelbine tartrate) and INX-076 (sphingosome-entrapped topotecan), both the liposomal preparations are categorized under sphingosomes, which are designed by Inex Pharmaceuticals and underwent phase I trials. Both the formulations comprise of sphingomyelin (SM) and cholesterol in molar ratio of 45:45 respectively and are suggested in case of solid tumors (Immordino et al. 2006). INX-0125 is designed for the treatment of Hodkin's disease as well as non-Hodkin's lymphoma which are capable of elevating tumor targeting of antineoplastic drugs (Torchilin 2014), whereas INX-076 entraps topotecan and is camptothecin derivative which helps to increase the efficiency and also reduces the dose in comparison to free drug (Bulbake et al. 2017).

10.4 Toxicity in Theranostics

Despite of advancements in the realm of drug delivery and nanomedicine due to the progress of NPs, it is still quite a newer concept to another form of pharmaceutical medication. There also exist specific-mechanisms through which these nano-systems are able to induce toxicities and are not characterized in a proper way, as very few nanomaterial drugs have succeeded in achieving FDA approval.

Toxicity is not only dependent upon drugs' physicochemical properties (size, shape, charge, chemical composition), but physiological status (disease condition, genetics) is also responsible. For example, patients suffering from "coronary artery disease" are susceptible to "heart attacks" when treated with NP therapy.

Nanomaterials like dendrimers, QD's, magnetic NPs, metallic NP, and carbon NPs are popularly utilized as bio-sensors, diverging agents, and drug-loading carriers. The toxic effects of these mentioned nano-materials are still an enigma in in vivo studies.

Polymer NPs are designed as non-viral vectors for gene delivery due to their safeness, and capability to transport larger genes, target specific, and lower immune response. As these non-viral vectors cause toxicities in gene vectors, it limits its use in gene therapy. To overcome this problem, degradable polymers in low pH conditions are designed. For an instance, toxic effects of lower molecular weight of polyethyleneimine polymers are observed to be reduced due to degradation of acid sensitive bridging.

The scenario of the utilization of heavy metals for in vivo use is disputed regarding its safety concerns. Cadmium and selenium are two metals that are employed in constructing the nucleus of QDs and cause moderate toxicities in vertebrates. Metallic NPs (e.g., gold NPs) have been frequently used for diagnosis and therapeutic purposes. They show numerous advantages, such as low toxicity in comparison to QDs, better potential as a contrast agent than organic dyes and better surface and refractive index spectroscopic properties. Although gold NPs are assumed to be non-toxic, specific forms, such as cationic gold nanospheres, may have considerable cell toxicity, but their anionic counterparts are not. AgNPs have also been demonstrated to cause oxidative stress in the brain and a few other organs, such as the liver, by lowering mitochondrial respiratory chain complexes (I, II, II, and IV). These particles have the ability to cause necrosis and apoptosis. These particles are also hazardous because they can provoke a large percentage of free radicals due to redox cycling at the particle's surface. ROS can cause oxidative stress and intervene with cellular activity.

When exposed for an extended period of time, MWCNT aggregation generates toxic effects inside the tissues due to the existence of larger particles, whereas SWCNTs, which are composed of smaller particles, undergo phagocytosis and are non-toxic. Because they can produce platelet aggregation, cationic carbon nanotubes may be more hazardous than neutral or negatively charged CNTs. Graphenes are the 2-D structures which induce blood clots or pleural edema as well as granuloma by accumulating inside the lungs after i.v. administration.

Mesoporous NPs are being utilized as theranostic agents in cancer treatment as they show the potential to target tumors and release the drug in a disciplined way. Even though they show lesser toxicity in comparison to colloidal silica, mesoporous silica NPs are still able to promote cytotoxicity related to oxidative stress by increasing ROS production and reducing GSH levels, which ultimately results in cell apoptosis. Silica NPs form protein agglomerates and inhibit gene expression, eventually interfering with the process of cell growth and replication (Kang et al. 2015; Barkat et al. 2021).

10.5 Regulations for Theranostics

Nano-medicines possess the power to accelerate the development of the pharmaceutical market and improve health outcomes. However, the large gap between scientific research and regulatory provisions is challenging to shorten. The marketing of nano-medicines is primarily reliant on a number of regulatory standards based on government policies governing manufacturing practices, quality control, safety, and patent protection. The lack of clear regulatory and safety ethics hampers NNM product development and clinical translation (Gaspar 2007; Tinkle et al. 2014; Sainz et al. 2015).

As an example, polymers are extensively researched as a viable stage for NNM methods; nevertheless, their safeness and efficacy rates are highly reliant on the polymer's molecular weight, its polydispersity, molecular structure, and conjugated chemistry (Gaspar and Duncan 2009; Diab et al. 2012). NNMs are presently managed under the traditional framework assisted by every single country's primary regulatory authority (for example, FDA, EMA and TGA). Despite the fact that NNMs have been in the market for nearly two decades, the very first generation of NNM drugs acquired regulatory clearance by simply meeting general medicinal ingredient guidelines. These rules are no longer sufficient for establishing NNM quality, safety, and efficacy in clinical settings (Gaspar 2007; Tinkle et al. 2014; Sainz et al. 2015). There is a necessity to validate regulatory standards and techniques, particularly for NPs that connect both medicine and medical device regulations. The complexity of an NNM means of administration, pharmacokinetics, pharmacodynamics, and safety profile, as well as providing suggestions on the optimum clinical trial design and patient selection, should be taken into account (Tinkle et al. 2014). International regulatory norms for NNMs should be designed in partnership with significant countries with financial interests. Despite the tremendous progress that has been made in the last 5 years, stronger collaboration between regulatory authorities, academia, research, and industry is essential (Gaspar 2007; Murday et al. 2009; Hafner et al. 2014). Complete review and documentation of the nanomedicine manufacturing process is required, including acceptable industry-oriented standards for quality control and preventing environmental concerns. Purity, sterility, stability, manufacturing procedures, and related industry-controlled standards will need to be met by manufactured NNMs (Gaspar 2007). These procedures will differ based on the nano-materials and nano-structures used. As a result, regulatory agencies need to collaborate to create testing methodologies and standard protocols for toxicity evaluation and regulatory provisions which guarantee the safety and efficacy of present and evolving NNMs (Hua et al. 2018).

10.6 Conclusion

Nanomedicines as a nano-theranostic agents are a boon to the world of medicine and have gained worldwide recognition and applicability as theranostic agents in various life-threatening disease conditions. These nanomaterials find their extensive application in today's world of medicine over conventional drug therapies available in the market for effective and target-specific drug delivery and minimizing their adverse effects. They provide certain advantages, like increased surface area, better stability, and decreased dosing frequency. Even though they help in minimizing and enhancing the bio-availability of drugs administered respectively, they themselves also contribute to show some side effects which are to be evaluated and studied while doing research in upcoming years and minimizing them will surely take the world of nanotheranostics to remarkable heights.

10.7 Future Prospects

Aside from the fact that a broad range of novel nanocarriers are developed over time, inherent complexities of biological environments show the substantial effects on nanomaterial activity and then protect their efficient use for therapeutic interventions. In order to develop novel treatment strategies and therapy based on the usage of smart nanocarriers, a thorough understanding of the true dynamics of malignant cells is required. The ideal nanoparticle for theranostic applications should possess a formulation approach with the least possible stages, be cost-efficient and practical, which will result in greater reproducibility and scale-up simplicity, and provide imaging and successful treatment.

Despite of several efforts to create peculiar customized nanocarriers, the FDA has approved only a few of them for medicinal use, requiring the development of more in vitro and in vivo models validating their targeting efficiency. As a result, NPs must be designed using multiscale models and computational models in order to enable more sophisticated targeted delivery of medicinal substances. There is an issue of dealing with multidrug resistance (MDR) (Muthu et al. 2014). Intriguingly, siRNA can be utilized in nano-medicine as a potential theranostic resistance suppressor (Akpa et al. 2023).

These is just a small overview of the current advancements and ongoing research in the field of nanotheranostics, its use, its potential in diagnosis and treatment of various types of cancers. Still there is need of gaining much more knowledge and perform research in order to dive deep-in into the ocean of Nanomaterials and Nanotheranostics for effective and efficient use of these technology for betterment and well-being of the mankind around the globe.

References

Abada E, Mashraqi A, Modafer Y, Al Abboud MA, El-Shabasy A (2024) Review green synthesis of silver nanoparticles by using plant extracts and their antimicrobial activity. Saudi J Biol Sci 31(1):103877. https://doi.org/10.1016/j.sjbs.2023.103877

Ahmad A, Wang YF, Ahmad I (2005) Separation of liposome-entrapped mitoxantrone from nonliposomal mitoxantrone in plasma: pharmacokinetics in mice. Methods Enzymol 391:176–185. https://doi.org/10.1016/S0076-6879(05)91010-0

Akpa PA, Peter IE, Onwuka AM et al (2023) Nanotheranostics: platforms, current applications, and mechanisms of targeting in breast and prostate cancers. J Nanotheranostics 4:346–383. https://doi.org/10.3390/jnt4030016

Bagalkot V, Zhang L, Levy-Nissenbaum E et al (2007) Quantum dot-aptamer conjugates for synchronous cancer imaging, therapy, and sensing of drug delivery based on bi-fluorescence resonance energy transfer. Nano Lett 7:3065–3070. https://doi.org/10.1021/nl071546n

Barkat MA, Harshita BS, Naim MJ, Pottoo FH, Singh SP, Ahmad FJ (2018) Current progress in synthesis, characterization and applications of silver nanoparticles: precepts and prospects. Recent Pat Antiinfect Drug Discov 13(1):53–69. https://doi.org/10.2174/1574891X12666171006102833

Barkat A, Beg S, Panda SK, Alharbi KS, Rahman M, Ahmed FJ (2021) Functionalized mesoporous silica nanoparticles in anticancer therapeutics. Semin Cancer Biol 69:365–375. https://doi.org/10.1016/j.semcancer.2019.08.022

Bulbake U, Doppalapudi S, Kommineni N, Khan W (2017) Liposomal formulations in clinical use: an updated review. Pharmaceutics 9:1–33. https://doi.org/10.3390/pharmaceutics9020012

Cano-Cortes MV, Navarro-Marchal SA, Ruiz-Blas MP et al (2020) A versatile theranostic nanodevice based on an orthogonal bioconjugation strategy for efficient targeted treatment and monitoring of triple negative breast cancer. Nanomedicine 24. https://doi.org/10.1016/j.nano.2019.102120

Chang Y, Meng X, Zhao Y et al (2011) Novel water-soluble and pH-responsive anticancer drug nanocarriers: doxorubicin-PAMAM dendrimer conjugates attached to superparamagnetic iron oxide nanoparticles (IONPs). J Colloid Interface Sci 363:403–409. https://doi.org/10.1016/j.jcis.2011.06.086

Choi JH, Nguyen FT, Barone PW et al (2007) Multimodal biomedical imaging with asymmetric single-walled carbon nanotube/iron oxide nanoparticle complexes. Nano Lett 7:861–867. https://doi.org/10.1021/nl062306v

Cline EN, Li MH, Choi SK et al (2013) Paclitaxel-conjugated PAMAM dendrimers adversely affect microtubule structure through two independent modes of action. Biomacromolecules 14:654–664. https://doi.org/10.1021/bm301719b

De Matteis V, Cascione M, Toma CC, Leporatti S (2018) Silver nanoparticles: synthetic routes, in vitro toxicity and theranostic applications for cancer disease. Nanomaterials 8. https://doi.org/10.3390/nano8050319

Diab R, Jaafar-Maalej C, Fessi H, Maincent P (2012) Engineered nanoparticulate drug delivery systems: the next frontier for oral administration. AAPS J 14:688–702. https://doi.org/10.1208/s12248-012-9377-y

Fang J-Y, Wen C-J, Zhang, et al (2012) Theranostic liposomes loaded with quantum dots and apomorphine for brain targeting and bioimaging. Int J Nanomedicine 1599. https://doi.org/10.2147/IJN.S29369

Fang Y, Peng C, Guo R et al (2013) Dendrimer-stabilized bismuth sulfide nanoparticles: synthesis, characterization, and potential computed tomography imaging applications. Analyst 138:3172. https://doi.org/10.1039/c3an00237c

Farrag NS, El-Sabagh HA, Al-mahallawi AM et al (2017) Comparative study on radiolabeling and biodistribution of core-shell silver/polymeric nanoparticles-based theranostics for tumor targeting. Int J Pharm 529:123–133. https://doi.org/10.1016/j.ijpharm.2017.06.044

Fu Y, Nitecki DE, Maltby D et al (2006) Dendritic iodinated contrast agents with PEG-cores for CT imaging: synthesis and preliminary characterization. Bioconjug Chem 17:1043–1056. https://doi.org/10.1021/bc060019c

Garriga R, Herrero-Continente T, Palos M et al (2020) Toxicity of carbon nanomaterials and their potential application as drug delivery systems: in vitro studies in Caco-2 and MCF-7 Cell lines. Nanomaterials 10:1617. https://doi.org/10.3390/nano10081617

Gaspar R (2007) Regulatory issues surrounding nanomedicines: setting the scene for the next generation of nanopharmaceuticals. Nanomedicine 2:143–147. https://doi.org/10.2217/17435889.2.2.143

Gaspar R, Duncan R (2009) Polymeric carriers: preclinical safety and the regulatory implications for design and development of polymer therapeutics. Adv Drug Deliv Rev 61:1220–1231. https://doi.org/10.1016/j.addr.2009.06.003

Gavas S, Quazi S, Karpiński TM (2021) Nanoparticles for cancer therapy: current progress and challenges. Nanoscale Res Lett 16. https://doi.org/10.1186/s11671-021-03628-6

Giepmans BNG, Adams SR, Ellisman MH, Tsien RY (2006) The fluorescent toolbox for assessing protein location and function. Science 312:217–224. https://doi.org/10.1126/science.1124618

Golubewa L, Timoshchenko I, Romanov O et al (2020) Single-walled carbon nanotubes as a photo-thermo-acoustic cancer theranostic agent: theory and proof of the concept experiment. Sci Rep 10. https://doi.org/10.1038/s41598-020-79238-6

Grünwald GK, Vetter A, Klutz K et al (2013a) Systemic image-guided liver cancer radiovirotherapy using dendrimer-coated adenovirus encoding the sodium iodide symporter as theranostic gene. J Nucl Med 54:1450–1457. https://doi.org/10.2967/jnumed.112.115493

Grünwald GK, Vetter A, Klutz K et al (2013b) EGFR-targeted adenovirus dendrimer coating for improved systemic delivery of the theranostic NIS gene. Mol Ther Nucleic Acids 2:e131. https://doi.org/10.1038/mtna.2013.58

Hafner A, Lovrić J, Lakŏ GP, Pepić I (2014) Nanotherapeutics in the EU: an overview on current state and future directions. Int J Nanomedicine 9:1005–1023. https://doi.org/10.2147/IJN.S55359

Hartman KB, Laus S, Bolskar RD et al (2008) Gadonanotubes as ultrasensitive pH-smart probes for magnetic resonance imaging. Nano Lett 8:415–419. https://doi.org/10.1021/nl0720408

Hua S, de Matos MBC, Metselaar JM, Storm G (2018) Current trends and challenges in the clinical translation of nanoparticulate nanomedicines: pathways for translational development and commercialization. Front Pharmacol 9:1–14. https://doi.org/10.3389/fphar.2018.00790

Huang B, Kukowska-Latallo JF, Tang S et al (2012) The facile synthesis of multifunctional PAMAM dendrimer conjugates through copper-free click chemistry. Bioorg Med Chem Lett 22:3152–3156. https://doi.org/10.1016/j.bmcl.2012.03.052

Huang P, Rong P, Lin J et al (2014) Triphase interface synthesis of plasmonic gold bellflowers as near-infrared light mediated acoustic and thermal theranostics. J Am Chem Soc 136:8307–8313. https://doi.org/10.1021/ja503115n

Immordino ML, Dosio F, Cattel L (2006) Stealth liposomes: review of the basic science, rationale, and clinical applications, existing and potential. Int J Nanomedicine 1:297–315

Kang H, Mintri S, Menon AV et al (2015) Pharmacokinetics, pharmacodynamics and toxicology of theranostic nanoparticles. Nanoscale 7:18848–18862. https://doi.org/10.1039/c5nr05264e

Kargozar S, Hoseini SJ, Milan PB et al (2020) Quantum dots: a review from concept to clinic. Biotechnol J 15:1–18. https://doi.org/10.1002/biot.202000117

Kim SC, Kim DW, Shim YH et al (2001) In vivo evaluation of polymeric micellar paclitaxel formulation: toxicity and efficacy. J Control Release 72:191–202. https://doi.org/10.1016/S0168-3659(01)00275-9

Kim D, Jeong YY, Jon S (2010) A drug-loaded aptamer—gold nanoparticle bioconjugate for combined ct imaging and therapy of prostate cancer. ACS Nano 4:3689–3696. https://doi.org/10.1021/nn901877h

Kim KS, Park SJ, Lee MY et al (2012) Gold half-shell coated hyaluronic acid-doxorubicin conjugate micelles for theranostic applications. Macromol Res 20:277–282. https://doi.org/10.1007/s13233-012-0062-x

Kim T, Braun GB, She Z et al (2016) Composite porous silicon–silver nanoparticles as theranostic antibacterial agents. ACS Appl Mater Interfaces 8:30449–30457. https://doi.org/10.1021/acsami.6b09518

Ko NR, Nafiujjaman M, Lee JS et al (2017) Graphene quantum dot-based theranostic agents for active targeting of breast cancer. RSC Adv 7:11420–11427. https://doi.org/10.1039/c6ra25949a

Kościk I, Jankowski D, Jagusiak A (2022) Carbon nanomaterials for theranostic use. J Carbon Res 8. https://doi.org/10.3390/c8010003

Kutty RV, Feng SS (2013) Cetuximab conjugated vitamin E TPGS micelles for targeted delivery of docetaxel for treatment of triple negative breast cancers. Biomaterials 34:10160–10171. https://doi.org/10.1016/j.biomaterials.2013.09.043

Lee H, Lee E, Kim DK et al (2006) Antibiofouling polymer-coated superparamagnetic iron oxide nanoparticles as potential magnetic resonance contrast agents for in vivo cancer imaging. J Am Chem Soc 128:7383–7389. https://doi.org/10.1021/ja061529k

Lee JH, Lee K, Moon SH et al (2009) All-in-one target-cell-specific magnetic nanoparticles for simultaneous molecular imaging and siRNA delivery. Angew Chem Int Ed 48:4174–4179. https://doi.org/10.1002/anie.200805998

Li T, Murphy S, Kiselev B et al (2015) A new interleukin-13 amino-coated gadolinium metallofullerene nanoparticle for targeted MRI detection of glioblastoma tumor cells. J Am Chem Soc 137:7881–7888. https://doi.org/10.1021/jacs.5b03991

Li Y, Chang Y, Lian X et al (2018) Silver nanoparticles for enhanced cancer theranostics: in vitro and in vivo perspectives. J Biomed Nanotechnol 14:1515–1542. https://doi.org/10.1166/jbn.2018.2614

Liu J, Ohta S, Sonoda A et al (2007) Preparation of PEG-conjugated fullerene containing Gd3+ ions for photodynamic therapy. J Control Release 117:104–110. https://doi.org/10.1016/j.jconrel.2006.10.008

Liu H, Xu Y, Wen S et al (2013) Targeted tumor computed tomography imaging using low-generation dendrimer-stabilized gold nanoparticles. Chemistry 19:6409–6416. https://doi.org/10.1002/chem.201204612

Liu H, Wang H, Xu Y et al (2014) Lactobionic acid-modified dendrimer-entrapped gold nanoparticles for targeted computed tomography imaging of human hepatocellular carcinoma. ACS Appl Mater Interfaces 6:6944–6953. https://doi.org/10.1021/am500761x

Liu R, Hu C, Yang Y et al (2019) Theranostic nanoparticles with tumor-specific enzyme-triggered size reduction and drug release to perform photothermal therapy for breast cancer treatment. Acta Pharm Sin B 9:410–420. https://doi.org/10.1016/j.apsb.2018.09.001

Ma M, Hao Y, Liu N et al (2012) A novel lipid-based nanomicelle of docetaxel: evaluation of antitumor activity and biodistribution. Int J Nanomedicine 7:3389–3398. https://doi.org/10.2147/IJN.S29827

Maheswaran T, Prabhunath Thuckanaickenpalayam Ragunathan Y, Ganapathy N et al (2018) Theranostics an emerging paradigm—a review. IOSR J Dent Med Sci 17:19–24. https://doi.org/10.9790/0853-1710071924

Majoros IJ, Thomas TP, Mehta CB, Baker JR (2005) Poly(amidoamine) dendrimer-based multifunctional engineered nanodevice for cancer therapy. J Med Chem 48:5892–5899. https://doi.org/10.1021/jm0401863

Mandal RP, Mandal G, Sarkar S et al (2017) "Theranostic" role of bile salt-capped silver nanoparticles—gall stone/pigment stone disruption and anticancer activity. J Photochem Photobiol B Biol 175:269–281. https://doi.org/10.1016/j.jphotobiol.2017.08.040

Morishita N, Nakagami H, Morishita R et al (2005) Magnetic nanoparticles with surface modification enhanced gene delivery of HVJ-E vector. Biochem Biophys Res Commun 334:1121–1126. https://doi.org/10.1016/j.bbrc.2005.06.204

Mukherjee S, Chowdhury D, Kotcherlakota R et al (2014) Potential theranostics application of bio-synthesized silver nanoparticles (4-in-1 system). Theranostics 4:316–335. https://doi.org/10.7150/thno.7819

Murday JS, Siegel RW, Stein J, Wright JF (2009) Translational nanomedicine: status assessment and opportunities. Nanomedicine 5:251–273. https://doi.org/10.1016/j.nano.2009.06.001

Muthu MS, Avinash Kulkarni S, Liu Y, Feng SS (2012) Development of docetaxel-loaded vitamin e TPGS micelles: formulation optimization, effects on brain cancer cells and biodistribution in rats. Nanomedicine 7:353–364. https://doi.org/10.2217/nnm.11.111

Muthu MS, Leong DT, Mei L, Feng SS (2014) Nanotheranostics—application and further development of nanomedicine strategies for advanced theranostics. Theranostics 4:660–677. https://doi.org/10.7150/thno.8698

Nanjwade BK, Bechra HM, Derkar GK et al (2009) Dendrimers: emerging polymers for drug-delivery systems. Eur J Pharm Sci 38:185–196. https://doi.org/10.1016/j.ejps.2009.07.008

Niederhafner P, Reiniš M, Šebestík J, Ježek J (2008) Glycopeptide dendrimers, part III—a review: use of glycopeptide dendrimers in immunotherapy and diagnosis of cancer and viral diseases. J Pept Sci 14:556–587. https://doi.org/10.1002/psc.1011

Pedrosa P, Vinhas R, Fernandes A, Baptista PV (2015) Gold nanotheranostics: proof-of-concept or clinical tool? Nanomaterials 5:1853–1879. https://doi.org/10.3390/nano5041853

Pui C-H, Jeha S (2007) New therapeutic strategies for the treatment of acute lymphoblastic leukaemia. Nat Rev Drug Discov 6:149–165. https://doi.org/10.1038/nrd2240

Rizzitelli S, Giustetto P, Cutrin JC et al (2015) Sonosensitive theranostic liposomes for preclinical in vivo MRI-guided visualization of doxorubicin release stimulated by pulsed low intensity non-focused ultrasound. J Control Release 202:21–30. https://doi.org/10.1016/j.jconrel.2015.01.028

Rizzitelli S, Giustetto P, Faletto D et al (2016) The release of doxorubicin from liposomes monitored by MRI and triggered by a combination of US stimuli led to a complete tumor regression in a breast cancer mouse model. J Control Release 230:57–63. https://doi.org/10.1016/j.jconrel.2016.03.040

Sainz V, Conniot J, Matos AI et al (2015) Regulatory aspects on nanomedicines. Biochem Biophys Res Commun 468:504–510. https://doi.org/10.1016/j.bbrc.2015.08.023

Sangiliyandi G (2010) Antitumor activity of silver nanoparticles in Dalton's lymphoma ascites tumor model. Int J Nanomedicine 753. https://doi.org/10.2147/IJN.S11727

Shao D, Zhang X, Liu W et al (2016) Janus silver-mesoporous silica nanocarriers for SERS traceable and pH-sensitive drug delivery in cancer therapy. ACS Appl Mater Interfaces 8:4303–4308. https://doi.org/10.1021/acsami.5b11310

Sheikhpour M, Naghinejad M, Kasaeian A et al (2020) The applications of carbon nanotubes in the diagnosis and treatment of lung cancer: a critical review. Int J Nanomedicine 15:7063–7078. https://doi.org/10.2147/IJN.S263238

Shi H, Ye X, He X et al (2014) Au@Ag/Au nanoparticles assembled with activatable aptamer probes as smart "nano-doctors" for image-guided cancer thermotherapy. Nanoscale 6:8754. https://doi.org/10.1039/C4NR01927J

Siafaka PI, Okur NÜ, Karantas ID et al (2021) Current update on nanoplatforms as therapeutic and diagnostic tools: a review for the materials used as nanotheranostics and imaging modalities. Asian J Pharm Sci 16:24–46. https://doi.org/10.1016/j.ajps.2020.03.003

Sims J, Miracle VA (2002) Phases of a clinical trial. Dimens Crit Care Nurs 21:152–153. https://doi.org/10.1097/00003465-200207000-00009

Sk UH, Kojima C (2015) Dendrimers for theranostic applications. Biomol Concepts 6:205–217. https://doi.org/10.1515/bmc-2015-0012

Sriram MI, Kanth SBM, Kalishwaralal K, Gurunathan S (2010) Antitumor activity of silver nanoparticles in Dalton's lymphoma ascites tumor model. Int J Nanomedicine 5:753–762. https://doi.org/10.2147/IJN.S11727

Tabata Y, Ikada Y (1999) Biological functions of fullerene. Pure Appl Chem 71:2047–2053. https://doi.org/10.1351/pac199971112047

Thorek DLJ, Chen AK, Czupryna J, Tsourkas A (2006) Superparamagnetic iron oxide nanoparticle probes for molecular imaging. Ann Biomed Eng 34:23–38. https://doi.org/10.1007/s10439-005-9002-7

Tinkle S, Mcneil SE, Mühlebach S et al (2014) Nanomedicines: addressing the scientific and regulatory gap. Ann N Y Acad Sci 1313:35–56. https://doi.org/10.1111/nyas.12403

Torchilin VP (2014) Multifunctional, stimuli-sensitive nanoparticulate systems for drug delivery. Nat Rev Drug Discov 13:813–827. https://doi.org/10.1038/nrd4333

Truskewycz A, Yin H, Halberg N et al (2022) Carbon dot therapeutic platforms: administration, distribution, metabolism, excretion, toxicity, and therapeutic potential. Small 18. https://doi.org/10.1002/smll.202106342

Wang Y, Zhang S, Xu T et al (2018) Ultra-sensitive and ultra-fast detection of whole unlabeled living cancer cell responses to paclitaxel with a graphene-based biosensor. Sensors Actuators B Chem 263:417–425. https://doi.org/10.1016/j.snb.2018.02.095

Wetzler M, Thomas DA, Wang ES et al (2013) Phase I/II trial of nanomolecular liposomal annamycin in adult patients with relapsed/refractory acute lymphoblastic leukemia. Clin Lymphoma Myeloma Leuk 13:430–434. https://doi.org/10.1016/j.clml.2013.03.015

Wicki A, Witzigmann D, Balasubramanian V, Huwyler J (2015) Nanomedicine in cancer therapy: challenges, opportunities, and clinical applications. J Control Release 200:138–157. https://doi.org/10.1016/j.jconrel.2014.12.030

Yallapu MM, Othman SF, Curtis ET et al (2012) Curcumin-loaded magnetic nanoparticles for breast cancer therapeutics and imaging applications. Int J Nanomedicine 7:1761–1779. https://doi.org/10.2147/IJN.S29290

Yuan X, Zhang X, Sun L et al (2019) Cellular toxicity and immunological effects of carbon-based nanomaterials. Part Fibre Toxicol 16:18. https://doi.org/10.1186/s12989-019-0299-z

Zhang L, Xia J, Zhao Q et al (2010) Functional graphene oxide as a nanocarrier for controlled loading and targeted delivery of mixed anticancer drugs. Small 6:537–544. https://doi.org/10.1002/smll.200901680

Zhang Y, Thomas TP, Lee KH et al (2011) Polyvalent saccharide-functionalized generation 3 poly(amidoamine) dendrimer-methotrexate conjugate as a potential anticancer agent. Bioorg Med Chem 19:2557–2564. https://doi.org/10.1016/j.bmc.2011.03.019

Zhang Z, Liu C, Bai J et al (2015) Silver nanoparticle gated, mesoporous silica coated gold nanorods (AuNR@MS@AgNPs): low premature release and multifunctional cancer theranostic platform. ACS Appl Mater Interfaces 7:6211–6219. https://doi.org/10.1021/acsami.5b00368

Zhao J, Mi Y, Feng SS (2013) Targeted co-delivery of docetaxel and siPlk1 by herceptin-conjugated vitamin E TPGS based immunomicelles. Biomaterials 34:3411–3421. https://doi.org/10.1016/j.biomaterials.2013.01.009

Zheng M, Liu S, Li J et al (2014) Integrating oxaliplatin with highly luminescent carbon dots: an unprecedented theranostic agent for personalized medicine. Adv Mater 26:3554–3560. https://doi.org/10.1002/adma.201306192

Nanotheranostics in Prostate Cancer: Current Status and Future Potential

11

Juveriya Israr, Rumana Ahmad, Harshita Barkat, and Sahabjada Siddiqui

Abstract

Prostate cancer is a common and serious disease that affects many men and unfortunately leads to a high number of deaths among male cancer patients. Early detection and monitoring of prostate cancer are being improved by continuous advancements in imaging and sensing approaches. A more targeted and personalized approach to disease management requires carefully categorizing patients. Hybrid imaging techniques provide complementing structural and morphological data to enable more accurate and efficient imaging. Precision and personalization in the diagnosis and treatment of prostate cancer could be greatly enhanced by the use of radiotracers, biomarkers, and nanoparticles in targeted imaging. By combining targeted delivery and diagnostics, nanotheranostics—which makes use of biocompatible nanoparticles—offers a viable strategy for managing cancer. These platforms offer a substantial advantage in cancer research and diagnostics due to their highly customizable molecular architectures and ability to interface with specific proteins. This chapter aims to examine the most recent functional and technological developments in prostate cancer

J. Israr
Institute of Biosciences and Technology, Shri Ramswaroop Memorial University, Barabanki, India

Department of Biotechnology, Era University, Lucknow, India

R. Ahmad
Department of Biochemistry, Era University, Lucknow, India

H. Barkat
Department of Pharmaceutics, College of Pharmacy, University of Hafr Al-Batin, Hafr Al-Batin, Saudi Arabia

S. Siddiqui (✉)
Department of Biotechnology, Era University, Lucknow, India

M. A. Barkat et al. (eds.), *Nanotheranostics for Diagnosis and Therapy*,
https://doi.org/10.1007/978-981-97-3115-2_11

diagnosis and therapy. The key problems with managing prostate cancer could be resolved through the implementation of these nanotheranostics advancements.

Keywords

Prevalent malignancy · Prostate cancer · Nanotheranostic · Sensing · Imaging

11.1 Introduction

Non-communicable diseases, including various autoimmune diseases, heart diseases, strokes, cancers, diabetes, chronic kidney disease, osteoporosis, Alzheimer's disease, and cataracts, are widespread globally and are a leading cause of premature death. These diseases are strongly influenced by certain lifestyle choices, including poor dietary habits, sedentary behavior, excessive alcohol consumption, and tobacco use. Prostate cancer is a significant concern, being the second most prevalent form of cancer worldwide. It particularly affects males, with around 29% of them being affected by this disease (Daar et al. 2007; World Health Organization 2021). By 2023, it is projected that there will be more than 1.9 million new cancer cases diagnosed in the United States. Prostate cancer is the most common form, impacting nearly 29% of males. It is followed by lung cancer, which accounts for about 12% of cases, and colorectal cancer, which represents approximately 8% of cases. Among women, breast cancer is the most common form of cancer, making up 31% of all cases (Singh et al. 2018; Parry et al. 2011; World Health Organization 2021; Rawla 2019). According to the Prostate Cancer United Kingdom, it is estimated that one in eight men will be diagnosed with prostate cancer at some point in their lives. There are around 400,000 men currently facing the difficulties of living with prostate cancer. It is crucial to closely monitor treatment outcomes and take proactive measures to prevent cancer recurrence. Utilizing simple and non-invasive diagnostic and monitoring tools would significantly improve these endeavors.

Customized treatment strategies are gaining popularity in the medical field, aiming to limit patients' exposure to unnecessary procedures and decrease healthcare expenses. When it comes to prostate cancer diagnosis, the main approach involves using prostate-specific antigen (PSA). However, this method has its drawbacks, including issues with accuracy and the potential for false positive or false negative results. Patients diagnosed with localized prostate cancer generally have a favorable 5-year survival rate of over 90%. However, those with distant metastases face considerably lower survival rates. The advancements in cancer diagnostics and therapeutics face challenges in terms of their applicability to human patients. While numerous theranostics have shown promise in controlled laboratory environments and animal studies, their limitations hinder their progress in the field (Siegel et al. 2021; Gupta et al. 2024). The prostate gland is often affected by conditions such as prostatitis, autoimmune diseases, and cancer, particularly non-communicable diseases (NCDs). PSA, PAP, and PSMA are markers that play a crucial role in evaluating and diagnosing prostate cancer.

The exact cause of prostate cancer is still unknown, but certain factors such as obesity, advanced age, and family history have been identified as major risk factors. Prostate cancer is more likely to occur in males over the age of 50. Various factors, such as genetic background, ethnicity, family history, and specific gene variants, can also play a role in determining the risk. Certain factors can potentially raise the risk of developing prostate cancer, such as dietary habits and infections with chlamydia, gonorrhea, syphilis, and prostatitis. Nevertheless, the evidence supporting these factors remains inconclusive. Timely identification of prostate cancer is essential for effective treatment, as the chances of survival and recovery decrease significantly in later stages. Early detection greatly improves the chances of a positive outcome, with a survival rate of over 95%. Nevertheless, for individuals experiencing local or inter-organ metastases, the chances of survival drop to around 50% and even less than 30%, respectively (Chavda et al. 2023). Figure 11.1 illustrates the different stages of prostate cancer during its development, Fig. 11.2 illustrates the diagnostics of prostate cancer, and Fig. 11.3 illustrates the different therapeutic strategies used in the treatment of prostate cancer.

Detecting early-stage prostate cancers can be quite challenging as they often don't show noticeable symptoms and the available screening methods are not widely accessible. Primary tumors in prostate cancer frequently present as symptomless growths, often referred to as "silent killers" because they can advance undetected. It is crucial to screen individuals who are at risk because of the intricate characteristics of this condition. For the diagnosis of prostate cancer, doctors typically start with the digital rectal examination (DRE) test. This involves a hands-on examination to detect any abnormal areas in the prostate gland, which can indicate the presence of cancer. The PSA test measures the levels of the enzyme PSA in the blood of males. It is recommended to have regular tests to determine if a biopsy is needed (Key

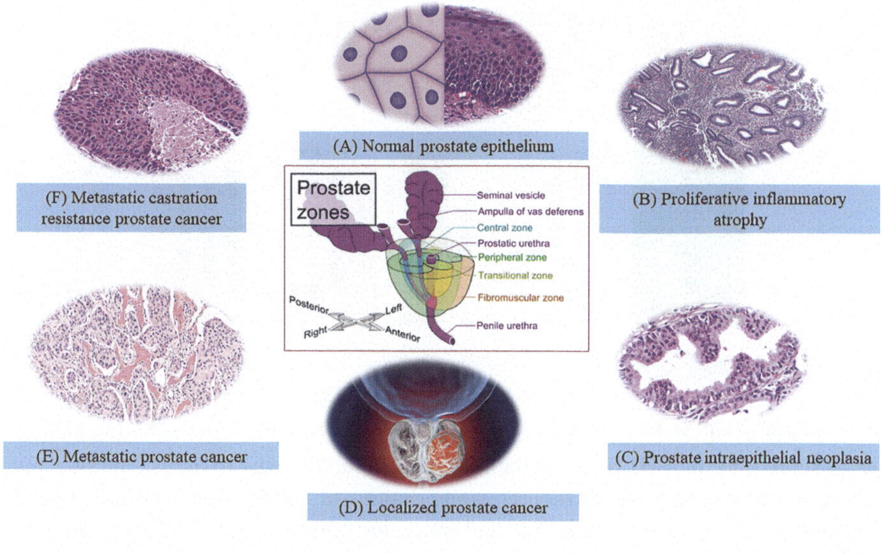

Fig. 11.1 Prostate cancer development

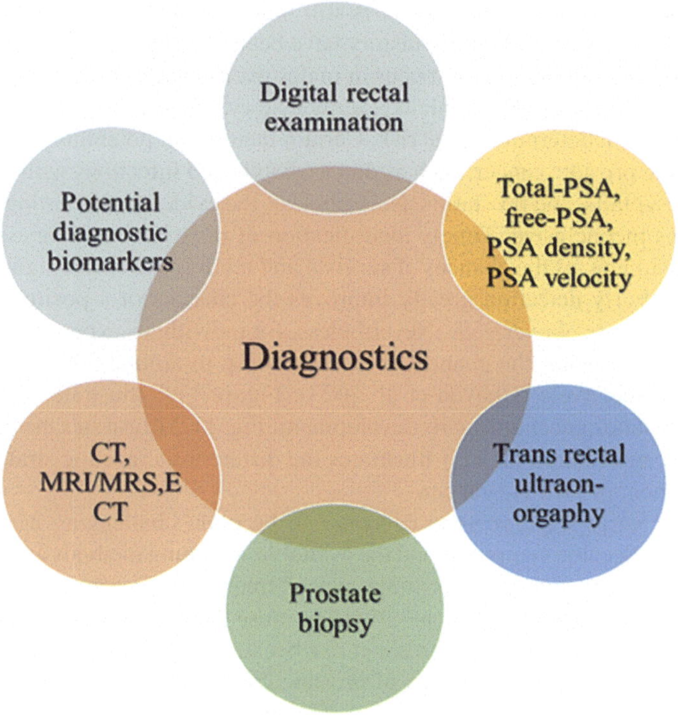

Fig. 11.2 Diagnostics of prostate cancer

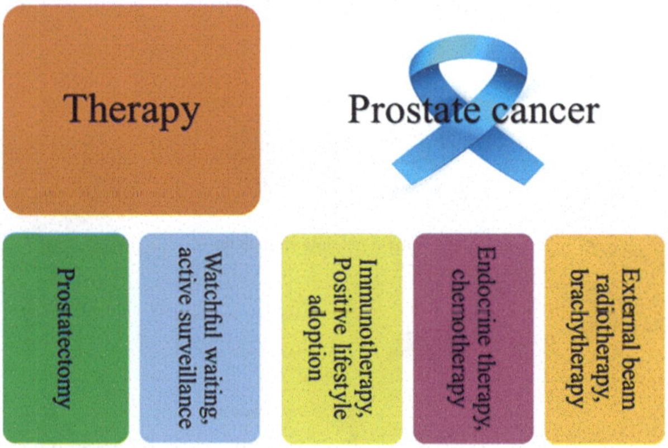

Fig. 11.3 Prostate cancer therapy

2011; Ledet et al. 2017). In addition, there are alternative diagnostic tools such as the 4Kscore and the PHI (prostate health index) that are used to assess the progression of prostate cancer. Nevertheless, some diagnostic techniques may have limitations when it comes to sensitivity and specificity, which can lead to inaccurate results. Furthermore, certain techniques may require invasive procedures, which can have a significant impact on the patient's overall experience (Eastham 2017). Improved detection and treatment methods are crucial in reducing the mortality rate among cancer patients. There are various conventional cancer treatments available, including hormone replacement treatment, immunotherapy, radiation, surgery, and chemotherapy (Melancon et al. 2012). It involves identifying individuals in need of medical care and providing them with a standard treatment. Many medical and pharmacological studies focus on categorizing diseases and developing treatments. These methods have not gained widespread use because of their limited selectivity, which can potentially harm normal tissue and the immune system. Potential outcomes of non-surgical cancer treatments include ineffectiveness, challenges with drug distribution, concerns about biological barriers, and the development of tumor cell resistance. Thus, progress in eradicating cancer is hindered by the lack of reliable methods for evaluating early therapy. Thus, the objective of anticancer research is to develop improved treatments (Falzone et al. 2018; Banthia et al. 2022).

Through the integration of various scientific disciplines, nanomedicine offers innovative approaches to diagnostics and treatment. The term "nano" refers to a unit of measurement that is equivalent to one billionth of a meter. Nanoparticles are tiny particles that range in size from 1 to 100 nm. "Theranostics" combines assessment and treatment. Nanotheranostics involves the utilization of nanoparticles for both diagnosis and treatment. This approach has the potential to revolutionize the diagnosis, treatment, and prognosis of cancer. In this chapter, we will explore the fascinating world of nano theranostics platforms, specifically examining their applications in the context of prostate cancer. This comprehensive discussion includes the benefits of using advanced treatment methods, the application of cutting-edge nanomedicines, challenges, and constraints, as well as the mechanisms and components involved in precise targeting.

11.2 Nanotheranostic Platforms/Sensing and Detection Platforms

Imaging and bio-sensing methodologies have become increasingly important in cancer research, offering valuable support in identifying and locating prostate cancer cells, regardless of the stage of the disease. Although there has been some improvement in reducing false-positive results, the current diagnostic methods for PCa can still be time-consuming and may not always detect small metastases or primary tumors. New screening approaches are being developed to improve patient stratification, disease tracking, and prognosis response data gathering. There have been proposals for clinical trials and potential applications of nanoparticle biosensors, including quantum dots, gold nanoparticles, and graphene oxide. Achieving success necessitates precise measurement and diligent monitoring of host biomarkers (Asem and Malmström 2018; Kröger et al. 2018).

11.2.1 Liposomes/Liposomes for PCa Biosensors

In the field of oncology, liposomes and lipid nanocapsules (LNCs) are two lipid-based nano-delivery platforms that are frequently used. Liposomes are commonly used as nanocarriers in drug delivery systems. They have several advantages, including their small size, compatibility with the body, ability to break down naturally, low toxicity, lack of immune response, and ability to hold different types of medications (Akbarzadeh et al. 2013). These platforms can be utilized as theranostic tools for different imaging techniques, such as fluorescence imaging, MRI contrast, and the use of cinobufagin, a potent anticancer medication (Rehman et al. 2020). Liposomal medications have shown significant effectiveness in combating cancer and display a sustained release pattern. Achieving tumor regression prediction has been made possible through the development of multifunctional RNA-loaded magnetic liposomes, which leverage magnetic resonance imaging (MRI) (Zhang et al. 2018; Karpuz et al. 2020). Studies have shown that liposomal drugs have demonstrated improved efficacy in models of prostate cancer. It is worth mentioning that liposomes containing curcumin have demonstrated impressive inhibition of cellular proliferation. Recent research findings suggest that there have been significant advancements in the development of engineered liposomes. These liposomes have been carefully formulated to contain daunorubicin and emodin, which have demonstrated encouraging outcomes in specifically targeting tumor sites. Liposomes have shown promise in the field of cancer nanotheranostics, specifically in diagnostic techniques like MRI, PET, and SPECT (Grippin et al. 2019; Akpa et al. 2023).

Studies have demonstrated the potential of nanocarriers such as nanoemulsions, liposomes, and polymeric nanoparticles to utilize Enhanced Permeability and Retention (EPR) effects. This indicates that these carriers could be effective in delivering anti-cancer medications (Chen et al. 2016). These innovative nanocarriers have been developed to perform multiple functions simultaneously in the diagnosis, scanning, and treatment of prostate cancer. This is made possible by their unique ability to attach chemicals that specifically target cancer cells. Liposomes have been used to improve drug delivery by taking advantage of their vesicular structure, which is surrounded by a lipid bilayer (Kim et al. 2010). Researchers have utilized polyethylene glycol (PGE) and other biodegradable polyethers to enhance lipid membranes, enabling more accurate active targeting of specific ligands (Malam et al. 2009). In the field of prostate cancer research, there has been a recent proposal to focus on self-assembling micelles. These micelles are based on elastin-like polypeptide (ELP) and are equipped with tethered gastrin-releasing peptide (GRP). In vitro experiments have shown that the use of hybrid ELP/liposome nanoparticles can effectively reduce cell viability in prostate cancer cells (Zhang et al. 2018). Researchers are currently developing liposomes that specifically target urotensin-II, to improve drug delivery to prostate and colon cancer cells. In the field of prostate cancer research, various synthetic nano platforms are employed to sense and detect the disease. These platforms encompass a range of materials such as QDs, AuNPs, liposomes, and CPs, which also include GONPs. The potential of these nanoparticles as revolutionary tools for cancer screening and early detection is truly remarkable (Zappavigna et al. 2019).

11.2.2 Solid Lipid Nanoparticles

These nano-sized particles, known as solid lipid nanoparticles (SLNs), consist of a solid lipid matrix and typically have a size ranging from 50 to 500 nm. These nanoparticles have numerous benefits in the realm of cancer treatment (Lima et al. 2013). They have shown great efficacy in preserving sensitive medications, improving the stability of drugs at a molecular level, and aiding in the formulation process. In addition, SLNs allow for the precise delivery of therapeutic agents to the specific areas affected by the tumor, resulting in enhanced treatment results (Wong et al. 2007). Through the reduction of systemic adverse effects, SLNs offer a more advantageous therapeutic profile. Research has shown that sentinel lymph nodes (SLNs) have proven to be well-tolerated and have minimal toxicity, making them a great option for incorporating pharmaceuticals into targeted nanosized carriers (Fathi et al. 2012). There are numerous advantages associated with the use of sentinel lymph nodes (SLNs) in cancer treatment. These benefits include the ability to combine both water-soluble and fat-soluble medications, a non-toxic nature, and excellent drug stability. Resveratrol solid lipid nanoparticles (RSV-SLNs) have shown impressive abilities to distribute within prostate cells, making them potential candidates for delivering chemotherapeutic agents directly to the target site. Various synthetic polymers, including chitosan, dextran, albumin, heparin, gelatin, and collagen, possess exceptional characteristics such as biodegradability, biocompatibility, non-immunogenicity, and non-toxicity (Sharma et al. 2022; Akanda et al. 2015; Akpa et al. 2023).

11.2.3 Deoxyribonucleic Acid-Based Nano Platforms for PCa Biosensors

Given its compact size, excellent biocompatibility, and customizable nature, deoxyribonucleic acid (DNA) holds great potential as a nano platform. It has a favorable impact on the immune system and can be transformed into beneficial nanostructures capable of drug delivery. The use of DNA nanotechnology, specifically self-assembled DNA structures, in the early detection and therapy of cancer, is becoming increasingly prominent. Fluorescent nanoparticles, known as upconversion nanoparticles (UCNPs), have been successfully utilized in biological detections by combining them with DNA (Chen et al. 2018; Dong et al. 2021; Lin et al. 2017).

11.2.4 Polymers

Polymers are a type of large molecules made up of repeating subunits known as monomers. Research has demonstrated that enhancing surface modifications in polymer-based drug delivery systems can greatly enhance the accuracy, precision, and efficacy of therapeutic and diagnostic approaches (Grossman and McNeil 2012). Various types of macromolecules, whether natural or synthetic, have been employed in nanotheranostics for their diverse applications. Considerable research has been carried out on peptide aptamer-targeted polymers, particularly in the field of tumor cell diagnosis and distribution. A polymeric core-shell nanosystem was developed by utilizing glycol chitosan, photosensitizer protoporphyrin IX, and PEG (Gardel 2013). This novel technique increased the fluorescence linked to cancer cells and made it easier for them to gather in specific areas. A specialized formulation of hyaluronic acid (HA)–iodixanol nanogels has been developed to meet the precise requirements of targeted X-ray computed tomography (CT) imaging and chemotherapeutic drug delivery (Zhu et al. 2016). Recent research has shown that nanocolloids measuring 3.6 nm in size have impressive photoluminescence emission characteristics. This makes them ideal for use as fluorescent nanoprobes and drug nanocarriers with strong inhibitory capabilities (Mansur et al. 2019). Regrettably, there are certain concerns closely linked to organic solvents when it comes to polymeric nanoparticles. In the field of prostate cancer research, the use of a powerful duo of chemotherapy agents, docetaxel and quercetin, in the creation of chemically modified polymeric nanocapsules (NCs), has shown promising results. These NCs not only exhibit increased cellular inhibitory activity but also possess improved absorption capabilities. The in vivo tests also showed encouraging results of nanoparticle carriers (NCs) that include docetaxel (DTX) and quercetin (QU). Recent studies have shown that certain types of nanoparticles, including superparamagnetic iron oxides and gold nanoparticles, may have the ability to enhance the effectiveness of radiation therapy and improve the detection of tumors (Yu et al. 2020; Shitole et al. 2020).

11.2.5 Metallic and Magnetic Nanoparticles/Core-Shell Metallic Nanoparticles for PCa Biosensors/Gold Nanoparticles for Bio-sensing Applications

Research on metallic and magnetic nanoparticles has garnered significant attention in the field of cancer studies. These nanoparticles have exceptional qualities that make them highly promising for a range of applications in the diagnosis and treatment of cancer. When it comes to studying prostate cancer, the field of nanotheranostics has made some exciting advancements. Metallic nanoparticles, like gold nanoparticles, have proven to be particularly interesting due to their unique shapes and physical properties. These characteristics have the potential to greatly impact their use in theranostic applications. Research has shown that the use of radioactive gold nanoparticles holds great potential in the field of targeted therapy. These nanoparticles release beta particles to specifically target and destroy tumor tissue, while gamma rays are utilized for scintigraphy or SPECT/CT imaging purposes (Xu et al. 2020). Research has shown the effectiveness of delivering drugs directly to prostate cancer cells. A recent study by Luo et al. (2020) has demonstrated the successful development of gold nanoparticles that are conjugated with gadolinium complexes and ligands. These nanoparticles have been specifically designed to target prostate-specific membrane antigen (PSMA) and show great potential for MR-guided therapy in prostate cancer. Utilizing nanoparticles in radiation therapy enables the precise delivery of radiation, minimizing harm to healthy tissues and maintaining a low irradiation dosage. Using an external magnetic field, nanotheranostic therapy can effectively target and eliminate cells in specific tissues. By incorporating various components into a PEGylated poly skeleton, the conversion of laser energy into thermal energy can be achieved. Combining chemotherapy and photothermal therapy has proven to be a successful method for treating prostate cancer, as demonstrated by studies conducted by Ferreira et al. (2020), Guo et al. (2018), and Barbezan et al. (2024).

Various types of nanoparticles made from metals like gold, copper, and silver are readily available in different sizes and shapes in the market. The properties of these particles vary based on the difference in thickness between the outer coat and the dielectric core at the center (Barkat et al. 2018). Gold nanoshells have proven to be valuable in Nanomedicine due to their efficient drug release and the capability to attach specific antibodies or proteins. This makes them effective biomarker recognition tools (Luo et al. 2015). Significant progress has been made in clinical experiments involving the use of superparamagnetic NPs for medication delivery. Additional research is needed to develop effective and secure metal-based nanoparticles, as their size can potentially elicit harmful immune reactions. We have developed luminescent nanomaterials with a spherical shape and precisely controlled dimensions for bioimaging purposes. The nanophosphors mentioned in the study by Sweet et al. (2012) have shells that are biocompatible and incorporate rare earth elements to prevent the degradation of fluorescence.

The unique properties of gold nanoparticles (AuNPs) have captured the attention of researchers due to their large surface areas, biocompatibility, strength, and optical qualities. Certain nano sensors utilize aptamers or antibodies, which offer the advantages of cost-effectiveness, flexibility, and long-term stability. It is crucial to employ a highly sensitive method for detecting the prostate cancer antigen (PSA), as it can be misleading when present in high levels. In a study conducted by researchers, they suggested utilizing the LSPR extinction spectrum of DNA aptamers functionalized Au nanodisk arrays (Berger 2003). This method offers a label-free approach that is efficient, straightforward, and extremely precise. While ELISA is frequently used to identify elevated levels of biomarkers in prostate cancer, its detection capabilities are restricted to later stages of the disease. There is a suggestion to use a highly sensitive fluorescence-activated probe to enhance cancer sensing, as it has a detection level significantly lower than standard fluorescent probes. In the field of prostate cancer research, scientists have developed a method to create Ab2-RBITC-AuNPs. This involves the process of conjugating RBITC to the surface of the AuNPs, as described by Fass (2008). Sarcosine, a methylated derivative of glycine, has shown promise as a potential biomarker for prostate cancer. Nevertheless, research has indicated that the molecular mass of alanine and sarcosine are the same, which raises doubts about the reliability of sarcosine as a marker for prostate cancer. Screen-printed carbon electrodes (SPCEs) with in situ modified AuNPs are proposed as a novel method for detecting and quantifying sarcosine (Tegler et al. 2012; Takayanagi et al. 2004).

11.2.6 Carbon Nanotubes/Carbon-Based Nanomaterials and Nanoparticles for PCa Biosensors

Carbon nanotubes (CNTs) demonstrate remarkable structural integrity and mechanical resilience thanks to their strong bonding properties. The use of these entities is highly advantageous due to their large surface area, ability to contain therapeutic agents, and impressive ability to be modified on the surface. Recent advancements have shown the successful use of nanotubes as carriers for chemotherapy agents and nucleic acids, enabling precise delivery and improved therapeutic capabilities (Fubini et al. 2010). Research is being conducted to explore the use of nanotubes as diagnostic agents in the field of oncology. A potential approach that has been studied is the use of a branching polyethylene glycol chain to attach paclitaxel (PTX) to water-soluble carbon nanotubes (Burke et al. 2012). This method showcases low levels of toxicity and shows a significant increase in tumor absorption, surpassing the conventional Taxol® treatment in the mouse model of 4T1 triple-negative breast cancer by ten times. Another interesting application involves the use of photothermal-induced ablation, which helps to make cell membranes more permeable and induce necrosis (Liu et al. 2008). This method shows potential in reducing the emergence of tumor resistance and the chances of tumor reoccurrence. A theranostic multi-walled carbon nanotube (MWCNT) was successfully developed by integrating acid-oxidized MWCNTs with functional moieties (Das et al. 2013; Shao et al. 2015).

Carbon nanotubes (CNTs) demonstrate remarkable structural integrity and mechanical resilience thanks to their strong bonding properties. The use of these entities is highly advantageous due to their large surface area, ability to contain therapeutic agents, and impressive ability to be modified on the surface. Recent advancements have shown the successful use of nanotubes as carriers for chemotherapy agents and nucleic acids, enabling precise delivery and improved therapeutic capabilities (Fubini et al. 2010). Research is being conducted to explore the use of nanotubes as diagnostic agents in the field of oncology. A potential approach that has been studied is the use of a branching polyethylene glycol chain to attach paclitaxel (PTX) to water-soluble carbon nanotubes (Burke et al. 2012). This method showcases low levels of toxicity and shows a significant increase in tumor absorption, surpassing the conventional Taxol® treatment in the mouse model of 4T1 triple-negative breast cancer by ten times. Another interesting application involves the use of photothermal-induced ablation, which helps to make cell membranes more permeable and induce necrosis (Liu et al. 2008). This method shows potential in reducing the emergence of tumor resistance and the chances of tumor reoccurrence. A theranostic multi-walled carbon nanotube (MWCNT) was successfully developed by integrating acid-oxidized MWCNTs with functional moieties (Das et al. 2013; Shao et al. 2015).

11.2.7 Dendrimers

Dendrimer conjugates are a type of drug-based product that has shown promising characteristics in the field of cancer treatment (Abbasi et al. 2014). These characteristics encompass minimal harm to healthy cells, excellent ability to dissolve in solutions, resistance to chemical degradation, and a tendency to specifically gather in cancerous cells. The effectiveness of these compounds depends on the type of polymer used, so it is crucial to create new biocompatible polymers that can be used as safer alternatives in the body (Yan et al. 2021). In the field of oncology, charged polymers like polyethyleneimine (PEI) and polyamidoamine (PAMAM) are widely used to enhance the delivery of nucleic acids and small molecules. Several clinical trials are currently underway to assess the capabilities of dendrimer-based products in a range of applications, including contrast agents, topical gels, transfection agents, and theranostic agents. Recent studies have demonstrated the potential of utilizing carbon quantum dots derived from sweet lemon peel and PAMAM dendrimers in the treatment of triple-negative breast cancer. In their study, Chittasupho et al. devised a novel therapeutic approach for cancer management. They utilized PAMAM dendrimers and doxorubicin to develop CXCR4-targeted dendrimers. The dendrimers were effectively internalized by breast carcinoma cells, resulting in increased cytotoxicity compared to the dendrimers that were not bound (Palmerston Mendes et al. 2017). In the research conducted by Wang et al., dendrimers were utilized to carry doxorubicin. The researchers noted that these dendrimers demonstrated improved antitumor activity via caveolae-mediated endocytosis. Dendrimers are essential in delivering medications to patients with various illnesses, including breast and prostate cancers. Quantum dots (QDs) are nanocrystals known for their remarkable optical and electrical properties, which make them ideal for continuous fluorescence monitoring and guiding treatment through imaging (Nottelet et al. 2015; Carvalho et al. 2020).

11.2.8 Quantum Dots/Quantum Dots as Bio-sensing Tools

Quantum dots are incredibly small semiconductor particles with exceptional optical and electronic properties. They are commonly utilized in a wide range of industries due to their promising capabilities. The nanoparticle needs to have certain characteristics such as being biocompatible, biodegradable, non-toxic, and non-immunogenic (Granada-Ramirez et al. 2018). In addition, it should possess the capability to package lipophobic and lipophilic medications, diagnostic substances, and targeting components. In the realm of cancer research, quantum dots, or QDs, have displayed immense promise. There have been notable breakthroughs in the development of nano theranostics platforms that employ QDs to detect and treat cancer. One method includes applying a layer of paclitaxel and lipoic acid to QDs, which boosts their effectiveness as a treatment. In addition, scientists have developed highly fluorescent QDs that are combined with gelatin/chondroitin, showing great potential in enhancing the efficacy of breast cancer therapy. These cutting-edge methods show potential for enhancing cancer treatment and enhancing patient results. In their study, Jiang et al. introduced a highly adaptable nanosystem that utilizes graphene oxide to administer enzalutamide intravenously, aiming to address the challenges of treating castrate-resistant prostate cancer. This nanosystem showcases exceptional targeting abilities specifically designed for prostate cancer. Quantum dots (QDs) have proven to be incredibly useful in nanodevices for detecting breast and prostate cancer due to their impressive photostability and brightness (Huang et al. 2017; Jahangir et al. 2019). The quinic acid-conjugated N-CQDs exhibit impressive luminous properties and a remarkable capacity to accumulate within tumors. These unique qualities make them extremely promising candidates for multifunctional theranostic agents (Samimi et al. 2021).

Quantum dots (QDs) possess exceptional optical properties that facilitate targeted and effective drug delivery to cancer cells. Fluorescent properties enable convenient tracking of various substances within the body, such as proteins, DNA, anions, small molecules, and metal ions. Developing sensors with high selectivity, sensitivity, and stability poses a significant challenge (Malam et al. 2009). Researchers have found that CdSSe/ZnS QDs have proven to be highly effective fluorophores. These fluorophores have a long fluorescence lifetime, which enables them to precisely target and locate individual cancer cells. This capability has the potential to significantly speed up the diagnosis process (Probst et al. 2013). Coating CdTe QDs with carboxymethyl chitosan can improve biocompatibility and reduce toxicity, addressing the limitations of their high toxicity and low sensitivity. This technique offers a robust binding surface that effectively activates the photoluminescence effect of the QDs (Puri et al. 2009). PSA (Prostate Specific Antigen) and ANXA3 (Annexin A3) are commonly used biomarkers in prostate cancer. PSA values can sometimes be misleading, but there is a specific and non-invasive biomarker called ANXA3 that can provide more accurate results. What's great about ANXA3 is that it remains stable in urine samples for over 48 h at room temperature (Bouchelouche et al. 2010). Using CdS QDs coupled to Au nanoimmunosensors on a Quartz Crystal Microbalance (QCM) provides a highly sensitive and label-free method for detecting ANXA3. Our CuInS2 QDs offer a range of benefits, including water solubility, near-IR emission, and high efficiency. These qualities result in lower toxicity, longer drug retention times in the body, improved solubility, and enhanced targeting capabilities. At certain wavelengths of light, there is a remarkable ability for it to easily penetrate through tissue, minimizing any scattering effects. These particular QDs utilize aptamers to attach drugs and transport them to the desired cells. A potential avenue for further exploration involves the development of a QD biosensor that utilizes the emission of H_2O_2 from cancerous cells. Electrochemical biosensors based on reduced Graphene Oxide QDs/ZnO hybrid nanofibers (rGO QDs/ZnO hybrid NFs) address the challenges of sensitivity, selectivity, stability, and repeatability associated with nanomaterials made from metals, metal compounds, and carbon. The rGO QDs/ZnO hybrid biosensors have demonstrated their effectiveness in drug screening for PCa cells treated with anticancer drugs (Imlimthan et al. 2018).

11.3 Prostate Cancer Detection Using Bioimaging and Biosensing Technologies

During the initial phases of prostate cancer, patients often do not exhibit any noticeable symptoms. However, as the tumor continues to grow, symptoms may start to appear in the lower urinary tract. Various clinical strategies are utilized for the detection of prostate cancer, including digital rectal examination (DRE), prostate-specific antigen (PSA) testing, transrectal ultrasonography, computed tomography (CT), emission computed tomography (ECT), and magnetic resonance imaging (MRI) (Lin et al. 2020). The main detection method revolves around the widely used and minimally invasive screening of prostate-specific antigen (PSA) levels in the bloodstream. However, the use of this approach has decreased due to its reduced accuracy. The use of PSA as the primary biomarker for evaluating the extent and implications of the disease is currently a subject of discussion, considering its unique tendency for tumor dormancy and increased levels linked to prostate abnormalities. There is a modern and advanced method used for screening prostate cancer, which includes the utilization of biosensing assays (Ahmed et al. 2017). These tests use enzyme-linked immunosorbent assays (ELISA) to identify prostate-specific antigen (PSA) in a patient's blood sample. However, the existing approaches encounter difficulties in achieving detection limits as low as several nanograms per milliliter (ng/mL) of prostate-specific antigen (PSA), which is the preferred receptor for diagnosis in this particular situation. It is crucial to prioritize the improvement of sensitivity levels in detecting biomarkers linked to circulating cancer cells in the body, even before symptoms of the disease appear. Utilizing cutting-edge biosensors with heightened sensitivity, there is potential to detect and diagnose prostate cancer at an earlier stage. This could allow for the screening of men at a younger age than what is currently recommended for PSA screening. In the realm of oncology, clinical testing is of utmost importance in categorizing patients based on different factors like age and disease severity. Emphasizing the importance of early detection in the field of oncology, it is crucial to utilize accurate measurement and monitoring techniques for emerging host biomarkers. In the field of prostate cancer, clinical imaging methods are crucial for identifying and accurately locating the disease. These methods provide invaluable insights that help guide treatment strategies (Hofmarcher et al. 2020). These imaging techniques are highly precise and accurate in identifying and pinpointing cancerous tissues, all while being minimally invasive:

11.3.1 Magnetic Resonance Imaging (MRI)

Utilizing nuclear magnetic resonance (NMR), magnetic resonance imaging (MRI) can detect cancer in soft tissues. Given the prevalence of hydrogen atom nuclei in both fat and water, they serve as useful probes (Rais-Bahrami et al. 2013). By manipulating the pulse sequence parameters, it is possible to generate images of the surrounding tissues with different levels of contrast. This method helps in assessing the extent of cancer beyond the prostate gland. There are methods available to differentiate between different types of cancer by utilizing radio waves to activate the proton elements within tumor cells (de Rooij et al. 2014). Scientists monitor the reorientation speed of proton nuclei to identify any abnormalities. When it comes to imaging soft tissues, MRI is a game-changer. It allows us to capture detailed, three-dimensional pictures with exceptional clarity and precision. Injecting an exogenous contrast agent, like paramagnetic nanoparticles (NPs), can improve tumor detection by reducing the relaxation rate of water protons. Choosing a contrast agent that exhibits a strong affinity for a specific type of tumor tissue is the main challenge (Han et al. 2017).

11.3.2 Single-Photon Emission Computed Tomography (SPECT)

Cutting-edge PCa diagnostic scans utilize long-lived radioisotopes and employ single-photon emission computed tomography (SPECT). These radionuclides can bind to macromolecules such as proteins and can be easily tracked using a gamma camera, serving as valuable biomarkers. By emitting a gamma ray of a specific energy upon decay, it enables the early detection and improved treatment of cancer. Thanks to the process of radioactive decay, there is now a wider range of radioactive materials that can be used in imaging methods. As an example, ^{235}U can undergo decay and produce daughter isotopes ^{134}Sn and ^{99}Mo through neutron bombardment. Through the generation of high-energy radiation, specifically-rays, which have a low wavelength and can be easily eliminated from the human body, the decay of ^{99}Mo to ^{99m}Mo has the potential to serve as an effective diagnostic agent. This could enable the precise localization of tumor sites, as suggested by Frangioni (2008) and Paterson et al. (2014).

11.3.3 Positron-Emission Tomography (PET) and PET Coupled with Computed Tomography (PET/CT)

Advanced imaging techniques such as positron-emission tomography (PET) and positron-emission tomography–computed tomography (PET CT) are utilized to study the intricate structure and function of tissues in living organisms. Substances that are radio-tagged are administered into the patient's bloodstream, allowing for their absorption and the subsequent recording of a scanned image of the affected area. The radioisotope undergoes decay, releasing a positron that can penetrate tissue, albeit within a restricted range. When the positron collides with an electron, a pair of photons (-photons) is generated, traveling in opposite directions. These photons generate a burst of light, which is then detected by a scintillator, converted into electrical signals, and processed into an image. The half-lives of PET radiotracers vary, with some lasting as short as a few minutes and others lasting over 12 h. The imaging data is derived from the biochemical changes that occur when the radiotracer interacts with tissues (Evans et al. 2018; Scher et al. 2008; Frangioni 2008).

11.3.4 Radiolabelled Tracers

Imaging probes with radiolabelled tracers have proven to be valuable tools for cancer diagnosis, thanks to their biodistribution, clearance rate, and site-specific binding. Choosing the most suitable radiotracer involves considering factors such as imaging quality and the half-life of the radioisotope. When it comes to prostate cancer, one important tool is the radiolabelled tracer 18-fluorodeoxyglucose ([18F]-FDG), which has the remarkable ability to distinguish between normal and malignant cells. As it is transported through glucose proteins, it conveniently accumulates at the tumor site and easily diffuses into the cell membrane of the tissue. PET/CT is an incredibly effective tool for detecting tumors and pinpointing cancer. The low metabolic rate and minimal absorption by tumor cells can pose challenges for [18F]-FDG in PCa. Considering the brief duration of its effectiveness and its capability to detect choline, a crucial element of cell membranes in tissues, PET tracers such as [11C]-Choline could be more favorable options compared to [18F]-FDG. Tracking the progression of cancer can be done by observing evidence of tumor cell proliferation, such as choline absorption. Due to its extended half-life of 109.8 min, the radioisotope 18F is commonly used in PET tracers. When it comes to targeted imaging using monoclonal antibodies (mAbs), the use of radioisotopes with a longer half-life, like 124I or 86Y, is necessary due to the size and clearance time of the antibodies. SPECT, PET, and CT all utilize ionising radiation, which has the potential to harm healthy tissue. However, MRI does not pose this risk, although it is a pricier alternative compared to the other options (Anderson and Welch 1999; Imlimthan et al. 2018).

11.3.5 PET and CT

Due to their versatility, PET and CT quickly gained popularity in the medical field in the early 2000s (Townsend 2008). CT scans offer precise and comprehensive images of the body's structure, while PET scans provide insights into metabolic processes. During CT imaging, a rotating tube emits arrow-shaped beams of X-rays toward the patient, generating 2D picture slices of the organ or tissue. A computer then processes these images, generating a three-dimensional representation of the scanned patient's anatomical structure. Research has shown that when combining these techniques with computed tomography (CT), a more comprehensive analysis can be achieved in terms of tumor tissue incidence, location, and stage determination. Unlike MRI, which only focuses on soft tissue, this approach proves to be more effective in detecting larger cancers. Through the use of cross-sectional imaging, CT scans provide a comprehensive view of tumor anatomy and shape. Imaging with exceptional accuracy and reliability of tumor location is achieved through the hybrid combination with CT (Berger 2003; Jadvar 2013).

11.3.6 Hybrid Combination of MRI/CT

Research has shown that the combination of MRI and CT imaging can significantly enhance the accuracy of cancer diagnosis and prognosis (Bouchelouche et al. 2010). When utilizing this approach, it is crucial to take into account the characteristics of radiotracer imaging probes, such as their half-life, decay mechanism, cost, and availability. To detect, pinpoint, and evaluate the stages and locations of prostate cancer, the use of MRI and PET/CT is necessary. Despite the progress made in improving accuracy, sensitivity, and precision, the high cost of these advancements often deters many patients from utilizing them. Additionally, access to these technologies for early-stage diagnosis remains limited to only a select few medical facilities. Thanks to advancements in imaging technology, hospitals are now able to detect and diagnose cancers at much earlier stages of development. The images provide a comprehensive analysis of the tumor cells' appearance and anatomy, enabling informed decisions regarding treatment options. When considering the most suitable imaging technique, it is important to prioritize health and safety considerations. Various industries are collaborating to improve early cancer detection and reduce mortality rates. As a result, imaging techniques are now being used to screen individuals with symptoms. Using nanoparticles as contrast agents offers significant research potential, as they enable more accurate imaging (Anderson and Welch 1999).

11.3.7 Point-of-Care Tests (POCTs)

For early disease detection, point-of-care tests (POCTs) are commonly used to detect specific biomarkers. These tests offer a cost-effective, efficient, and user-friendly solution that can be utilized in any setting, without requiring specialized equipment or medical expertise. The PSA test is a commonly used method employed by doctors to identify prostate cancer in its early stages (Barry 2001). Nevertheless, there are concerns about the reliability of PSA testing in detecting prostate cancer. This is because the presence of other non-cancerous prostate conditions, like benign prostatic hyperplasia (BPH), can cause an increase in PSA levels in the blood, potentially leading to false-positive results for prostate cancer (Verhamme et al. 2002). Research in pancreatic cancer aims to discover early biomarkers that are produced by cancer cells in a specific manner and at low levels. Biosensors are commonly utilized to interact with specific biomarkers that arise during the progression of a disease. In the field of biosensors, three key components are commonly found: a bio-recognition element, a transducing component, and a readout system (Devillers et al. 2017). Transduced signals can include electrochemical signals or fluctuations in fluorescence intensity. Electrochemical biosensors rely on alterations in electrical properties to identify binding events with the target analytes (Pohanka and Skládal 2008). Achieving high sensitivity and specificity at low analyte concentrations has posed a significant challenge in the realm of biosensor design and manufacturing. In recent decades, there has been a growing use of new electrochemical biosensors. One particular tool, electrical impedance spectroscopy (EIS), has proven to be effective in detecting prostate cancer (Topkaya et al. 2016). In EIS, the detection method used does not require any labels. The binding event between the biorecognition element and analytes is sensed directly, with high sensitivity, and in real time. In recent years, there have been modifications to create smaller electronic devices used for point-of-care diagnosis in detecting biomarkers for prostate cancer. Recent studies have demonstrated that a miniaturized EIS-based biosensor has achieved responses comparable to the clinical threshold of 4 ng/mL for detecting PSA in human serum. Thanks to its compact design, affordability, high sensitivity, consistent results, and reliability, this biosensor can be used as a blueprint for creating cutting-edge devices that can detect biomarkers for early-stage prostate cancer (Mehrabani et al. 2014; Ibau et al. 2019).

11.4 Synthetic Scaffolds for Bio-sensing Applications in Living Cells

In the field of medical research, a crucial component is the connection between the bio-recognition element and the generation of a physical characteristic. This characteristic is then transformed into a signal using either an electrochemical or optical transducer, as discussed by Damborský et al. (2016). NDIs have been extensively researched as synthetic scaffolding for biosensing probes in PCa for three decades. These compounds have been extensively researched in various fields such as electronics, material science, biology, and medicinal chemistry. They have shown promise in connecting different components in molecular devices and aiding in the development of sensors (Al Kobaisi et al. 2016). NDIs, the smallest homolog of rylenediimides, are synthesized by combining NDA with an amine. They were initially utilized as synthetic pigments and colors (Thalacker et al. 2006). NDI derivatives display a versatile electron-deficient aromatic π system that can be absorbed and emit fluorescence at various wavelengths. Basic NDI derivatives typically have absorption maxima around 350 nm, but through derivatization, this range can be extended to 500–760 nm. Their major characteristics include a high fluorescence quantum yield and fluorescence emissions in the range of 650–780 nm. Modulating the optical properties of NDIs through easy functionalization with amino acids and peptides is being utilized for optical imaging purposes. Recent research indicates that they can enter prostate cancer cells. Modifying the aromatic core alters the electrochemical and optical properties of NDIs, which is their primary distinguishing feature. Roxane-based mechanically interlocked molecules (MIMs) have been extensively researched for their applications in catalysts, chemical sensors, polymers, molecular switches, and motors. In the field of biology, self-organized molecules have proven to be valuable tools for drug delivery, targeting bacterial proteins, and releasing bioactive peptides (Griffiths and Stoddart 2008). NDI-based conjugates are currently under study as imaging agents or building blocks for more complex supramolecular constructions with sensing and imaging applications. Water solubility and biological media use of amino acids on the NDI core have been tested (Pengo et al. 2006). The examination has been conducted on amino acid-tagged NDIs to study their aggregation and behavior in aqueous solutions. Furthermore, the fluorescence and aggregation properties of amino acid-tagged NDIs have been extensively investigated, along with their compatibility in living cells. The NDI core of these compounds, when combined with peptide moieties, exhibits fluorescence in living cells. The peptide-tagged NDI demonstrates excellent uptake and distribution within living prostate cancer cells, making it an ideal choice for fluorescence spectroscopy imaging probes. Research has been conducted on the potential of G-quadruplexes as anticancer therapeutics, thanks to their biocompatibility, water solubility, and ability to interact with electron-rich species of amino acid-functionalized NDIs. These compounds have a strong affinity for G-quadruplexes, thanks to their ability to stack aromatically and interact hydrophobically, leading to the formation of supramolecular aggregates (Răsădean et al. 2017).

11.4.1 Molecular Imaging Approaches to Drug Discovery Processes

Imaging is crucial in the search for new medications, as it provides valuable evidence through scans and images to predict their effectiveness. Developing and introducing a new medicine to the market requires both time and financial resources. Collaboration between research institutions and pharmaceutical companies is crucial in determining the priority of drug development projects that have the potential to revolutionize medical care (Murphy 2010). To effectively pursue advancements in the field of prostate cancer, it is crucial to thoroughly understand the molecular mechanisms at play and then confirm the validity of potential targets through molecular and structural imaging techniques. When a target is found, the success of a potential drug depends on the presence of a molecule that has a strong affinity for the target, ensuring its effectiveness, safety, and compatibility. Conducting thorough experiments on animals is crucial for optimizing leads (Rudin 2005). Clinical advance trials have provided evidence of tumor shrinkage and the detection of potential safety concerns, which enhances the overall process. It is crucial to showcase the drug's behavior in the body during the pre-clinical phase and detect any potential safety hazards. One of the challenges in current research trials is the absence of standardized protocols. Enhancing focused imaging methods to enable more accurate therapy predictions is currently a top priority in the pharmaceutical sector. The remarkable potential of nanotechnology to tackle current challenges in the field, along with the utilization of molecules possessing exceptional imaging capabilities, is emphasized (Takayanagi et al. 2004; Yankeelov et al. 2016).

11.4.2 Small-Size Peptides as Targeting Agents

Since the late 1990s, monoclonal antibody-based therapy has been focused on receptors that are overexpressed, mutant, or selectively expressed in human cancer cells (Aina et al. 2007). Nevertheless, the antibodies tend to be absorbed by the liver or bone marrow, struggle to penetrate tumoral cell membranes, and face difficulties reaching tumoral cells due to their large sizes. In the field of cancer research, there has been a shift towards utilizing small peptides to enhance cancer cell receptors, mimicking the effects of antibodies. These peptides have the advantage of being easily produced and able to penetrate tumoral tissues due to their small size. Additionally, they specifically target cancer cell receptors, which helps to minimize damage to the liver and bone marrow (Zhang et al. 2012). Research has also been conducted on synthetic drug-free treatments using small peptides for cancer. Anticancer peptides (ACPs) have shown remarkable efficacy in targeting cancer cells while sparing healthy cells and tissues, making them a promising option for cancer treatment. Various peptides with anticancer properties disrupt the cell membrane and infiltrate. Nevertheless, the design of these peptides must be meticulously executed to achieve the desired functions (Chen et al. 2016; Gaspar et al. 2013). There is a novel group of peptides that specifically focus on identifying and attacking markers found on the membrane of cancer cells. These targeted peptides expose specific and highly affinity cancer cell receptor regions. Targeting peptides have shown promising potential for clinical imaging probe design and manufacturing. They possess desirable characteristics such as retention in the target, fast clearance from healthy tissues or cells, and great stability in vivo (Marqus et al. 2017).

Utilizing various imaging techniques such as PET/MRI and PET/CT is crucial in detecting cancer. This approach allows for the use of a single contrast agent across different imaging modalities, ensuring consistent signal at the target area while preserving biodistribution. There have been some recent studies that have discussed various peptide-based imaging probes, both in vitro and in vivo. Due to its extended half-life and ability to form stable complexes with various ligands, 64Cu has remained a widely used PET imaging radioisotope for an extended period. When it comes to imaging and therapy, the effectiveness of this metal radioisotope is greatly enhanced by ensuring that the radiolabelling conditions are biocompatible. Additionally, selecting a suitable bifunctional chelator simplifies the process of conjugating it to the targeting peptide or protein (Key et al. 2016). Although peptides have improved the safety of imaging probes, concerns about radiation exposure persist. Fluorescence imaging offers a range of benefits, including non-invasiveness, real-time results, high resolution, and cost-effectiveness. The near-infrared (NIR) wavelength range (700–1000 nm) has become increasingly significant in fluorescence imaging due to its advantages such as low cell and tissue absorption and autofluorescence, deep tissue penetration, non-invasiveness, and high picture sensitivity. The focus on small-sized peptides that target malignant cells, the ability to adjust the fluorescence of the scaffolds they are attached to, and the utilization of NIR wavelengths have led to a rise in the number of publications exploring imaging probes for cancer diagnosis (Cai et al. 2018; Luo et al. 2011).

11.4.3 Bombesin-Related Peptides as Potential Targeting Groups for the Detection of PCa

GRPR is found in solid tumors such as breast, colon, lung, and prostate cancer. In prostate cancer, these cells express them at a rate of 63–100% and increase the levels of angiogenesis promoters, which play a crucial role in the spread of the disease. Research is currently being conducted on the study of GRPRs for prostate cancer, as they have a high affinity for GRP, a 27-amino-acid chain, and bombesin, a 14-amino-acid chain. The strong receptor affinity of Bombesin is attributed to its amino acid sequence, which bears resemblance to human GRP (Beer et al. 2012). There are three different bombesin receptor subtypes found in mammals: neuromedin B (NMB), gastrin-releasing peptide (GRP), and BB3. Various subtypes play important roles in immune defense, thyroid and adrenal function, swallowing, weight regulation, and cognitive processes. The smooth muscle contraction of GRP facilitates the release of gastrointestinal hormones, as well as the secretion of pancreatic and gastric secretions. Through the utilization of BBN's affinity and binding, researchers have successfully created therapeutic and imaging compounds specifically designed to target prostate cancer (Pooja et al. 2019). Exploring the affinity of bombesin fragments 7–13 or 7–14 to the GRPR, researchers have developed a range of imaging probes for optical imaging, SPECT, and radiotherapy, as well as MRI and PET/CT imaging. These probes have shown promise in both in vitro and in vivo studies. In previous preclinical studies, researchers utilized 'cold' or radiolabelled 64Cu-bombesin conjugates (Sancho et al. 2011). Various radiolabelled targeting-peptide compounds have been examined for their potential to impede the proliferation of prostate cancer cells. Scintigraphy has been utilized in imaging patients with the Gastrin-releasing peptide (GRP) receptor to anticipate the effectiveness of GRP receptor-targeted treatments, detect GRP receptor-positive tumors at an early stage, and classify them in vivo (Elshafae et al. 2016). In a pilot study conducted in 2000 and approved by the University Hospital Ghent Ethical Board, researchers investigated the safety, imaging properties, and tumor detection efficacy of the GRP analog 99mTc-RP527 for GRP-R scintigraphy. Technetium-99m RP527, a bombesin-derived 99mTc chelated targeting peptide, has a strong affinity for GRP receptors and is commonly employed in tumor detection. Research findings indicate that 99mTc-RP527 shows promising results in detecting prostate tumors that have spread to other parts of the body in male patients. This imaging technique has been found to have good imaging features and a favorable T/N ratio, which can be further improved through the use of SPECT (Van de Wiele et al. 2000).

11.4.4 Radiolabelled Biomarkers for the Imaging of PCa

In cancer detection and treatment planning, biomarkers play a crucial role. Various biomarkers, such as cells, proteins, enzymes, and membrane receptors, have been studied extensively (Shi et al. 2016). The rapid multiplication of tumor cells, known as angiogenesis, plays a crucial role in detecting malignancy. Integrin $\alpha v\beta 3$, a transmembrane receptor found in the extracellular matrix, has been identified as an angiogenic biomarker (Wang et al. 2013). Using a peptide chain commonly found in cancer sites, it is possible to enhance imaging clarity and sensitivity by combining it with a drug or carrier molecule. Combining cyclic RGD (cRGD) with radioisotopes has the potential to produce more effective anti-cancer drugs (Bernsen et al. 2015). Although the first PET radiotracer, [cRGD]Fgalacto-RGD, showed promise in detecting cancer earlier, it did not perform well in the liver, spleen, and kidneys. This was due to low RGD uptake and a high background-to-noise ratio. Repeating tests using radioactively labeled multimeric cRGD peptides could potentially enhance binding affinity, as suggested by Chen et al. (2016). PSA, a serine protein enzyme, is produced by endothelial cells of the prostate gland to monitor the progression of cancer in the prostate. Monoclonal antibodies can identify this specific pattern on tumor tissue, either as a standalone molecule or as part of a larger complex. The potential of this area of research is limited due to its volatility and the need for further study (Ulmert et al. 2012). PSMA targeting is an additional biomarker for PCa. PSMA, a transmembrane glycoprotein associated with prostate cancer stages, shows great potential as a diagnostic target for prostate cancer. Before fully exploring and harnessing its therapeutic potential to halt the spread of cancer, certain obstacles need to be addressed and overcome (Chang 2004). The PET imaging results showed a noticeable increase in absorption in the PSMA-expressing xenografts compared to the isogenic control tumor and background tissue. PSMA is commonly associated with prostate cancer, but it can also be present in other cancerous tissues. Recent advancements in imaging technology indicate that the use of low-molecular-weight radiopharmaceuticals, such as [68Ga] GaPSMA-11, may have superior performance compared to radiolabelled monoclonal antibodies. Urea-based gallium-68 short peptide compounds are utilized to specifically target the prostate-specific membrane antigen (PSMA) in PET imaging. Studies have shown that the use of 177Lu-PSMA has demonstrated promising results in treating metastatic prostate cancer that has developed resistance to conventional therapies (von Eyben et al. 2018; Plichta et al. 2021).

11.4.5 Thiosemicarbazone-Based Metal Complexes as Cancer Theranostics

Thiosemicarbazone-based metal complexes (TSCs) play a crucial role in biosensors, imaging probes, and medicinal medicines. Researchers worldwide have been drawn to their ease of synthesis and functionalization for biological applications (Yildirim et al. 2014). TSCs can bind metal cations, which can enhance their cytotoxicity, fluorescence emission, and biocompatibility. [Cu(ATSM)] has been used to address copper imbalance caused by neurological conditions such as Alzheimer's disease. Since the 1990s, researchers have been studying mono- and bis-TSCs as free ligands or as metal complexes (Paterson and Donnelly 2011). TSC metal complexes have gained popularity as medicinal agents due to their cytotoxicity and biocompatibility. TSC derivatives with Zn(II) have been suggested as cytotoxic agents, while nickel(II) complexes have been employed for HIV antiretroviral, antibacterial, and cell line apoptosis (Lobana et al. 2009). TSC metal complexes can specifically target areas with extremely low oxygen levels, known as "hypoxia," which are commonly found in malignant tissues. In hypoxic conditions, the decrease of Cu(II) to Cu(I) is irreversible. This leads to the selection of cells by bis-TSCs such as [Cu(ATSM)] (Lobana et al. 2009).

Radioisotopes in radiopharmaceuticals play a crucial role in delivering ionizing radiation to cancerous cells, making them indispensable for both diagnosing and treating the disease. These radiopharmaceuticals have the potential to be utilized for PET imaging. In vivo, imaging of hypoxic tissues is increasingly utilizing copper(II) bis-thiosemicarbazone (60Cu, 62Cu, and 64Cu) due to its convenient coordination and redox chemistry (Dearling et al. 2002). The extended half-life and restricted accessibility of Copper-64 have contributed to its popularity in PET imaging. Gallium-68 is a promising choice due to its shorter half-life and its potential for both therapeutic and imaging purposes in the context of prostate cancer (Alam et al. 2016). Additional fluorescence and PET radioisotopes are indium111, zirconium-89, and technetium-99m. Generators are required for these radioisotopes and they may be located quite a distance from the imaging center. To address these concerns, an alternative approach has been developed using fluorescent TSC metal complexes instead of radiolabelled compounds. Fluorophores linked to or integrated into the backbone can regulate the fluorescence emission of these metal complexes at various wavelengths (Cortezon-Tamarit et al. 2016). Fluorescence imaging has emerged as a convenient and cost-effective alternative to PET imaging, offering rapid and sensitive results with readily available equipment. Advancements in fluorescence methods have led to the development of fluorescent compounds that specifically target hypoxic tissues or indicators of malignant cells. These advancements have shown promise in the field of prostate cancer research (Sturzu et al. 2014; Tseng et al. 2017).

11.4.6 Applications of Nanoparticles in the Molecular Imaging of Prostate Cancer

Nanotheranostics have shown great potential in the field of prostate cancer, offering a personalized and precise approach to diagnostics and therapy. By combining the advantages of NPs with well-established imaging techniques, highly targeted and precise imaging approaches can be achieved (Cai et al. 2008). Because of their unique physical and optical properties, gold nanoparticles (AuNPs) have been widely used due to their non-toxic nature. Utilizing AuPs as a contrast agent in PET/

SPECT and CT scans can lead to enhanced contrast images while reducing radiation exposure (Bulte and Modo 2017). In addition, they can specifically target and bind to cancerous tissue through their interaction with trans-endothelial pores. Quantum dots (QDs) have the potential to revolutionize imaging and detection techniques. The fluorescent nanoparticles have been specifically designed to emit light at a single wavelength within the UV to IR spectrum. This is due to their unique optical and chemical characteristics (Mirzaei et al. 2017). QDs possess unique features such as single-wavelength excitation, strong light absorbance, and easy adaptability to be compatible with human body tissues. These qualities make them highly selective and sensitive when used in NP-based therapy to target tumor sites. Unfortunately, safety concerns have hindered their advancement to clinical trials. While further clinical trials are necessary to fully understand the potential risks and benefits of QDs to human health, it is encouraging to see that advances in technology offer hope in addressing these issues (Peng and Li 2010; Azzazy et al. 2006).

11.4.7 Targeting Prostate Cancer Using Nanotechnology Tools

As a biological platform, nanotechnology enables the molecular-level interaction of nanomaterials with biomolecules. Various fields have experienced significant advancements, as evidenced by the contributions of Saini et al. (2010). Recent advancements in cancer nanotechnology have led to the development of improved imaging technologies that allow for more precise identification and targeted therapy of prostate cancer. There is a technique called nanotheragnosis that combines diagnostic and therapeutic capabilities, which has the potential to lower the mortality rate from PCa (Satalkar et al. 2016). There are two ways to target tumor cells: passively and actively. Using biomolecules, active targeting modifies the surface of the nanocarrier agent to enable specific binding. On the other hand, passive targeting involves the accumulation of agents on abnormal blood vessels that are permeable. Active targeting is the preferred approach due to its ability to tightly regulate pharmacokinetics and biodistribution. This allows for increased cellular binding while minimizing uptake by healthy cells. Extensive research has been dedicated to the study of targeted nanoparticles for drug delivery systems, as they hold great potential in the field of Nanomedicine (Nune et al. 2009; McKittrick and Shea-Rohwer 2014).

11.5 Considerations of the Challenges of Prostate Cancer Treatment for the Future of Nanomedicine

Clinical trials play a crucial role in evaluating the effectiveness of potential anticancer drugs and assessing any potential adverse effects in patients with cancer. There are four stages in each of these trials, labeled I–IV. In phase I, the main objectives involve establishing the drug's safety profile and determining its maximum tolerated dose (MTD) (Guan 2012). The maximum tolerated dose (MTD) of a drug's anti-tumor efficacy is determined through two stages of phase II testing, referred to as IIA and IIB. In phase III, the drug's effectiveness is compared to existing treatments, while phase IV focuses on uncovering the potential risks and benefits (Sedgwick 2011). Clinical trials offer new opportunities for studying and improving the well-being of cancer patients, by enhancing our understanding of risks and reducing mortality rates. During clinical trials, patients are not required to cover the cost of the drugs, and the findings from these studies can contribute to the progress of future research. The journey from chemical discovery to clinical practice can be quite lengthy and complex. Over the past three decades, remarkable progress has been made in the field of nanotechnology and Nanomedicine, resulting in a wide range of groundbreaking methods for combating cancer using nanoparticles. Theranostics utilizing nanoparticles show immense potential in combating various types of cancers, such as those affecting the prostate, breast, and esophagus. Studies have shown that green tea extracts, such as epigallocatechin gallate (EGCG), exhibit promising anti-cancer effects in laboratory settings (Lymperopoulos et al. 2017; Tian et al. 2018).

Over the past two decades, nanotechnology has garnered significant attention for its ability to meet clinical needs in cancer diagnosis and therapy. Unfortunately, only a limited number of Nanomedicines have been approved for use in clinical trials for cancer. These factors contribute to the significance of the medical industry in the economy, the time-consuming nature of clinical trials, and the strict regulatory framework. Despite the potential advantages, there are ongoing concerns regarding the in vivo behavior of Nanomedicine, particularly the toxicity of cadmium-containing QDs on organic frameworks. For a regulatory approach to Nanomedicine to be successful, it is crucial to address regulatory obstacles and establish a precise definition (Etheridge et al. 2013). Continuous evaluation is necessary to determine the efficacy of nanomaterials in healthcare settings. Current metrics place a higher value on predicting the approval and payment of new applications while neglecting the importance of considering the preferences of patients and their loved ones. Instead of focusing solely on the impact of emerging Nanomedicines, researchers should consider the ethical, legal, and social implications of their work (Moghimi et al., 2005). Understanding coordination chemistry, synthesis, and encapsulating methods is crucial for the development of nanomaterials with theranostics applications. It is crucial to ensure strong compatibility among the core, targeting agent, diagnostic moiety, and active medication in the development of new nanomaterials. Through the improvement of diagnostic and therapeutic outcomes, the utilization of targeted imaging agents can assist in determining the most effective treatment approach for individual patients (Sarkar et al. 2015; Hua et al. 2018).

11.6 Conclusion

Nanotheranostic agents have revolutionized the diagnosis and treatment of various diseases. These compounds can be modified to specifically target certain reactions, which could potentially enhance the effectiveness of molecular imaging and treatment for cancer. However, due to the presence of nanoparticles in living systems, concerns about their long-term suitability persist. The healthcare industry and the electronics sector are actively seeking solutions to the challenges involved in designing Nanomedicine, by integrating electronic devices with cellular environments and biological materials. There is potential for the advancement of rapid and reliable technologies using nanotheranostic tools to enable early diagnosis and non-invasive post-diagnostic monitoring of prostate cancer. This could be a promising solution to the challenges faced in managing prostate cancer. Additional efforts are needed to tackle the significant obstacles, especially when it comes to implementing nanotheranostics in the clinical management of prostate cancer.

References

Abbasi E, Aval SF, Akbarzadeh A, Milani M, Nasrabadi HT, Joo SW, Hanifehpour Y, Nejati-Koshki K, Pashaei-Asl R (2014) Dendrimers: synthesis, applications, and properties. Nanoscale Res Lett 9:1–10

Ahmed HU, El-Shater Bosaily A, Brown LC, Gabe R, Kaplan R, Parmar MK et al (2017) Diagnostic accuracy of multi-parametric MRI and TRUS biopsy in prostate cancer (PROMIS): a paired validating confirmatory study. Lancet 389:815–822. https://doi.org/10.1016/S0140-6736(16)32401-1

Aina OH, Liu R, Sutcliffe JL, Marik J, Pan C-X, Lam KS (2007) From combinatorial chemistry to cancer-targeting peptides. Mol Pharm 4:631–651. https://doi.org/10.1021/mp700073y

Akanda MH, Rai R, Slipper IJ, Chowdhry BZ, Lamprou D, Getti G, Douroumis D (2015) Delivery of retinoic acid to LNCap human prostate cancer cells using solid lipid nanoparticles. Int J Pharm 493:161–171

Akbarzadeh A, Rezaei-Sadabady R, Davaran S, Joo SW, Zarghami N, Hanifehpour Y, Samiei M, Kouhi M, Nejati-Koshki K (2013) Liposome: classification, preparation, and applications. Nanoscale Res Lett 8:102

Akpa PA, Peter IE, Onwuka AM, Obi BC, Akunne MO, Nworu CS, Ejikeme PM, Akunne TC, Attama AA, Akah PA (2023) Nanotheranostics: platforms, current applications, and mechanisms of targeting in breast and prostate cancers. J Nanotheranostics 4(3):346–383. https://doi.org/10.3390/jnt4030016

Al Kobaisi M, Bhosale SV, Latham K, Raynor AM, Bhosale SV (2016) Functional naphthalene diimides: synthesis, properties, and applications. Chem Rev 116:11685–11796. https://doi.org/10.1021/acs.chemrev.6b00160

Alam IS, Arrowsmith RL, Cortezon-Tamarit F, Twyman F, Kociok-Köhn G, Botchway SW et al (2016) Microwave gallium-68 radiochemistry for kinetically stable bis(thiosemicarbazone) complexes: structural investigations and cellular uptake under hypoxia. Dalton Trans 45:144–155. https://doi.org/10.1039/C5DT02537K

Anderson CJ, Welch MJ (1999) Radiometal-labeled agents (non-technetium) for diagnostic imaging. Chem Rev 99:2219–2234. https://doi.org/10.1021/cr980451q

Asem H, Malmström E (2018) Polymeric nanoparticles explored for drug-delivery applications. In: ACS symposium series. American Chemical Society, Washington, pp 315–331

Azzazy HM, Mansour MM, Kazmierczak SC (2006) Nano diagnostics: a new frontier for clinical laboratory medicine. Clin Chem 52:1238–1246. https://doi.org/10.1373/clinchem.2006.066654

Banthia P, Gambhir L, Sharma A, Daga D, Kapoor N, Chaudhary R, Sharma G (2022) Nano to rescue: repository of nanocarriers for targeted drug delivery to curb breast cancer. 3 Biotech 12:1–23

Barbezan AB, Rosero WA, Perez DV, Rigo ME, Da Silva GD, Rodrigues AA, De Almeida LF, Da Silva FF, Rivera AG, Da Silva NG, Bernardes ES, Zeituni CA, Rostelato ME (2024)

Radioactive gold nanoparticles coated with BSA: a promising approach for prostate cancer treatment. Nanotheranostics 8(1):112–126. https://doi.org/10.7150/ntno.91507

Barkat MA, Harshita BS, Naim MJ, Pottoo FH, Singh SP et al (2018) Current progress in synthesis, characterization and applications of silver nanoparticles: precepts and prospects. Recent Pat Antiinfect Drug Discov 13(1):53–69. https://doi.org/10.2174/1574891X12666171006102833

Barry MJ (2001) Prostate-specific-antigen testing for early diagnosis of prostate cancer. N Engl J Med 344:1373–1377. https://doi.org/10.1056/NEJM200105033441806

Beer M, Montani M, Gerhardt J, Wild PJ, Hany TF, Hermanns T et al (2012) Profiling gastrin-releasing peptide receptor in prostate tissues: clinical implications and molecular correlates. Prostate 72:318–325. https://doi.org/10.1002/pros.21434

Berger A (2003) Positron emission tomography. BMJ 326:1449. https://doi.org/10.1136/bmj.326.7404.1449

Bernsen MR, Kooiman K, Segbers M, van Leeuwen FWB, de Jong M (2015) Biomarkers in pre-clinical cancer imaging. Eur J Nucl Med Mol Imaging 42:579–596. https://doi.org/10.1007/s00259-014-2980-7

Bouchelouche K, Turkbey B, Choyke P, Capala J (2010) Imaging prostate cancer: an update on positron emission tomography and magnetic resonance imaging. Curr Urol Rep 11:180–190. https://doi.org/10.1007/s11934-010-0105-9

Bulte JWM, Modo MMJ (2017) Design and applications of nanoparticles in biomedical imaging. Springer, Berlin

Burke AR, Singh RN, Carroll DL, Wood JCS, D'Agostino RB, Ajayan PM, Torti FM, Torti SV (2012) The resistance of breast cancer stem cells to conventional hyperthermia and their sensitivity to nanoparticle-mediated photothermal therapy. Biomaterials 33:2961–2970

Cai W, Gao T, Hong H, Sun J (2008) Applications of gold nanoparticles in cancer nanotechnology. Nanotechnol Sci Appl 1:17–32. https://doi.org/10.2147/NSA.S3788

Cai H, Xie F, Mulgaonkar A, Chen L, Sun X, Hsieh J-T et al (2018) Bombesin functionalized Cu-copper sulfide nanoparticles for targeted imaging of orthotopic prostate cancer. Nanomedicine 13(14). https://doi.org/10.2217/nnm-2018-0062

Carvalho MR, Reis RL, Oliveira JM (2020) Dendrimer nanoparticles for colorectal cancer applications. J Mater Chem B 8:1128–1138

Chang SS (2004) Overview of prostate-specific membrane antigen. Rev Urol 6:S13–S18

Chavda VP, Khadela A, Shah Y, Postwala H, Balar P, Vora L (2023) Current status of cancer nanotheranostics: emerging strategies for cancer management. Nanotheranostics 7(4):368–379. https://doi.org/10.7150/ntno.82263

Chen G, Roy I, Yang C, Prasad PN (2016) Nanochemistry and nanomedicine for nanoparticle-based diagnostics and therapy. Chem Rev 116:2826–2885. https://doi.org/10.1021/acs.chemrev.5b00148

Chen T, Ren L, Liu X, Zhou M, Li L, Xu J et al (2018) DNA nanotechnology for cancer diagnosis and therapy. I J Mol Sci 19:1671. https://doi.org/10.3390/ijms19061671

Cortezon-Tamarit F, Sarpaki S, Calatayud DG, Mirabello V, Pascu SI (2016) Applications of "hot" and "cold" bis(thiosemicarbazonato) metal complexes in multimodal imaging. Chem Rec 16:1380–1397. https://doi.org/10.1002/tcr.201500292

Daar AS, Singer PA, Leah Persad D, Pramming SK, Matthews DR, Beaglehole R et al (2007) Grand challenges in chronic non-communicable diseases. Nature 450:494–496. https://doi.org/10.1038/450494a

Damborský P, Švitel J, Katrlík J (2016) Optical biosensors. Essays Biochem 60:91–100. https://doi.org/10.1042/EBC20150010

Das M, Datir SR, Singh RP, Jain S (2013) Augmented anticancer activity of a targeted, intracellularly activatable, theranostic nanomedicine based a fluorescent and radiolabeled, methotrxate-folic acid-multiwalled carbon nanotube conjugate. Mol Pharm 10:2543–2557

de Rooij M, Hamoen EHJ, Fütterer JJ, Barentsz JO, Rovers MM (2014) Accuracy of multiparametric MRI for prostate cancer detection: a meta-analysis. Am J Roentgenol 202:343–351. https://doi.org/10.2214/AJR.13.11046

Dearling JL, Lewis JS, Mullen GE, Welch MJ, Blower PJ (2002) Copper bis(thiosemicarbazone) complexes as hypoxia imaging agents: structure-activity relationships. J Biol Inorg Chem 7:249–259. https://doi.org/10.1007/s007750100291

Devillers M, Ahmad L, Korri-Youssoufi H, Salmon L (2017) Carbohydrate-based electrochemical biosensor for detection of a cancer biomarker in human plasma. Biosens Bioelectron 96:178–185. https://doi.org/10.1016/j.bios.2017.04.031

Dong Z, Xue X, Liang H, Guan J, Chang L (2021) DNA nanomachines for identifying cancer biomarkers in body fluids and cells. Anal Chem 93:1855–1865. https://doi.org/10.1021/acs.analchem.0c03518

Eastham J (2017) Prostate cancer screening. Investig Clin Urol 58:217–219. https://doi.org/10.4111/icu.2017.58.4.217

Elshafae SM, Hassan BB, Supsavhad W, Dirksen WP, Camiener RY, Ding H et al (2016) Gastrin-releasing peptide receptor (GRPr) promotes EMT, growth, and invasion in canine prostate cancer. Prostate 76:796–809. https://doi.org/10.1002/pros.23154

Etheridge ML, Campbell SA, Erdman AG, Haynes CL, Wolf SM, McCullough J (2013) The big picture on nanomedicine: the state of investigational and approved nanomedicine products. Nanomedicine 9:1–14. https://doi.org/10.1016/j.nano.2012.05.013

Evans JD, Jethwa KR, Ost P, Williams S, Kwon ED, Lowe VJ et al (2018) Prostate cancer-specific PET radiotracers: a review on the clinical utility in recurrent disease. Pract Radiat Oncol 8:28–39. https://doi.org/10.1016/j.prro.2017.07.011

Falzone L, Salomone S, Libra M (2018) Evolution of cancer pharmacological treatments at the turn of the third millennium. Front Pharmacol 9:1300

Fass L (2008) Imaging and cancer: a review. Mol Oncol 2:115–152. https://doi.org/10.1016/j.molonc.2008.04.001

Fathi M, Mozafari M, Mohebbi M (2012) Nanoencapsulation of food ingredients using lipid based delivery systems. Trends Food Sci Technol 23:13–27

Ferreira M, Sousa J, Pais A, Vitorino C (2020) The role of magnetic nanoparticles in cancer nanotheranostics. Materials 13:266

Frangioni JV (2008) New technologies for human cancer imaging. J Clin Oncol 26:4012–4021. https://doi.org/10.1200/JCO.2007.14.3065

Fubini B, Ghiazza M, Fenoglio I (2010) Physio-chemical features of engineered nanoparticles relevant to their toxicity. Nanotoxicology 4:347–363

Gardel ML (2013) Synthetic polymers with biological rigidity. Nature 493:619

Gaspar D, Veiga AS, Castanho MARB (2013) From antimicrobial to anticancer peptides. A review. Front Microbiol 4. https://doi.org/10.3389/fmicb.2013.00294

Granada-Ramirez DA, Arias-Ceron JS, Rodriguez-Fragoso P, Vazquez-Hernandez F, Luna-Arias JP, Herrera-Perez JL, Mendoza-Alvarez JG (2018) Quantum dots for biomedical applications. In: Narayan R (ed) Nanobiomaterials. Woodhead Publishing, Sawston, pp 411–436

Griffiths KE, Stoddart JF (2008) Template-directed synthesis of donor/acceptor [2] catenanes and [2] rotaxanes. Pure Appl Chem 80:485–506. https://doi.org/10.1351/pac200880030485

Grippin AJ, Wummer B, Wildes T, Dyson K, Trivedi V, Yang C, Sebastian M, Mendez-Gomez HR, Padala S, Grubb M et al (2019) Dendritic cell-activating magnetic nanoparticles enable early prediction of antitumor response with magnetic resonance imaging. ACS Nano 13:13884–13898

Grossman JH, McNeil SE (2012) Nanotechnology in cancer medicine. Phys Today 65:38–42

Guan S (2012) Statistical designs for early phases of cancer clinical trials. J Biopharm Stat 22:1109–1126. https://doi.org/10.1080/10543406.2011.571801

Guo Y, Wang X-Y, Chen Y-L, Liu F-Q, Tan M-X, Ao M, Yu J-H, Ran H-T, Wang Z-X (2018) A light-controllable specific drug delivery nanoplatform for targeted bimodal imaging-guided photothermal/chemo synergistic cancer therapy. Acta Biomater 80:308–326

Gupta D, Roy P, Sharma R, Kasana R, Rathore P, Gupta TK (2024) Recent nanotheranostic approaches in cancer research. Clin Exp Med 24(1):8. https://doi.org/10.1007/s10238-023-01262-3

Han Z, Li Y, Roelle S, Zhou Z, Liu Y, Sabatelle R et al (2017) Targeted contrast agent specific to an oncoprotein in tumor microenvironment with the potential for detection and risk stratification of prostate cancer with MRI. Bioconjug Chem 28:1031–1040. https://doi.org/10.1021/acs.bioconjchem.6b00719

Hofmarcher T, Lindgren P, Wilking N, Jönsson B (2020) The cost of cancer in Europe 2018. Eur J Cancer 129:41–49. https://doi.org/10.1016/j.ejca.2020.01.011

Hua S, de Matos MBC, Metselaar JM, Storm G (2018) Current trends and challenges in the clinical translation of nanoparticulate nanomedicines: pathways for translational development and commercialization. Front Pharmacol 9. https://doi.org/10.3389/fphar.2018.00790

Huang HK, Yan J, Liu P, Zhao BY, Cao Y, Zhang XF (2017) A novel cancer nanotheranostics system based on quantum dots encapsulated by a polymer-prodrug with controlled release behavior. Aust J Chem 70:1302–1311

Ibau C, Arshad MK, Gopinath SCB, Nuzaihan MNM, Fathil MF, Estrela P (2019) Gold interdigitated triple-microelectrodes for label-free prognosticative aptasensing of prostate cancer biomarker in serum. Biosens Bioelectron 136:118–127. https://doi.org/10.1016/j.bios.2019.04.048

Imlimthan S, Otaru S, Keinänen O, Correia A, Lintinen K, Santos HA et al (2018) Radiolabeled molecular imaging probes for the in vivo evaluation of cellulose nanocrystals for biomedical applications. Biomacromolecules 20:674–683. https://doi.org/10.1021/acs.biomac.8b01313

Jadvar H (2013) Molecular imaging of prostate cancer with PET. J Nucl Med 54:1685–1688. https://doi.org/10.2967/jnumed.113.126094

Jahangir MA, Gilani SJ, Muheem A, Jafar M, Aslam M, Ansari MT et al (2019) Quantum dots: next generation of smart nano-systems. Pharm Nanotechnol 7(3):234–245. https://doi.org/10.2174/2211738507666190429113906

Karpuz M, Silindir-Gunay M, Ozer AY, Ozturk SC, Yanik H, Tuncel M, Aydin C, Esendagli G (2020) Diagnostic and therapeutic evaluation of folate-targeted paclitaxel and vinorelbine encapsulating theranostic liposomes for non-small cell lung cancer. Eur J Pharm Sci 156:105576

Key TJ (2011) Fruit and vegetables and cancer risk. Br J Cancer 104:6–11. https://doi.org/10.1038/sj.bjc.6606032

Key J, Dhawan D, Cooper CL, Knapp DW, Kim K, Kwon IC et al (2016) Multicomponent, peptide-targeted glycol chitosan nanoparticles containing ferrimagnetic iron oxide nanocubes for bladder cancer multimodal imaging. Int J Nanomedicine 11:4141–4155. https://doi.org/10.2147/IJN.S109494

Kim D, Jeong YY, Jon S (2010) A drug-loaded aptamer−gold nanoparticle bioconjugate for combined CT imaging and therapy of prostate cancer. ACS Nano 4:3689–3696. https://doi.org/10.1021/nn901877h

Kröger APP, Hamelmann NM, Juan A, Lindhoud S, Paulusse JMJ (2018) Biocompatible single-chain polymer nanoparticles for drug delivery—a dual approach. ACS Appl Mater Interfaces 10:30946–30951

Ledet EM, Ernst EM, Schiff J, Lin S, Lewis BE, Sartor AO (2017) Germline variants and family history in caucasian and African-American prostate cancer. J Clin Oncol 35:e16548. https://doi.org/10.1200/JCO.2017.35.15_suppl.e16548

Lima AM, Pizzol CD, Monteiro FB, Creczynski-Pasa TB, Andrade GP, Ribeiro AO, Perussi JR (2013) Hypericin encapsulated in solid lipid nanoparticles: phototoxicity and photodynamic efficiency. J Photochem Photobiol B Biol 125:146–154

Lin M, Gao Y, Diefenbach TJ, Shen JK, Hornicek FJ, Park YI et al (2017) Facial layer-by-layer engineering of upconversion nanoparticles for gene delivery: near-infrared-initiated fluorescence resonance energy transfer tracking and overcoming drug resistance in ovarian cancer. ACS Appl Mater Interfaces 9:7941–7949. https://doi.org/10.1021/acsami.6b15321

Lin Y, Zhao X, Miao Z, Ling Z, Wei X, Pu J et al (2020) Data-driven translational prostate cancer research: from biomarker discovery to clinical decision. J Transl Med 18:119. https://doi.org/10.1186/s12967-020-02281-4

Liu Z, Chen K, Davis C, Sherlock S, Cao Q, Chen X, Dai H (2008) Drug delivery with carbon nanotubes for in vivo cancer treatment. Cancer Res 68:6652–6660

Lobana TS, Sharma R, Bawa G, Khanna S (2009) Bonding and structure trends of thiosemicarbazone derivatives of metals-an overview. Coord Chem Rev 253:977–1055. https://doi.org/10.1016/j.ccr.2008.07.004

Luo S, Zhang E, Su Y, Cheng T, Shi C (2011) A review of NIR dyes in cancer targeting and imaging. Biomaterials 32:7127–7138. https://doi.org/10.1016/j.biomaterials.2011.06.024

Luo Y-H, Chang LW, Lin P (2015) Metal-based nanoparticles and the immune system: activation, inflammation, and potential applications. Biomed Res Int 2015:143720. https://doi.org/10.1155/2015/143720

Luo D, Johnson A, Wang X, Li H, Erokwu BO, Springer S, Lou J, Ramamurthy G, Flask CA, Burda C et al (2020) Targeted radiosensitizers for MR-guided radiation therapy of prostate cancer. Nano Lett 20:7159–7167

Lymperopoulos G, Lymperopoulos P, Alikari V, Dafogianni C, Zyga S, Margari N (2017) Application of theranostics in oncology. Adv Exp Med Biol 989:119–128. https://doi.org/10.1007/978-3-319-57348-9_10

Malam Y, Loizidou M, Seifalian AM (2009) Liposomes and nanoparticles: nanosized vehicles for drug delivery in cancer. Trends Pharmacol Sci 30:592–599. https://doi.org/10.1016/j.tips.2009.08.004

Mansur AA, Caires AJ, Carvalho SM, Capanema NS, Carvalho IC, Mansur HS (2019) Dual-functional supramolecular nanohybrids of quantum dot/biopolymer/chemotherapeutic drug for bioimaging and killing brain cancer cells in vitro. Colloids Surf B Biointerfaces 184:110507

Marqus S, Pirogova E, Piva TJ (2017) Evaluation of the use of therapeutic peptides for cancer treatment. J Biomed Sci 24:21. https://doi.org/10.1186/s12929-017-0328-x

McKittrick J, Shea-Rohwer LE (2014) Review: Down conversion materials for solid-state lighting. J Am Ceram Soc 97:1327–1352. https://doi.org/10.1111/jace.12943

Mehrabani S, Maker A, Armani A (2014) Hybrid integrated label-free chemical and biological sensors. Sensors 14:5890–5928. https://doi.org/10.3390/s140405890

Melancon MP, Stafford RJ, Li C (2012) Challenges to effective cancer nanotheranostics. J Control Release 164:177–182

Mirzaei H, Shakeri A, Rashidi B, Jalili A, Banikazemi Z, Sahebkar A (2017) Phytosomal curcumin: a review of pharmacokinetic, experimental and clinical studies. Biomed Pharmacother 85:102–112. https://doi.org/10.1016/j.biopha.2016.11.098

Moghimi SM, Hunter AC, Murray JC (2005) Nanomedicine: current status and future prospects. FASEB J 19:311–330

Murphy P (2010) Imaging in clinical trials. Cancer Imaging 10:S74–S82. https://doi.org/10.1102/1470-7330.2010.9027

Nottelet B, Darcos V, Coudane J (2015) Aliphatic polyesters for medical imaging and theranostic applications. Eur J Pharm Biopharm 97:350–370

Nune SK, Gunda P, Thallapally PK, Lin Y-Y, Laird Forrest M, Berkland CJ (2009) Nanoparticles for biomedical imaging. Expert Opin Drug Deliv 6:1175–1194. https://doi.org/10.1517/17425240903229031

Palmerston Mendes L, Pan J, Torchilin VP (2017) Dendrimers as nanocarriers for nucleic acid and drug delivery in cancer therapy. Molecules 22:1401

Parry CD, Patra J, Rehm J (2011) Alcohol consumption and non-communicable diseases: epidemiology and policy implications. Addiction 106:1718–1724. https://doi.org/10.1111/j.1360-0443.2011.03605.x

Paterson BM, Donnelly PS (2011) Copper complexes of bis(thiosemicarbazones): from chemotherapeutics to diagnostic and therapeutic radiopharmaceuticals. Chem Soc Rev 40:3005–3018. https://doi.org/10.1039/C0CS00215A

Paterson BM, Roselt P, Denoyer D, Cullinane C, Binns D, Noonan W et al (2014) PET imaging of tumours with a 64Cu labeled macrobicyclic cage amine ligand tethered to Tyr3-octreotate. Dalton Trans 43:1386–1396. https://doi.org/10.1039/C3DT52647J

Peng C-W, Li Y (2010) Application of quantum dots-based biotechnology in cancer diagnosis: current status and future perspectives. J Nanomater 1–11. https://doi.org/10.1155/2010/676839

Pengo P, Pantoş GD, Otto S, Sanders JKM (2006) Efficient and mild microwave-assisted stepwise functionalization of naphthalenediimide with α-amino acids. J Org Chem 71:7063–7066. https://doi.org/10.1021/jo061195h

Plichta KA, Graves SA, Buatti JM (2021) Prostate-specific membrane antigen (PSMA) theranostics for treatment of oligometastatic prostate cancer. Int J Mol Sci 22:12095. https://doi.org/10.3390/ijms222212095

Pohanka M, Skládal P (2008) Electrochemical biosensors—principles and applications. J Appl Biomed 6:57–64. https://doi.org/10.32725/jab.2008.008

Pooja D, Gunukula A, Gupta N, Adams DJ, Kulhari H (2019) Bombesin receptors as potential targets for anticancer drug delivery and imaging. Int J Biochem Cell Biol 114:105567. https://doi.org/10.1016/j.biocel.2019.105567

Probst CE, Zrazhevskiy P, Bagalkot V, Gao X (2013) Quantum dots as a platform for nanoparticle drug delivery vehicle design. Adv Drug Deliv Rev 65:703–718. https://doi.org/10.1016/j.addr.2012.09.036

Puri A, Loomis K, Smith B, Lee J-H, Yavlovich A, Heldman E et al (2009) Lipid-based nanoparticles as pharmaceutical drug carriers: from concepts to clinic. Crit Rev Ther Drug Carrier Syst 26:523–580. https://doi.org/10.1615/critrevtherdrugcarriersyst.v26.i6.10

Rais-Bahrami S, Siddiqui MM, Turkbey B, Stamatakis L, Logan J, Hoang AN et al (2013) Utility of multiparametric magnetic resonance imaging suspicion levels for detecting prostate cancer. J Urol 190:1721–1727. https://doi.org/10.1016/j.juro.2013.05.052

Răsădean DM, Sheng B, Dash J, Pantoş GD (2017) Amino-acid-derived naphthalenediimides as versatile G-quadruplex binders. Chem Eur J 23:8491–8499. https://doi.org/10.1002/chem.201700957

Rawla P (2019) Epidemiology of prostate cancer. World J Oncol 10:63–89. https://doi.org/10.14740/wjon1191

Rehman AU, Akram S, Seralin A, Vandamme T, Anton N (2020) Lipid nanocarriers: formulation, properties, and applications. In: Smart nanocontainers. Elsevier, Amsterdam, pp 355–382

Rudin M (2005) Molecular imaging. In: Basic principles and applications in biomedical research. University of Zürich, pp 1–541. https://doi.org/10.1142/P379

Saini R, Saini S, Sharma S (2010) Nanotechnology: the future medicine. J Cutan Aesthet Surg 3:32–33. https://doi.org/10.4103/0974-2077.63301

Samimi S, Ardestani MS, Dorkoosh FA (2021) Preparation of carbon quantum dots-quinic acid for drug delivery of gemcitabine to breast cancer cells. J Drug Deliv Sci Technol 61:102287

Sancho V, Di Florio A, Moody TW, Jensen RT (2011) Bombesin receptor-mediated imaging and cytotoxicity: review and current status. Curr Drug Deliv 8:79–134. https://doi.org/10.2174/156720111793663624

Sarkar S, Sautier L, Schilling G, Bokemeyer C, Koch U, Mehnert A (2015) Anxiety and fear of cancer recurrence and its association with supportive care needs and health-care service utilization in cancer patients. J Cancer Surviv 9:567–575. https://doi.org/10.1007/s11764-015-0434-2

Satalkar P, Elger BS, Shaw DM (2016) Defining nano, nanotechnology and nanomedicine: why should it matter? Sci Eng Ethics 22:1255–1276. https://doi.org/10.1007/s11948-015-9705-6

Scher HI, Halabi S, Tannock I, Morris M, Sternberg CN, Carducci MA et al (2008) Design and end points of clinical trials for patients with progressive prostate cancer and castrate levels of testosterone: recommendations of the prostate cancer clinical trials working group. J Clin Oncol 26:1148–1159. https://doi.org/10.1200/JCO.2007.12.4487

Sedgwick P (2011) Survival (time to event) data: censored observations. BMJ 343:d4816. https://doi.org/10.1136/bmj.d4816

Shao W, Paul A, Rodes L, Prakash S (2015) A new carbon nanotube-based breast cancer drug delivery system: preparation and in vitro analysis using paclitaxel. Cell Biochem Biophys 71:1405–1414

Sharma AN, Upadhyay PK, Dewangan HK (2022) Development, evaluation, pharmacokinetic and biodistribution estimation of resveratrol-loaded solid lipid nanoparticles for prostate cancer targeting. J Microencapsul 39:563–574

Shi J, Kantoff PW, Wooster R, Farokhzad OC (2016) Cancer nanomedicine: progress, challenges and opportunities. Nat Rev Cancer 17:20–37

Shitole AA, Sharma N, Giram P, Khandwekar A, Baruah M, Garnaik B, Koratkar S (2020) LHRH-conjugated, PEGylated, poly-lactide-co-glycolide nanocapsules for targeted delivery of combinational chemotherapeutic drugs Docetaxel and Quercetin for prostate cancer. Mater Sci Eng C 114:111035

Siegel RL, Miller KD, Fuchs HE, Jemal A (2021) Cancer statistics, 2021. CA Cancer J Clin 71(1):7–33. https://doi.org/10.3322/caac.21654

Singh P, Pandit S, Mokkapati VRSS, Garg A, Ravikumar V, Mijakovic I (2018) Gold nanoparticles in diagnostics and therapeutics for human cancer. Int J Mol Sci 19:1979. https://doi.org/10.3390/ijms19071979

Sturzu A, Sheikh S, Echner H, Nägele T, Deeg M, Amin B et al (2014) Rhodamine-marked bombesin: a novel means for prostate cancer fluorescence imaging. Investig New Drugs 32:37–46. https://doi.org/10.1007/s10637-013-9975-2

Sweet MJ, Chessher A, Singleton I (2012) Review: Metal-based nanoparticles; size, function, and areas for advancement in applied microbiology. Adv Appl Microbiol 80:113–142. https://doi.org/10.1016/B978-0-12-394381-1.00005-2

Takayanagi R, Nakamura Y, Nakajima Y, Shimizu A, Nakamura H, Yamada Y et al (2004) Analysis of information submitted by clinical trial sponsors regarding the safety of investigational drugs. Yakugaku Zasshi 124:225–229. https://doi.org/10.1248/yakushi.124.225

Tegler G, Ericson K, Sörensen J, Björck M, Wanhainen A (2012) Inflammation in the walls of asymptomatic abdominal aortic aneurysms is not associated with increased metabolic activity detectable by 18-fluorodeoxglucose positron-emission tomography. J Vasc Surg 56:802–807. https://doi.org/10.1016/j.jvs.2012.02.024

Thalacker C, Röger C, Würthner F (2006) Synthesis and optical and redox properties of core-substituted naphthalene diimide dyes. J Org Chem 71:8098–8105. https://doi.org/10.1021/jo0612269

Tian J-Y, Guo F-J, Zheng G-Y, Ahmad A (2018) Prostate cancer: updates on current strategies for screening, diagnosis and clinical implications of treatment modalities. Carcinogenesis 39:307–317. https://doi.org/10.1093/carcin/bgx141

Topkaya SN, Azimzadeh M, Ozsoz M (2016) Electrochemical biosensors for cancer biomarkers detection: recent advances and challenges. Electroanalysis 28:1402–1419. https://doi.org/10.1002/elan.201501174

Townsend DW (2008) Combined positron emission tomography-computed tomography: the historical perspective. Semin Ultrasound CT MRI 29:232–235. https://doi.org/10.1053/j.sult.2008.05.006

Tseng J-C, Narayanan N, Ho G, Groves K, Delaney J, Bao B et al (2017) Fluorescence imaging of bombesin and transferrin receptor expression is comparable to 18F-FDG PET in early detection of sorafenib-induced changes in tumor metabolism. PLoS One 12:e0182689. https://doi.org/10.1371/journal.pone.0182689

Ulmert D, Evans MJ, Holland JP, Rice SL, Wongvipat J, Pettersson K et al (2012) Imaging androgen receptor signaling with a radiotracer targeting free prostate-specific antigen. Cancer Discov 2:320–327. https://doi.org/10.1158/2159-8290.cd-11-0316

Van de Wiele C, Dumont F, Vanden Broecke R, Oosterlinck W, Cocquyt V, Serreyn R et al (2000) Technetium-99m RP527, a GRP analogue for visualisation of GRP receptor-expressing malignancies: a feasibility study. Eur J Nucl Med 27:1694–1699. https://doi.org/10.1007/s002590000355

Verhamme KMC, Dieleman JP, Bleumink GS, van der Lei J, Sturkenboom MCJM (2002) Incidence and prevalence of lower urinary tract symptoms suggestive of benign prostatic hyperplasia in primary care-the triumph project. Eur Urol 42:323–328. https://doi.org/10.1016/S0302-2838(02)00354-8

von Eyben FE, Baumann GS, Baum RP (2018) PSMA diagnostics and treatments of prostate cancer become mature. Clin Translat Imaging 6:145–148. https://doi.org/10.1007/s40336-018-0270-2

Wang F, Li Y, Shen Y, Wang A, Wang S, Xie T (2013) The functions and applications of RGD in tumor therapy and tissue engineering. Int J Mol Sci 14:13447–13462. https://doi.org/10.3390/ijms140713447

Wong HL, Bendayan R, Rauth AM, Li Y, Wu XY (2007) Chemotherapy with anticancer drugs encapsulated in solid lipid nanoparticles. Adv Drug Deliv Rev 59:491–504

World Health Organization (2021) Noncommunicable diseases: mortality. Available at: https://www.who.int/data/gho/data/themes/topics/topic-details/GHO/ncd-mortality. Accessed 24 Nov 2021

Xu H, Jiang S, Wang J, Li X, Wu T, Xu P, Santos-Oliveira R, Zhang A (2020) Radioactive gold nanoparticle in two forms (19879Au GNPs and 99mTc-GNPs) for lung cancer antiproliferative induction and intralesional imaging: a proof of concept. Anti Cancer Agents Med Chem 20:1648–1653

Yan X, Yang Y, Sun Y (2021) Dendrimer applications for cancer therapies. J Phys Conf Ser 1948:012205

Yankeelov TE, Mankoff DA, Schwartz LH, Lieberman FS, Buatti JM, Mountz JM et al (2016) Quantitative imaging in cancer clinical trials. Clin Cancer Res 22:284–290. https://doi.org/10.1158/1078-0432.ccr-14-3336

Yildirim H, Guler E, Yavuz M, Ozturk N, Kose Yaman P, Subasi E et al (2014) Ruthenium (II) complexes of thiosemicarbazone: synthesis, biosensor applications and evaluation as antimicrobial agents. Mater Sci Eng C 44:1–8. https://doi.org/10.1016/j.msec.2014.08.007

Yu F, Zhu M, Li N, Ao M, Li Y, Zhong M, Yuan Q, Chen H, Fan Z, Wang Y et al (2020) Imaging-guided synergistic targeting-promoted photo-chemotherapy against cancers by methotrexate-conjugated hyaluronic acid nanoparticles. Chem Eng J 380:122426

Zappavigna S, Abate M, Cossu AM, Lusa S, Campani V, Scotti L et al (2019) Urotensin-II-targeted liposomes as a new drug delivery system towards prostate and colon cancer cells. J Oncol 2019:9293560. https://doi.org/10.1155/2019/9293560

Zhang X-X, Eden HS, Chen X (2012) Peptides in cancer nano medicine: drug carriers, targeting ligands and protease substrates. J Control Release 159:2–13. https://doi.org/10.1016/j.jconrel.2011.10.023

Zhang W, Song Y, Eldi P, Guo X, Hayball J, Garg S et al (2018) Targeting prostate cancer cells with hybrid elastin-like polypeptide/liposome nanoparticles. Int J Nanomedicine 13:293–305. https://doi.org/10.2147/IJN.S152485

Zhu Y, Wang X, Chen J, Zhang J, Meng F, Deng C, Cheng R, Feijen J, Zhong Z (2016) Bio responsive and fluorescent hyaluronic acid-iodixanol nanogels for targeted X-ray computed tomography imaging and chemotherapy of breast tumors. J Control Release 244:229–239

Theranostics as an Emerging Integrated Imaging and Therapies in Precision Oncology

12

Mohamad Taleuzzaman, Sanjay Chauhan, Himanshu Singh, and Sumayya Khan

Abstract

Theranostics is the combination of screening biomarkers and curative agents that share a common target in cancerous cells or tissues. Scanning has progressed from an important clinical tool to a current and influential reference for improving disease categorization, patient selection, forecasting of therapeutic efficacy and tissue toxicity, and diagnosis as a core component of care delivery, often enabling the development of new medications and therapies. The nuclear approach for the treatment refers to use of radioactive molecules to photo pharmacologic occurrences by expressing specific cancer targets, including cell surface receptors or cellular carriers, and then using specifically designed agents to deliver ionizing radioactivity to the tissues that express these targets. The nuclear theranostic initiative has also ignited increase attention and significance in conjunction with the development of molecular diagnostics and personalised medicine. It assists in providing customized monitoring for a range of diseases, enhancing health choices, forecasts of response, toxicological, disease prediction determination, attempting to avoid unproductive and expensive diagnostic evaluations for a wide range of diseases. This chapter aims to provide an overview of theranostic approaches in nuclear medicine for cancer diagnosis and treatment.

M. Taleuzzaman (✉)
Department of Pharmaceutical Chemistry, Faculty of Pharmacy, Maulana Azad University, Jodhpur, Rajasthan, India

S. Chauhan · H. Singh
Institute of Pharmacy, Ram-Eesh Institute of Vocational and Technical Education, Greater Noida, Uttar Pradesh, India

S. Khan
Department of Pharmacology, Faculty of Pharmacy, Maulana Azad University, Jodhpur, Rajasthan, India

M. A. Barkat et al. (eds.), *Nanotheranostics for Diagnosis and Therapy*,
https://doi.org/10.1007/978-981-97-3115-2_12

Keywords

Theranostics · Oncology · Targeted therapy · Diagnostic · Toxicological

12.1 Introduction

A new area of precision oncology called "theranostics" combines therapeutic and diagnostic techniques to offer cancer patients individualized treatment plans. It involves the use of imaging technologies to diagnose and monitor cancer, as well as use of targeted therapies to treat the disease. It also involves the use of imaging modalities to identify specific molecular targets or biomarkers in cancer cells, and then using targeted therapies to attack these specific targets (Xue et al. 2021; Kashyap et al. 2023). In traditional oncology, the treatment of cancer usually requires chemotherapy, radiation therapy, or surgery, it might not be beneficial for all patients and can have serious adverse effects (Liu et al. 2023). However, in precision oncology, theranostics is becoming an increasingly important tool. Doctors can determine the exact spot and scale of a patient's cancer by employing imaging procedures like magnetic resonance imaging (MRI) and positron emission tomography (PET). This information can then be used to determine the most appropriate treatment options (Krzyszczyk et al. 2018).

For example, theranostics can be used to identify specific biomarkers on cancer cells that can be targeted by specific therapies. By adopting imaging technique to confirm the presence of these biomarkers, doctors can then select the most appropriate therapy for the patient. One theranostic proceed towards the use of radiolabelled molecules which combine to specific biomarkers on cancer cells. These radiolabelled molecules can then be visualized using PET imaging to pick out position and level of the cancer. Furthermore, the same radiolabelled chemical can be applied as a targeted therapy to kill cancer cells just while leaving healthy tissue unharmed (Chiu and Yen 2023). One example of theranostics in action is use of radioactive tracers in PET imaging. These tracers can bind to specific proteins on cancer cells, allowing doctors to identify the location and extent of disease. Targeted treatments can be directly administered to cancer cells using the same tracers. The use of medicine delivery systems based on nanotechnology that can target particular cancer cells is another example. Physicians can use these devices to load both therapeutic and diagnostic chemicals, and they can track the efficacy of the treatment in real time (Marin et al. 2020).

In targeted theranostics, imaging techniques are used to locate the diseased cells or tissues, and then therapeutic agents are delivered to these specific areas to treat the diseased cells. By using this approach minimizes the side effects of treatment by focusing on only the diseased cells, rather than affecting healthy cells as well. Diseases like cancer, cardiovascular and neurological are being treated with targeted theranostics. It has capacity to revolutionize the way diseases are diagnosed and treated by giving more accurate and personalized treatment options (Kang et al. 2018).

Biological pretargeting, also known as pretargeting theranostics, is a technique that used in cancer treatment by delivering a targeting molecule (usually an antibody or aptamer) to the tumor site, followed by the administration of a radiolabelled or fluorescently labelled probe that binds to the targeting molecule. This two-step approach allows for the delivery of high doses of therapeutic or imaging agents to the tumor while minimizing their exposure to healthy tissues (Bai et al. 2023). The pretargeting approach works by exploiting the high individual and affinity of antibodies or aptamers for their target molecules. The targeting molecule is designed to bind to a specific receptor or antigen exposed on the surface of cancer cells, while the probe is designed to bind specifically to the targeting molecule (Bailly et al. 2017; Hapuarachchige and Artemov 2020, Fig. 12.1).

1. Pretargeting using images.
2. Step for delivering medication or radionuclides.
3. Internalization of the complex of components involved in pretargeting receptor delivery.
4. Therapeutic or radioactive material release within cells.

For ideal pretargeting theranostic approaches, the following conditions are required: (a) strong binding ability pretargeting element with elevated target expression of receptors. (b) The pretargeting component-receptor molecular complex's ideal uptake is a complex characteristic that depends on the circumstances of the tumor's microenvironment and must offer a sufficient window of opportunity for in situ interactions with a second delivery component. (c) Two components' affinity for adhering to the targeted cell surface. (d) Appropriate pairing of medicines and imaging agents/modalities for the medicinal aspects (Hapuarachchige and Artemov 2020).

Theranostics offers several advantages over traditional diagnostic and therapeutic approaches. Personalized therapy regimens that consider each patient's distinct

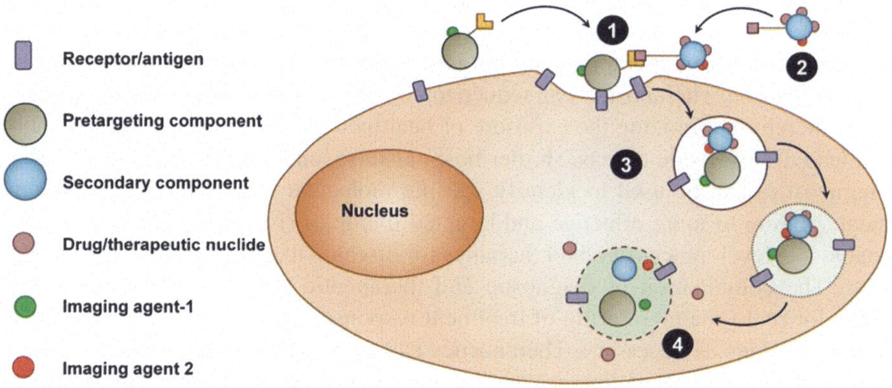

Fig. 12.1 An illustration of the pretargeting theranostic strategy (Hapuarachchige and Artemov 2020)

genetic and molecular composition are made possible by theranostics. By this idea can improve treatment outcomes and decrease the chance of adverse effects. Theranostics combines imaging techniques with targeted therapy, allowing for precise and accurate targeting of disease cells. Theranostics can reduce toxicity by targeting diseased cells specifically, thereby minimizing the open up of healthy cells to therapeutic agents. The use of imaging techniques in theranostics can identify disorder at an early stage, allowing for earlier intervention and improved outcomes. Theranostics can reduce healthcare costs by minimizing the need for repeat procedures and reducing the risk of adverse effects, resulting in shorter hospital stays and fewer complications (Chehelgerdi et al. 2023). While theranostics offers several advantages over traditional diagnostic and therapeutic approaches, there are also some limitations and challenges to consider. The success of theranostics depends on the specificity of the targeting agent used. If the targeting agent does not bind specifically to the intended target, the therapy may be ineffective or may harm healthy tissues (Jeelani et al. 2014). The personalized treatment is a key advantage of theranostics; patient heterogeneity can make it challenging to develop effective therapies for everyone. Differences in disease progression, genetics, and environmental factors can all affect treatment outcomes. The availability of imaging agents that can accurately detect specific disease markers is limited. This can limit the effectiveness of theranostics in some cases. The development and production of theranostic agents can be expensive, which can limit their availability and accessibility to patients. The regulatory approval process for theranostics can be challenging due to the combination of diagnostic and therapeutic agents. This can delay the availability of new therapies to patients (Kevadiya et al. 2019).

Theranostics has capacity to improve cancer treatment outcomes by providing a more precise and targeted approach to cancer therapy. It can also reduce the side effects associated with traditional cancer treatments and improve the quality patient's life. As the field of precision oncology continues to evolve, it is anticipated that theranostics will become more crucial in the creation of individualized cancer treatment plans.

Theranostics has significant implications and future possibilities for the field of medicine. Theranostics makes personalized and precision medicine possible by enabling customized therapies according to each patient's own genetic and molecular composition. Theranostics can reduce toxicity by targeting diseased cells specifically, thereby minimizing the exposure of healthy cells to therapeutic agents. This can lead to fewer side effects, shorter hospital stays, and reduced healthcare costs. Theranostics can be used to identify specific molecular targets for drug development, leading to more effective and targeted therapies (Fan et al. 2023). This can enable the development of novel therapies for diseases that are currently difficult to treat. The combination of diagnostic and therapeutic agents in theranostics can allow for real-time monitoring of treatment response, enabling clinicians to adjust treatment plans as necessary. Theranostics can be integrated with other technologies, such as artificial intelligence and robotics, to further enhance the precision and accuracy of disease diagnosis and therapies (Krzyszczyk et al. 2018; Kong et al. 2023). Figure 12.2 represents the theranostic imaging strategies.

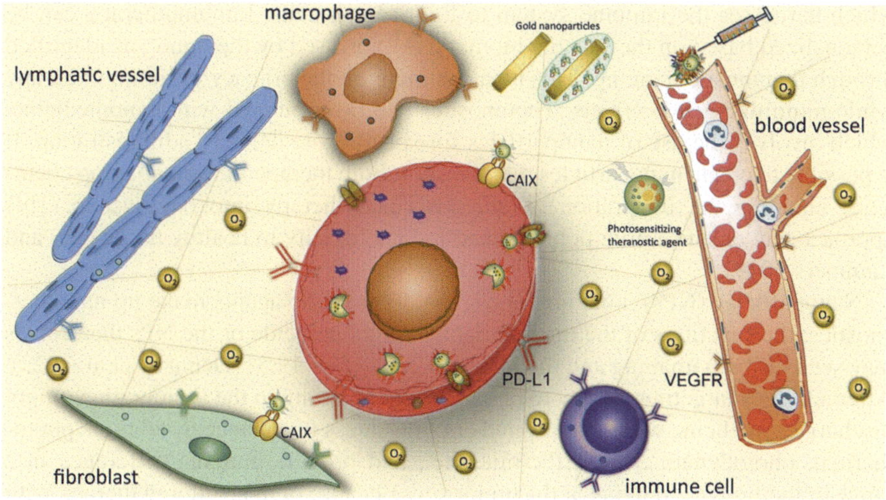

Fig. 12.2 Theranostic imaging strategies with a schematic representation of the tumor ecosystem (Bhujwalla et al. 2018; Kakkad et al. 2019)

In conclusion, theranostics is an emerging field that holds great promise for the future of precision oncology. By combining diagnostic and therapeutic approaches, doctors can provide more personalized and effective treatments for cancer patients. Overall, theranostics is a promising approach to disease diagnosis and treatment, but it is not without its limitations and challenges. As the field continues to evolve, researchers will need to address these challenges to ensure that theranostics can reach its full potential in improving patient outcomes.

12.2 Integrating Theranostics to Additional Methods of Diagnosis and Care

Theranostics can be combined with additional forms of diagnosis and therapy to further improve cancer care. Like, it can be combined with molecular imaging techniques, such as PET or SPECT, to provide a more accurate diagnosis and treatment monitoring. Molecular imaging allows clinicians to visualize specific biological processes within cells and tissues, which can help to identify early-stage tumors and track the response to therapy (Hosseini et al. 2023).

The theranostic techniques, which have been evolved lately, are multidisciplinary in nature. Integration with other diagnostic techniques and methods, like liquid biopsies and genetic testing, falls under this category. It makes sense to combine theranostic techniques with immunotherapy, chemotherapy, and external-beam radiation therapy; it is likely to be long until these combined applications are implemented (Marin et al. 2020). The theranostics can be combined with immunotherapy, which harnesses the immune system to fight cancer cells. Immunotherapy can be personalized based on the specific biomarkers expressed by the tumor, as identified through theranostic imaging. This idea can increase the efficacy of immunotherapy while minimizing side effects. Theranostics can be combined with nanomedicine, which involves the use of nanoparticles for specific site delivery administration. It is possible to create nanoparticles that transport both therapeutic and imaging chemicals, enabling the integration of diagnosis and therapy into one system. This approach can enhance drug potential and reduce toxicity to healthy tissue (Shi and Lammers 2019).

Nanoparticles (NPs), imaging labels, and therapeutic agents make up nanotheranostics; the functions of the three variables might coincide or the NPs themselves may serve as both imaging and therapeutic agents (Fig. 12.3) (Dennahy et al. 2022). After adding a targeting moiety to enable active targeting to the tumour, they might discharge a medicine within controlled circumstances to enhance irradiation, photothermal, photodynamic, hyperthermia, etc., and provide imaging difference in a range of methods of imaging at the tumor site. Because of this, tailored therapy with nanotheranostics might be directed in an extremely adaptable manner before successful cancer treatment (Sun et al. 2023). Chemotherapy, that employs high-energy radiation to kill cancer cells, is frequently paired with theranostics. This theranostics imaging can identify areas of the tumour that are most susceptible to radiation, allowing for more precise and effective targeting of the tumour. Clinicians can create more individualized and efficient treatment regimens for cancer patients by integrating theranostics with other diagnostic and therapeutic methods. This approach can lead to better outcomes and improved patients life standard (Kashyap et al. 2023).

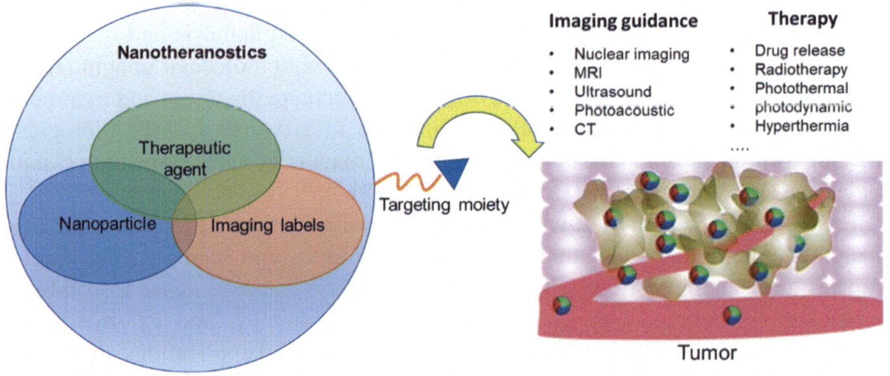

Fig. 12.3 An overview of nanotheranostic technology's diverse imaging and therapeutic capabilities (Dennahy et al. 2022)

12.3 Theranostics in Nuclear Medicine

Nuclear theranostics, also known as theranostics in nuclear medicine, is a fastly growing field that combines diagnostic and therapeutic techniques using radiopharmaceuticals to treat cancer and other diseases. It involves the use of nuclear medicine imaging to diagnose and stage diseases, as well as targeted radiotherapy to treat them. Utilizing a single radiopharmaceutical agent to perform therapeutic and diagnostic imaging applications is the fundamental principle of nuclear theranostics. This agent is designed to specifically target disease-specific biomarkers or receptors, allowing for highly accurate diagnostic imaging and targeted treatment. Along with the development of molecular imaging and personalized medicine, the nuclear theranostic approach has attracted growing attention and importance. It assists in providing tailored management for a variety of diseases, as well as improving patient selection, predicting response and toxicity, determining prognosis, and avoiding pointless and expensive diagnostic tests and treatments for many diseases (Marin et al. 2020). In nuclear theranostics, the radiopharmaceutical agent is usually a molecule that binds specifically to a target receptor or biomarker expressed by the cancer cells. The agent is labelled with a radioactive isotope that emits radiation, which can be detected by nuclear imaging modalities, such as PET or SPECT (Benfante et al. 2023).

Diagnostic nuclear imaging allows for the precise localization of cancer and the determination of its stage, size, and extent of spread. Once the target is identified, the same radiopharmaceutical agent can be used as a therapeutic agent to selectively deliver high-energy radiation to the tumor, killing cancer cells while sparing healthy tissue (Weber et al. 2020). Important applications of nuclear theranostics are for the treatment of neuroendocrine tumours (NETs) and prostate cancer. For example, in the treatment of metastatic prostate cancer, theranostic agents such as Lutetium-177 PSMA can be used to both diagnose and treat the cancer, resulting in high response rates and improved survival. Overall, nuclear theranostics represents a highly promising approach to personalized cancer treatment that holds great potential for enhancing outcomes and decreasing the burden of disease in population (Barca et al. 2021).

12.4 Metabolotheranostics in Oncology for Precision Medicine

Metabolotherapeutics is the term for methods that specifically target disease-based alterations in metabolic processes identified by MRS using image-guided platforms for administration with the goal to achieve disease-specific therapy (Bhujwalla et al. 2018).

In current scenario, metabolotheranostics is a growing area in precision medicine that combines metabolomics with theranostic application for the diagnosis, treatment, and monitoring of cancer. It involves the use of metabolomics to identify specific metabolic alterations in cancer cells, which can be used as targets for theranostic approaches.

Metabolomics is the study of the metabolic processes within cells and tissues and can be used to identify metabolic pathways that are deregulated in cancer cells. By identifying these metabolic pathways, researchers can develop theranostic approaches that target specific metabolic pathways in cancer cells, while sparing healthy tissue.

A particular strategy of metabolotheranostics is metabolic imaging, which includes ultrasound (US), positron emission tomography (PET), magnetic resonance spectroscopy (MRS), single photon emission computer tomography (SPECT), and optical imaging. These techniques can be used to detect specific metabolic alterations in cancer cells (Bhujwalla et al. 2018). In cancer, the metabolic imaging capabilities of MRS open up different possibilities for "metabolotheranostics," in which, alteration in particular cancer metabolic alterations observed by MRS can be normalized by administering drugs, siRNA, or cDNA to tumors under image guidance to downregulate, and then evaluating their efficacy with MRS (Bhujwalla et al. 2018).

These imaging techniques can give knowledge regarding the metabolic activity of tumors, including the uptake and utilization of specific metabolites, such as glucose or amino acids. Once the metabolic alterations are identified, theranostic approaches can be developed that target these metabolic pathways. For example, targeted therapies can be developed that selectively inhibit enzymes involved in specific metabolic pathways (Wang et al. 2023).

Unlike other diseases, which have pathologies resulting from cell death, cancer is a disease of maintaining cells and tissue colonization (Babar et al. 2023). Hypoxia, extracellular acidosis, and an aberrant metabolism are caused by the dysfunctional vasculature in tumors, which are added to the inherently disturbed metabolism of cancer cells. Imaging allows for the visualization of these tumor ecosystem complexities and their transformation into opportunities for theranostic imaging (Bhujwalla et al. 2018).

Metabolotheranostics has the potential to improve cancer diagnosis and treatment outcomes by providing a more precise and personalized approach to cancer therapy. By targeting specific metabolic pathways in cancer cells can improve the efficacy of cancer therapies while reducing toxicity to healthy tissue. Additionally, metabolotheranostics can also be used to monitor the treatment response and disease progression, allowing for timely adjustments in treatment plans (Debela et al. 2021).

12.5 Nuclear Theranostics with Molecular Imaging and Personalized Medicine

Nuclear theranostics is the application of radioactive substances to understand biologic phenomena through the expression of particular disease targets, such as membrane transporters or cell surface receptors, and then using specially formulated agents to deliver ionizing radiation to the tissues that express these targets (Marin et al. 2020).

Nuclear theranostics is a rapidly evolving field that combine molecular imaging with targeted therapy for personalized cancer treatment. It involves the use of radiopharmaceuticals, where the molecules are labelled with a radioactive isotope that can be detected by molecular imaging techniques, such as PET or SPECT (Aboagye et al. 2023). The goal of nuclear theranostics is to provide an accurate diagnosis, treatment monitoring, and targeted therapy by using a single radiopharmaceutical agent. The agent is designed to specifically target tumour-specific biomarkers, such as receptors or antigens, which are over expressed on cancer cells but are absent in normal cells.

Molecular imaging allows clinicians to visualize the specific biomarkers expressed by the tumour, which can help to identify early-stage tumors and track the response to therapy. The radiopharmaceutical agent is used both for diagnostic imaging and targeted therapy, as the radioactive isotope delivers high-energy radiation to the tumor, destroying cancer cells while sparing healthy tissue. Nuclear theranostics is a promising approach to personalized medicine, as it enables clinicians to select the most appropriate treatment for individual patients based on their tumor characteristics. By tailoring treatment to the specific molecular profile of each patient's tumor, nuclear theranostics can improve treatment efficacy while minimizing toxicity and adverse effects. One of the most promising applications of nuclear theranostics is in the treatment of neuroendocrine tumors (NETs). Radiolabeled somatostatin analogs, such as Gallium-68 DOTATATE, can be used for both diagnosis and therapy of NETs. The diagnostic imaging allows for the identification of tumor lesions and the measurement of somatostatin receptor expression levels, while the therapeutic agent selectively delivers high-energy radiation to the tumor (Gervasoni et al. 2023).

In summary, nuclear theranostics with molecular imaging is highly promising approach to personalized cancer treatment that holds great potential for improving patient outcomes and reducing the burden of disease. By combining diagnostic imaging with targeted therapy, clinicians can develop personalized treatment plans that are tailored to the specific molecular profile of each patient's tumor.

12.6 Theranostics in Prognosis and Treatment of Diseases

Theranostics is a rapidly developing field that combines diagnostic and therapeutic strategies to improve the prognosis and treatment of diseases. It involves the use of diagnostic tools to identify specific biomarkers or molecular targets associated with a disease, followed by targeted therapy using a single agent. By using a single agent that can both diagnose and treat the disease, theranostics provides a personalized approach for treatment that is tailored to the unique characteristics of each patient's disease. This approach can improve treatment efficacy while minimizing the risk of side effects and toxicity. Theranostics has been applied to a variety of diseases, including cancer, cardiovascular disease, and neurodegenerative diseases. In cancer, theranostics has shown great promise in the treatment of tumors that express specific biomarkers or molecular targets. For example, in prostate cancer, theranostics involves the use of radiolabeled prostate-specific membrane antigen (PSMA) to identify tumour cells and selectively deliver high-energy radiation to destroy them. In addition to its diagnostic applications, PSMA theranostics has also shown promising results in the treatment of metastatic castration-resistant prostate cancer. Theranostics has also been applied in the treatment of cardiovascular disease, where it can be used to target specific molecular pathways associated with atherosclerosis or thrombosis. For example, radiolabeled agents targeting the integrin αβ receptor that have been used for both diagnosis and treatment of atherosclerotic plaques in animal models. In neurodegenerative diseases, theranostics has the potential to improve diagnosis and treatment by identifying specific molecular targets associated with disease progression. For example, radiolabeled amyloid-beta (Aβ) and tau protein imaging agents can be used to diagnose and monitor Alzheimer's disease, while theranostic agents targeting Aβ or tau protein could be used to deliver disease-modifying therapies. Overall, theranostics has the potential to transform the diagnosis and treatment of diseases by providing a personalized approach that is tailored to the unique characteristics of each patient's disease. As research in this field continues to advance, we can expect to see further applications of theranostics in a variety of diseases, leading to improved patient outcomes and a better quality of life (Pini et al. 2022; Hertz and Greenspan 2023).

12.7 Radioligands in Dosimetry, Imaging, and Treatment

Radioligands are a type of radiopharmaceutical that is used in dosimetry, imaging, and treatment of various diseases. They are composed of a radioactive isotope (radionuclide) attached to a targeting molecule (ligand) that specifically binds to a target molecule or receptor in the body (Mahmoudi et al. 2021). In dosimetry, radioligands are used to estimate the absorbed radiation dose delivered to the target tissue or organ during therapy. This is important to ensure that the therapeutic dose is sufficient to destroy the target cells while minimizing radiation exposure to healthy tissues. The dosimetry can be calculated using imaging data obtained from PET or SPECT scans after administering the radioligand. In imaging, radioligands are used to visualize the distribution and activity of specific receptors or biomarkers in the body. This allows clinicians to identify and monitor disease progression, select appropriate treatment options, and evaluate treatment response. PET and SPECT imaging are commonly used in combination with radioligands for imaging applications. In treatment, radioligands can deliver a targeted dose of radiation to cancer cells or other disease targets. The radioligand is designed to specifically bind to a target molecule or receptor expressed on the surface of cancer cells, delivering high-energy radiation to destroy the cells while minimizing damage to healthy tissue. Radioligand therapy has been used to treat various types of cancer, including neuroendocrine tumours, prostate cancer, and lymphoma. The selection of the targeting molecule or ligand is critical for the effectiveness of radioligand-based therapies.

The ligand must have high affinity and specificity for the target molecule or receptor, while the radionuclide must have the appropriate half-life, decay mode, and energy for effective therapy. Radioligands have the potential to revolutionize the field of precision medicine by providing a personalized approach to diagnosis and treatment. By using radioligands for imaging, clinicians can identify specific targets for therapy and monitor the response to treatment. Radioligand therapy can provide a targeted approach to cancer treatment that delivers high doses of radiation to cancer cells while minimizing toxicity to healthy tissue (Rosar et al. 2021; Hoshi et al. 2023).

12.8 Current Developments and Future Perspectives

Theranostics is a rapidly evolving field with significant potential to revolutionize the diagnosis and treatment of various diseases. Current developments in theranostics include the development of novel radiopharmaceuticals, advances in molecular imaging technologies, and the expansion of theranostic applications to new disease targets. One of the most exciting recent developments in theranostics is the use of radiolabeled PSMA ligands for the diagnosis and treatment of prostate cancer. PSMA theranostics has shown remarkable success in the treatment of metastatic castration-resistant prostate cancer, leading to improved outcomes and prolonged survival in patients. Another area of development in theranostics is the use of nanotechnology-based approaches to deliver theranostic agents more effectively to target tissues. This includes the development of nanoparticles and other nanocarriers that can encapsulate radioligands and other therapeutic agents, enabling targeted delivery to specific tissues or cells (Langbein et al. 2019).

In addition, there is a growing interest in the use of theranostics for the treatment of neurodegenerative diseases such as Alzheimer's disease. Recent advances in molecular imaging technologies have led to the development of novel imaging agents that can detect the earliest pathological changes associated with Alzheimer's disease, providing opportunities for early diagnosis and targeted treatment. Preclinical and early clinical results from a number of novel therapeutic strategies based on antibodies, peptidomimetics, and small-molecule compounds have been encouraging (Langbein et al. 2019). Looking to the future, theranostics has significant potential to transform the field of precision medicine, providing personalized and targeted approaches to disease diagnosis and treatment. Advances in radiopharmaceuticals, molecular imaging, and nanotechnology are likely to continue to drive progress in the field, leading to new applications and improved outcomes for patients. There is also significant interest in expanding theranostic applications beyond cancer and neurodegenerative diseases, with researchers exploring the potential of theranostics for a wide range of conditions, including cardiovascular disease, infectious diseases, and autoimmune disorders.

Overall, the rapid pace of development in the field of theranostics is generating significant excitement and enthusiasm among researchers and clinicians, with the potential to transform the way we diagnose and treat diseases in the coming years.

References

Aboagye EO, Barwick TD, Haberkorn U (2023) Radiotheranostics in oncology: making precision medicine possible. CA Cancer J Clin 73(3):255–274

Babar Q, Saeed A, Tabish TA, Sarwar M, Thorat ND (2023) Targeting the tumor microenvironment: potential strategy for cancer therapeutics. Biochim Biophys Acta Mol Basis Dis 1869:166746

Bai JW, Qiu SQ, Zhang GJ (2023) Molecular and functional imaging in cancer-targeted therapy: current applications and future directions. Sig Transduct Target Ther 8:89. https://doi.org/10.1038/s41392-023-01366-y

Bailly C, Bodet-Milin C, Rousseau C, Faivre-Chauvet A, Kraeber-Bodéré F, Barbet J (2017) Pretargeting for imaging and therapy in oncological nuclear medicine. EJNMMI Radiopharm Chem 2(1):6. https://doi.org/10.1186/s41181-017-0026-8

Barca C, Griessinger CM, Faust A, Depke D, Essler M et al (2021) Expanding theranostic radiopharmaceuticals for tumor diagnosis and therapy. Pharmaceuticals (Basel, Switzerland) 15(1):13. https://doi.org/10.3390/ph15010013

Benfante V, Stefano A, Ali M, Laudicella R, Arancio W, Cucchiara A et al (2023) An overview of in vitro assays of 64Cu-, 68Ga-, 125I-, and 99mTc-labelled radiopharmaceuticals using radiometric counters in the era of radiotheranostics. Diagnostics 13(7):1210. https://doi.org/10.3390/diagnostics13071210

Bhujwalla ZM et al (2018) Theranostics and metabolotheranostics for precision medicine in oncology. J Magn Reson 291:141–151. https://doi.org/10.1016/j.jmr.2018.03.004

Chehelgerdi M, Chehelgerdi M, Allela OQB et al (2023) Progressing nanotechnology to improve targeted cancer treatment: overcoming hurdles in its clinical implementation. Mol Cancer 22:169. https://doi.org/10.1186/s12943-023-01865-0

Chiu FY, Yen Y (2023) Imaging biomarkers for clinical applications in neuro-oncology: current status and future perspectives. Biomark Res 11:35. https://doi.org/10.1186/s40364-023-00476-7

Debela DT, Muzazu SG, Heraro KD, Ndalama MT, Mesele BW, Haile DC, Kitui SK, Manyazewal T (2021) New approaches and procedures for cancer treatment: current perspectives. SAGE Open Med 9:20503121211034366. https://doi.org/10.1177/20503121211034366

Dennahy IS et al (2022) Nanotheranostics for image-guided cancer treatment. Pharmaceutics 14(5). https://doi.org/10.3390/pharmaceutics14050917

Fan D, Cao Y, Cao M et al (2023) Nanomedicine in cancer therapy. Sig Transduct Target Ther 8:293. https://doi.org/10.1038/s41392-023-01536-y

Gervasoni S, Öztürk I, Guccione C, Bosin A, Ruggerone P, Malloci G (2023) Interaction of radiopharmaceuticals with somatostatin receptor 2 revealed by molecular dynamics simulations. J Chem Inf Model 63(15):4924–4933

Hapuarachchige S, Artemov D (2020) Theranostic pretargeting drug delivery and imaging platforms in cancer precision medicine. Front Oncol 10:1–8. https://doi.org/10.3389/fonc.2020.01131

Hertz B, Greenspan BS (2023) Present and future of target therapies and theranostics: refining traditions and exploring new frontiers. Eur J Nucl Med Mol Imaging 50:1–1

Hoshi S, Yaginuma K, Meguro S, Onagi A, Matsuoka K, Hata J, Kojima Y (2023) PSMA targeted molecular imaging and radioligand therapy for prostate cancer: optimal patient and treatment issues. Curr Oncol 30(8):7286–7302

Hosseini SM, Mohammadnejad J, Salamat S, Zadeh ZB, Tanhaei M, Ramakrishna S (2023) Theranostic polymeric nanoparticles as a new approach in cancer therapy and diagnosis: a review. Mater Today Chem 29:101400

Jeelani S, Reddy RC, Maheswaran T, Asokan GS, Dany A, Anand B (2014) Theranostics: a treasured tailor for tomorrow. J Pharm Bioallied Sci 6(Suppl 1):S6–S8. https://doi.org/10.4103/0975-7406.137249

Kakkad S, Krishnamachary B, Jacob D, Pacheco-Torres J, Goggins E, Bharti SK et al (2019) Molecular and functional imaging insights into the role of hypoxia in cancer aggression. Cancer Metastasis Rev 38(1-2):51–64. https://doi.org/10.1007/s10555-019-09788-3

Kang H, Hu S, Cho MH, Hong SH, Choi Y, Choi HS (2018) Theranostic nanosystems for targeted cancer therapy. Nano Today 23:59–72. https://doi.org/10.1016/j.nantod.2018.11.001
Kashyap BK, Singh VV, Solanki MK, Kumar A, Ruokolainen J, Kesari KK (2023) Smart nanomaterials in cancer theranostics: challenges and opportunities. ACS Omega 8(16):14290–14320
Kevadiya BD, Ottemann BM, Thomas MB, Mukadam I, Nigam S, McMillan J, Gendelman HE (2019) Neurotheranostics as personalized medicines. Adv Drug Deliv Rev 148:252–289
Kong X, Gao P, Wang J et al (2023) Advances of medical nanorobots for future cancer treatments. J Hematol Oncol 16:74. https://doi.org/10.1186/s13045-023-01463-z
Krzyszczyk P, Acevedo A, Davidoff EJ, Timmins LM, Marrero-Berrios I et al (2018) The growing role of precision and personalized medicine for cancer treatment. Technology 6(3–4):79–100. https://doi.org/10.1142/S2339547818300020
Langbein T, Weber WA, Eiber M (2019) Future of theranostics: an outlook on precision oncology in nuclear medicine. J Nucl Med 60(9):13S–19S. https://doi.org/10.2967/jnumed.118.220566
Liu Y, Wu W, Cai C et al (2023) Patient-derived xenograft models in cancer therapy: technologies and applications. Sig Transduct Target Ther 8:160. https://doi.org/10.1038/s41392-023-01419-2
Mahmoudi E, Pirayesh E, Deevband MR, Amoui M, Rad MG, Ghorbani M (2021) Patient-specific dosimetry in radioligand therapy (RLT) for metastatic prostate cancer using ^{177}Lu-DKFZ-PSMA-617. Nucl Med Mol Imaging 55(5):237–244. https://doi.org/10.1007/s13139-021-00713-7
Marin JFG, Nunes RF, Coutinho AM, Zaniboni EC, Costa LB, Barbosa FG, Buchpiguel CA (2020) Theranostics in nuclear medicine: emerging and re-emerging integrated imaging and therapies in the era of precision oncology. Radiographics 40(6):1715–1740
Pini C, Gelardi F, Sollini M (2022) Present and future of target therapies and theranostics: refining traditions and exploring new frontiers—highlights from annals of nuclear medicine 2021. Eur J Nucl Med Mol Imaging 49(11):3613–3621
Rosar F, Schön N, Bohnenberger H, Bartholomä M, Stemler T, Maus S, Schaefer-Schuler A (2021) Comparison of different methods for post-therapeutic dosimetry in [177Lu] Lu-PSMA-617 radioligand therapy. EJNMMI Phys 8(1):40
Shi Y, Lammers T (2019) Combining nanomedicine and immunotherapy. Acc Chem Res 52(6):1543–1554. https://doi.org/10.1021/acs.accounts.9b00148
Sun L, Liu H, Ye Y et al (2023) Smart nanoparticles for cancer therapy. Sig Transduct Target Ther 8:418. https://doi.org/10.1038/s41392-023-01642-x
Wang X, Zhang J, Zheng K, Du Q, Wang G, Huang J, He J (2023) Discovering metabolic vulnerability using spatially resolved metabolomics for antitumor small molecule-drug conjugates development as a precise cancer therapy strategy. J Pharm Anal 13:776
Weber WA, Czernin J, Anderson CJ, Badawi RD, Barthel H, Bengel F et al (2020) The future of nuclear medicine, molecular imaging, and theranostics. J Nucl Med 61(Suppl 2):263S–272S
Xue Y, Gao Y, Meng F, Luo L (2021) Recent progress of nanotechnology-based theranostic systems in cancer treatments. Cancer Biol Med 18(2):336–351. Advance online publication. https://doi.org/10.20892/j.issn.2095-3941.2020.0510

GPSR Compliance

The European Union's (EU) General Product Safety Regulation (GPSR) is a set of rules that requires consumer products to be safe and our obligations to ensure this.

If you have any concerns about our products, you can contact us on ProductSafety@springernature.com

In case Publisher is established outside the EU, the EU authorized representative is:

Springer Nature Customer Service Center GmbH
Europaplatz 3
69115 Heidelberg, Germany

The manufacturer's authorised representative in the EU is Springer
Nature Customer Service Centre GmbH, Europaplatz 3, 69115 Heidelberg,
Germany. If you have any concerns regarding our products, please
contact ProductSafety@springernature.com

Printed and bound by CPI Group (UK) Ltd, Croydon, CR0 4YY
01/10/2026
02225863-0007